Gene Editing by CRISPR-Cas

CRISPR-Cas systems have redefined the landscape of genetic engineering, offering an unprecedented level of precision, efficiency, and versatility in genome editing. This book, titled *Gene Editing by CRISPR-Cas*, provides a comprehensive and well-structured approach on the historical context of genetics and gene-editing techniques leading to the groundbreaking discovery of CRISPR-Cas systems.

Distinguished by its multidisciplinary scope, this volume not only delves into CRISPR-Cas systems but also examines alternative and next-generation gene-editing technologies. The book aims to inform and engage both general audience with a keen interest in science and biology, and professionals and students in related fields. By offering a balanced blend of scientific explanations and real-world applications, the book seeks to empower readers to understand the potential of CRISPR technology and its impact across diverse domains, such as medicine, agriculture, biotechnology, and diagnostics, while addressing the ethical, regulatory, and biosafety considerations essential for responsible implementation. Furthermore, this book engages in a discussion on the future directions and broader impacts of CRISPR technologies, providing valuable insights into the evolving landscape of genome engineering and its implications for scientific advancements.

Key Features:

- This book is written in a clear and accessible manner, making complex scientific concepts understandable to a general audience without overwhelming them with technical jargon.
- This book provides a comprehensive exploration of the discovery and evolution of CRISPR technology, delving into the underlying science and its transformative capabilities. It serves as an excellent introduction to both the technical intricacies and the human implications of this groundbreaking technology.
- This book provides an in-depth exploration of the potential uses of CRISPR across multiple domains, with a particular focus on genetic disease treatment, cancer studies, agricultural advancements, HIV, blood disorders, neurological conditions, insect biology, microbial engineering, and both infectious and non-infectious diseases.
- Furthermore, this book also provides critical analysis of bioethics, safety concerns, and regulatory frameworks, providing a responsible perspective on genome editing.

Gene Editing by CRISPR-Cas

Edited by
Aarif Ali and
Farheen Shafique

CRC Press
Taylor & Francis Group
Boca Raton London New York

CRC Press is an imprint of the
Taylor & Francis Group, an **informa** business

Designed cover image: Shutterstock

First edition published 2026
by CRC Press
2385 NW Executive Center Drive, Suite 320, Boca Raton FL 33431

and by CRC Press
4 Park Square, Milton Park, Abingdon, Oxon, OX14 4RN

CRC Press is an imprint of Taylor & Francis Group, LLC

ISBN: 978-1-032-88312-0 (hbk)
ISBN: 978-1-032-88316-8 (pbk)
ISBN: 978-1-003-53715-1 (ebk)

DOI: 10.1201/9781003537151

Typeset in Times
by Apex CoVantage, LLC

Contents

About the Editors

Dr. Aarif Ali holds a doctorate of philosophy in clinical biochemistry from the University of Kashmir, Srinagar, Jammu, and Kashmir, India, and has obtained his masters in clinical biochemistry from the University of Kashmir, Srinagar, Jammu, and Kashmir, India. Dr. Aarif Ali serves as a reviewer of several high-impact national as well as international scientific journals. He has published the books *Pharmacogenomics: From Discovery to Clinical Implementation*, *Biological Insights of Multi-Omics Technologies in Human Diseases*, and *Fundamentals of Biochemistry.*

Dr. Ali has collaboratively worked on and published more than 100 research publications in different reputed national as well as international journals. He has been awarded with the Young Scientist Award at the AI-CIAS–2024, with his contributions to the scientific community extending to numerous research publications covering various aspects of human health. Dr. Ali's dedication to research and his passion for contributing to scientific knowledge make him a valuable asset to the academic and scientific community.

Dr. Farheen Shafique is a highly accomplished zoologist and molecular biologist with a strong academic background and extensive research experience. She holds a PhD in zoology and molecular biology, obtained jointly from the University of Azad Jammu and Kashmir, Pakistan, and the University of Sheffield, UK. Prior to that, she completed her master's degree in zoology/molecular biology. Farheen's dedication to research has earned her prestigious awards and scholarships, including the Commonwealth PhD Split-Site Scholarship in the United Kingdom and the International Research Support Initiative Program (IRSIP) PhD Fellowship by the Higher Education Commission of Pakistan. Farheen's contributions extend to numerous research publications covering topics ranging from infectious diseases, vaccination, and human health to ecological impacts of climate change and biodiversity preservation.

Contributors

Noreen Aasim
IBGE
The University of Agriculture
Peshawar, Pakistan

Sadhna Aggarwal
MD Anderson Cancer Center
University of Texas
Houston, Texas, United States

Shazia Akram Ghumman
College of Pharmacy
University of Sargodha
Sargodha, Pakistan

Sheikh Bilal Ahmad
Division of Veterinary Biochemistry, Faculty of Veterinary Sciences and Animal Husbandry
SKUAST-K
Shuhama Alusteng, Srinagar, Jammu and Kashmir, India

Syed Mudasir Ahmad
Division of Animal Biotechnology
Sher-e-Kashmir University of Agricultural Sciences and Technology of Kashmir
Shalimar, Srinagar, Jammu and Kashmir, India

Aarif Ali
Division of Basic Sciences and Humanities, Faculty of Horticulture
SKUAST-K
Shalimar, Srinagar, Jammu and Kashmir, India

Nisar ul Ashraf
Government College for Women
Nawakadal, Srinagar, Jammu and Kashmir, India

Javeed Iqbal Ahmad Bhat
Division of Basic Sciences and Humanities, Faculty of Horticulture
SKUAST-K
Shalimar, Srinagar, Jammu and Kashmir, India

Kaisar Ahmad Bhat
Department of Biotechnology
Baba Ghulam Shah Badshah University
Rajouri, Jammu and Kashmir, India

Mohammad Saleem Bhat
Hassan Khoyihami Memorial Degree College
Bandipora, Jammu and Kashmir, India

Lateef Ahmad Dar
Division of Animal Biotechnology
Sher-e-Kashmir University of Agricultural Sciences and Technology of Kashmir
Shalimar, Srinagar, Jammu and Kashmir, India

Nelofer Ereej
Faculty of Pharmaceutical Sciences
University of Central Punjab (UCP)
Lahore, Pakistan

Saleha Faheem
Faculty of Pharmaceutical Sciences
University of Central Punjab (UCP)
Lahore, Pakistan

Ubaid Gani
Division of Animal Biotechnology
Sher-e-Kashmir University of Agricultural Sciences and Technology of Kashmir
Shalimar, Srinagar, Jammu and Kashmir, India
Lovely Professional University
Punjab, India

Showkat Ahmad Ganie
Department of Clinical Biochemistry, School of Biological Sciences
University of Kashmir
Srinagar, Jammu and Kashmir, India

Meena Godha
Jaipur National University
Jagatpura, Jaipur

Anam Hameed
Department of Human Nutrition and Dietetics, Faculty of Rehabilitation and Allied Health Sciences
Riphah International University
Gulberg III, Lahore, Pakistan

Huma Hameed
Faculty of Pharmaceutical Sciences
University of Central Punjab (UCP)
Lahore, Pakistan

Rabia Hamid
Department of Nanotechnology
University of Kashmir
Hazratbal, Srinagar, Jammu and Kashmir, India

Irfan Hussain
Division of Animal Biotechnology
Sher-e-Kashmir University of Agricultural Sciences and Technology of Kashmir
Shalimar, Srinagar, Jammu and Kashmir, India

Ali Irfan
Department of Chemistry
Government College University Faisalabad
Faisalabad, Pakistan

Farhat Jabeen
Department of Clinical Biochemistry, School of Biological Sciences
University of Kashmir
Srinagar, Jammu and Kashmir, India

Farkhanda Jabeen
Pir Mehr Ali Shah Arid Agriculture University
Rawalpindi, Islamabad, Punjab, Pakistan

Priyanka Jha
Lovely Professional University
Punjab, India

Arjumand John
School of Bioengineering and Bioscience
Lovely Professional University
Phagwara, Jalandhar Punjab

Mahtab Ahmad Khan
Faculty of Pharmaceutical Sciences
University of Central Punjab (UCP)
Lahore, Pakistan

Nuzhat Khursheed
Department of Clinical Biochemistry, School of Biological Sciences
University of Kashmir
Srinagar, Jammu and Kashmir, India

Imtiaz Majeed
Faculty of Pharmaceutical Sciences
University of Central Punjab (UCP)
Lahore, Pakistan

Bashir Ahmad Malla
Department of Biochemistry, School of Biological Sciences
University of Kashmir
Hazratbal, Srinagar, Jammu and Kashmir, India

Tasaduq Manzoor
Division of Animal Biotechnology
Sher-e-Kashmir University of Agricultural Sciences and Technology of Kashmir
Shalimar, Srinagar, Jammu and Kashmir, India
Jaipur National University
Jagatpura Jaipur

Zainab Manzoor
Department of Clinical Biochemistry
University of Kashmir
Srinagar, Jammu and Kashmir, India

Tooba Mehboob
Faculty of Pharmaceutical Sciences
University of Central Punjab (UCP)
Lahore, Pakistan

Gh Jeelani Mir
Department of Biochemistry, School of Biological Sciences
University of Kashmir
Hazratbal, Srinagar, Jammu and Kashmir, India

Yaawar Bashir Mir
Division of Animal Biotechnology
Sher-e-Kashmir University of Agricultural Sciences and Technology of Kashmir
Shalimar, Srinagar, Jammu and Kashmir, India

Junaid Nazir
Division of Animal Biotechnology
Sher-e-Kashmir University of Agricultural Sciences and Technology of Kashmir
Lovely Professional University
Punjab, India

Umar Rasool
Department of Biotechnology
Central University of Kashmir
Ganderbal, Jammu and Kashmir, India

Arshid Reshi
Department of Clinical Biochemistry
University of Kashmir
Srinagar, Jammu and Kashmir, India

Amaninder Kaur Riat
School of Bioengineering and Bioscience
Lovely Professional University
Phagwara, Jalandhar Punjab

Tehseen Riaz
Faculty of Pharmaceutical Sciences
University of Central Punjab (UCP)
Lahore, Pakistan

Nausheen Shafi
Department of Clinical Biochemistry
University of Kashmir
Srinagar, Jammu and Kashmir, India

Farheen Shafique
King Abdullah Campus
University of Azad Jammu and Kashmir Muzaffarabad
Azad Kashmir, Pakistan

Naila Tabassam
Faculty of Pharmaceutical Sciences
University of Central Punjab (UCP)
Lahore, Pakistan

Muhammad Zaman
Faculty of Pharmaceutical Sciences
University of Central Punjab (UCP)
Lahore, Pakistan

1 CRISPR-Cas Systems
The Molecular Scissors

Bashir Ahmad Malla, Aarif Ali, Nuzhat Khursheed, Mohammad Saleem Bhat, Farhat Jabeen, and Javeed Iqbal Ahmad Bhat

The completion of the human genome sequencing represents a significant milestone in the realm of genetic research focused on diseases. Nevertheless, there exists a notable gap in our understanding of the roles of genes and their variants, primarily as we currently lack the know-hows necessary for precise gene sequence modifications, which are vital for experimentally assessing their molecular functions. The term *genome editing* denotes the targeted alteration of genomic DNA across a range of cell types and organisms. This technique includes the substitution, insertion, and deletion of DNA, which can result in the silencing of target genes, acquisition of new genetic traits, and the correction of deleterious gene mutations (Charpentier and Marraffini, 2014). In recent years, a remarkable progress in life sciences has been witnessed, establishing genome editing as the primary method for exploring gene function, shedding light on the pathogenesis of genetic disorders, identifying innovative targets for gene therapy, and advancing agricultural crop varieties (Cox et al., 2015). Molecular scissors that have been engineered for specific genomic modifications hold substantial potential for conducting functional analysis. These advancements could significantly enhance our understanding about the molecular basis of diseases and lead to the development of novel therapeutic approaches. A variety of platforms for these molecular scissors have been created, including zinc finger nucleases (ZFNs), transcription activator–like effector nucleases (TALENs), and the recently introduced clustered regularly interspaced short palindromic repeats (CRISPR), along with CRISPR-associated protein 9 (Cas9). Earlier, scientists had discovered the CRISPR-Cas protein system as an innovative microbial defense system against bacteria and archaea (Bolotin et al., 2005). In bacteria, it protects from bacteriophages (bacterial viruses) and mobile genetic elements (transposons); hence, it works as an adaptive immune system in various prokaryotes (Mojica et al., 2009; Jinek et al., 2012). Additionally, in bacteria it was found that after phage infection, a spacer sequence was present in the genome, and alteration in these sequences affects phage growth within the target cell (Barrangou et al., 2007).

The CRISPR-Cas protein system is a ribonucleic acid (RNA)–mediated selective deoxyribonucleic acid (DNA) editing or degradation system; that is, it is an RNA-mediated deoxyribonuclease (DNase). The CRISPR system consists of two foremost parts: CRISPR and Cas. The CRISPR is a short-sized guide RNA (gRNA),

DOI: 10.1201/9781003537151-1

approximately 20–30 nucleotides, and is complementary to the target DNA, whereas Cas is an endo-DNase protein. The gRNA, also known as CRISPR-RNA (crRNA), guides the Cas protein to degrade or edit the target nucleic acid (DNA). In CRISPR-Cas systems, there are two main parts, the crRNA and the Cas, working interdependently in a target-specific manner. In this system, the Cas protein works like a molecular scissor, cutting/breaking the target DNA at specific locations that exactly matches the sequence of the crRNA. Hence, in CRISPR-Cas systems, the target DNA sequence is the focus of the crRNA; therefore, it is designed by utilizing bioinformatic techniques to decrease the non-specificity. After the double-strand DNA break is triggered by Cas, multiple changes can be introduced at the site with the help of cellular repair processes using either homology-directed repair (HDR) or non-homologous end joining (NHEJ) pathways (Fu et al., 2021). In the beginning, the genome editing functions of CRISPR-Cas systems were unknown because of the vast defense systems in bacteria and archaea (Bolotin et al., 2005; Mojica et al., 2009; Jinek et al., 2012). Hence, earlier research scientists were focused on the reason behind the bacterial defense system. However, after the discovery of the CRISPR-Cas system, the genome editing process has witnessed a revolution in the molecular science field, and it is now becoming a leading technique garnering significant interest (Rao et al., 2022). A revolutionary age in genome (DNA) editing has been ushered in by the CRISPR-Cas system because it is fast, cheap, and has extraordinary precision and versatility. It is a major breakthrough in the field of molecular science because it performs better in terms of cost-efficiency and is relatively easier to use than previous genome editing methods, like ZFNs and TALENs. Further, it also demonstrates that this system improves editing outcomes due to its high efficacy and less off-target effects. Thus, due to these benefits, CRISPR-Cas has been widely used and has a more significant impact when compared to previous ZFNs and TALENs (Gaj et al., 2013). Thus, the changes that are introduced at the target site by the CRISPR-Cas system are incredibly useful for basic research, medicine, and crop improvement because of its precise and programmable genome editing capabilities (Li et al., 2023). Hence, it is used in diverse fields, including drug development science, human gene therapy, medical research, plant science, and in agricultural development (Liu et al., 2021).

1.1 STRUCTURE OF CRISPR

The chromosomes of prokaryotes generally contain one CRISPR array, while some other species contain about eight arrays (Jansen et al., 2002; Mojica et al., 2000, 2005; Koonin and Makarova, 2019). Three nucleotide sequences make up the CRISPR array.

1.1.1 Direct Repeats (DRs) Bordered with Spacer Sequence

DRs are short identical nucleotide sequences in the genome of an organism. The DR is normally of 32 base pairs (bp) in size, but in some species, the size varies from 21 to 47 bp. Spacer sequences are unique DNA segments that match the length of nearby repeats. They come from DNA pieces taken from invading viruses or plasmids, known as protospacers. The size of a spacer sequence is fixed at about 20–70 bp, but they show high variation in sequence composition (Karginov and Hannon, 2010).

1.1.2 Leader Sequence

The leader sequence in CRISPR systems is a non-coding DNA segment that is positioned before the CRISPR array. This sequence is essential for the proper functioning and regulation of the system. It is typically rich in adenine (A) and thymine (T) nucleotides, contributing to its AT-rich nature. This sequence is found directly in front of the first direct repeat within the CRISPR array. The leader sequence exhibits limited sequence conservation and is involved in the transcription and adaptation processes. It was found that a locus without a leader sequence is inactive in the adaptation process but active in the interference process (Alkhnbashi et al., 2016).

1.1.3 CRISPR-Associated Genes

CRISPR-associated genes represent the cluster of nucleotide sequence with different order and orientation. The Cas genes linked to CRISPR are crucial for encoding proteins that engage with CRISPR loci, offering adaptive immunity to bacteria and archaea against foreign genetic elements, like phages and plasmids. These genes and their protein products are essential in the acquisition, processing, and interference stages of the CRISPR-Cas immune mechanism. These sequences code for specific Cas proteins (about 93 proteins) and are classified into different classes, types, or subtypes (Makarova et al., 2020).

1.2 CRISPR-CAS MOLECULAR MECHANISM

Utilizing crRNA, the CRISPR-Cas system functions in a sequence-specific way to detect and cleave foreign DNA. The defense mechanism comprises three essential functional stages:

1. Adaptation stage, also called spacer acquisition or insertion stage
2. Expression stage, also called crRNA biogenesis or processing or maturation stage
3. Interference stage, also called target interference stage

1.2.1 Adaptation Stage

Adaptation is a process of insertion of a distinct sequence of foreign DNA called protospacer into CRISPR arrays to yield a new duplicated spacer (Barrangou et al., 2007; Yosef et al., 2012; Wei et al., 2015a). This process helps the target organism memorize this intruder nucleic acid sequence so that it can develop adaptive immunity (Barrangou et al., 2007). In *Escherichia coli* (*E. coli*), the combination of newfangled spacers at the leader sequence of the CRISPR array is done with the help of a hetero-hexameric protein complex composed of two Cas proteins: $[(Cas1)_2\text{-}Cas2]_2$ or $(Cas1)_2\text{-}(Cas2)_2\text{-}(Cas1)_2$ (Nunez et al., 2014). In the hexameric complex, both Cas proteins act as a nuclease (Wiedenheft et al., 2009), but the catalytic activity of Cas2 is dispensable for the process of insertion/adaptation (Nunez et al., 2014). The adaptation process is further divided into two types based on CRISPR-Cas types I and II. In type I, the integration factor of the target/host with the leader sequence is used, and it causes bending in the target DNA. This bending helps the hexameric complex to produce an initial break in the target DNA for the insertion of the spacer sequence. Meanwhile, in type II, for the insertion of polar spacer sequences, the anchoring sequence of the leader is used, as it causes the recognition of Cas1 protein of the hexameric complex. The hexameric complex is sufficient to promote the adaptation process in most CRISPR-Cas types, but type I-B requires Cas4 (Li et al., 2014), type I-F requires the interference machinery (Vorontsova et al., 2015), and type II-A requires Csn2, Cas9, and tra-crRNA (trans-activating crRNA) for spacer acquisition (Barrangou et al., 2007; Wei et al., 2015b). In CRISPR-Cas systems, the random selection of target nucleic acid to be integrated into the CRISPR locus can also be done. But this non-specificity can be neglected with the help of some specific nucleotide sequences, as in types I and V, where the short protospacer adjacent motif (PAM) sequence located near the protospacer is important for the adaptation and interference processes (Zetsche et al., 2015), while in type II systems, the PAM-interacting (PI) domain of Cas9 protein is important for the protospacer section (Wei et al., 2015b).

1.2.2 Expression Stage

This stage signifies the development of the long precursor crRNA (pre-crRNA) originating from the CRISPR locus. The processing (cleavage) of pre-crRNA is done within DR sequences to get small and mature crRNAs (guide RNAs) containing the invader nucleic acid sequence for memorization (Carte et al., 2008). The enzymes used for crRNA processing are different in CRISPR-Cas types and subtypes. In types I and III, Cas6 members perform the processing step and form an intermediate crRNA flanked with short sequences. The type I-C system does not produce Cas6 and, hence, uses the Cas5d protein for processing (Nam et al., 2012; Sashital et al., 2011), while the processing of crRNAs in type II systems involve tra-crRNA. During crRNA processing, the anti-repeat sequence of tra-crRNA hybridizes with the repeat sequence of pre-crRNA and forms a double-stranded RNA (tra-crRNA:crRNA). The formed double stranded RNA is then processed by the host's RNase-III and forms an intermediate crRNA, which further undergoes processing by an unknown mechanism and yields mature crRNA (Deltcheva et al., 2011). Moreover, some intermediate crRNAs can further undergo nucleotide (nt) trimming at their 5' or 3' end. The trimming of intermediate crRNA is done by an unknown nuclease yielding the mature crRNA. The mature crRNA is composed of the 5' sequence (spacer sequence) and the 3' sequence (repeat sequence), which forms a hairpin-like (loop-like) structure in most of the type I systems (Charpentier et al., 2015).

1.2.3 Interference Stage

In this stage, the mature crRNA acts as a guide RNA to specifically interfere with the invading nucleic acids with the help of the effector complex made up of single or multiple proteins. In the class 1 system, for the target interference, the multi-protein effector complex is used, and it contains multiple Cas proteins along with crRNA. Meanwhile, in class 2 systems, the single-protein effector is used for the target interference, and it contains a single multidomain Cas protein along with crRNA (Zetsche et al., 2015; Wang et al., 2011). The functions of both the complexes are to recognize the targeted nucleic acid and then to degrade it with the help of Cas3 (type I) or Cas9 (types II and V) (Makarova et al., 2020). During this process, the system can also cause self-destruction; thus, to avoid self-destruction, the PAM sequence is used. The used PAM site is present upstream (types I and V) or downstream (type II) of the specific spacer (Barrangou et al., 2007; Fonfara et al., 2016; Zetsche et al., 2015; Jiang et al., 2013). This PAM site is essential for Cas protein binding and is followed by the specific target nucleic acid recognition and degradation. The PAM selection occurs through the hexameric complex in type I systems and the Cas9-PI domain in type II systems (Xiao et al., 2017).

1.3 CRISPR-ASSOCIATED SYSTEMS AND THEIR TYPES

Understanding CRISPR-Cas systems' classification is crucial for comprehending their origins and in facilitating new research developments. The classification of CRISPR-Cas systems is based on different factors, like the composition of Cas proteins and effector complexes. Makarova and his colleagues first classified CRISPR-Cas systems (Makarova et al., 2017, 2015, 2020).

1.3.1 CRISPR-Cas System Classification

The CRISPR-Cas system consists of two classes (class 1 and class 2), six types (types I through VI), and a total of 33 subtypes, each presenting various variants (Makarova et al., 2020).

1.3.1.1 Class 1 System

In the class 1 system, the RNA-guided DNA cleavage is performed with the help of multiple protein cascade effector complexes. There are three types of class 1 systems, namely, type I, III, and IV, respectively (Makarova et al., 2017).

1.3.1.1.1 Type I System

In this type, the CRISPR-Cas array mainly consists of the Cas3 protein encoding gene, which degrades the target DNA (Brouns et al., 2008). This type is further divided into seven subtypes (I-A to I-G). In this type, pe-crRNAs have palindromic sequences that are unstructured (I-A and I-B) or form loop-like structures (I-C to I-F). In this system, crRNA maturation is done by the Cas6 protein, except in the I-C subtype (Cas5), and the effector complex contains multiple Cas proteins, including Cas3, Cas5, Cas7, and Cas8. Among all these Cas proteins, the Cas3 protein is an important component of the complex, as it cleaves the target DNA, while Cas3 is associated with Cas2 protein in some subtypes (Makarova et al., 2017; Charpentier et al., 2015).

1.3.1.1.2 Type III System

In this type, the CRISPR locus mainly contains Cas10 multidomain protein encoding genes, which cleaves the target DNA with the help of the palm domain (Marraffini and Sontheimer, 2008). Within type III, there are six subcategories, labeled from III-A to III-F. In this type, some subtypes involve reverse transcriptase for the spacer acquisition process. In this system, the pre-crRNA processing is done with the help of the Cas6 protein, while the subtypes which lack Cas6 encoding genes uses the Cas6 protein of other CRISPR arrays. Among the six subtypes, the Csm cascade effector complex (Csm-Cas protein, along with the crRNA) is used by four subtypes (III-A, III-D, III-E, and III-F). Similarly, the Cmr cascade effector complex (Cmr-Cas protein, along with the CrRNA) is used by two subtypes (III-B and III-C). The effector complex of subtypes III-A, III-B, and III-C cleaves foreign DNA/RNA. However, III-D and III-E complexes may cleave the RNA, while the III-F complex may cleave the DNA (Makarova et al., 2020, 2015; Charpentier et al., 2015).

1.3.1.1.3 Type IV System

This type contains genes which encode a ribonucleoprotein, called the CRISPR-associated splicing factor 1 (Csf1), with an unknown function (Koonin and Makarova, 2019). Type IV is further classified into three subtypes (IV-A to IV-C). In this system, pre-crRNA processing is performed with the help of the Cas6 protein. The cascade effector complex of this type contains Csf1, Cas5, and Cas7 proteins.

1.3.1.2 Class 2 System

In the class 2 system, the RNA-guided DNA degradation is mediated by a single multidomain Cas protein. The class 2 system is further classified into three types: types II, V, and VI systems (Makarova et al., 2020).

1.3.1.2.1 Type II System

In type II, the CRISPR locus encodes a multidomain protein to degrade the foreign DNA (Jinek et al., 2012). II-A, II-B, and II-C are the three subtypes within the type II system. In this type, the pre-crRNA processing is done with the help of the Cas9 protein, RNase-III (non-Cas protein), and tra-crRNA, but in subtype II-C, an alternative pre-crRNA processing was found. The Cas9 protein, which has two nuclease domains, is used in type II systems for cleaving the strands of the target DNA molecule (Makarova et al., 2020; Charpentier et al., 2015).

1.3.1.2.2 Type V System

This type contains Cas12 protein encoding genes, which are also known as C2c1 or C2c3 or Cpf1 (CRISPR from *Prevotella* and *Francisella* 1), with the RuvC endonuclease domain involved in single-stranded DNA (ssDNA) or double-stranded DNA (dsDNA) degradation (Zetsche et al., 2015). Ten subtypes have been identified within the type V system, which are V-A through V-I, along with V-K. In this type, Cas12 is used as an effector protein to degrade the foreign DNA molecule, and the processing of pre-crRNA is done with the help of RNase-III, except in subtype V-A (with the help of an effector).

1.3.1.2.3 Type VI System

This type contains Cas13 (C2c2) protein encoding genes, with two nucleotide-binding RNase domains, also known as higher eukaryotes and prokaryotes nucleotide-binding (HEPN) domains, which degrade the foreign single-stranded RNA (ssRNA) molecule (Abudayyeh et al., 2017). The classification of the type VI system includes four subtypes, specifically, VI-A through VI-D. In this type, the Cas13 protein is used as an effector molecule for foreign RNA degradation, and the processing of pre-crRNA is done with the help of an effector containing HEPN domains (Makarova et al., 2020).

1.4 CLASSIFICATION OF CAS PROTEINS

Cas proteins are the nuclease proteins that degrade nucleic acids, like DNA. Cas protein discovery has increased in popularity among scientists for their applications in diverse research fields, including agriculture, gene editing, and medicine. The classification, mechanism, and applications of Cas proteins are described in what follows.

1.4.1 Cas1 and Cas2 Proteins

In the prokaryotic adaptive immune system, Cas1 and Cas2 proteins are highly conserved (Barrangou et al., 2007). The CRISPR array contains Cas1 and Cas2–protein encoding genes, and they belong to the type II system (Wiedenheft et al., 2009). The Cas1 and Cas2 form a stable complex in *E. coli*–mediated *in vivo* spacer acquisition with an unknown mechanism (Babu et al., 2011). Cas1 and Cas2 are found in the integrase complex containing two distal dimers of Cas1 protein linked by Cas2 dimer, like $(Cas1)_2$-$(Cas2)_2$-$(Cas1)_2$ integrase (Yosef et al., 2012). They bind to the pre-spacer sequences and cause their integration into the leader region of the CRISPR locus with the help of the leader sequence (Wright and Doudna, 2016). The proteins Cas1 and Cas2 exhibit a high degree of conservation across all CRISPR-Cas systems. They are used for the integration process, but the exact mechanism is unknown.

1.4.2 Cas9 Protein

Cas9 is also called the *spy*Cas9 protein because it is involved in the CRISPR-mediated adaptive immunity of *Streptococcus pyogenes*. The Cas9 protein is a multidomain protein with different functions, and in the CRISPR system, it mostly acts as a DNA endonuclease. Among all Cas proteins, the Cas9 mechanism is studied extensively. There are six domains in Cas9 proteins, and these include the following: (1) recognition lobe (REC) I, (2) REC II, (3) arginine-rich bridge helix, (4) PAM-interacting (PI) domain, (5) HNH (histidine-asparagine-histidine) domain, and (6) RuvC domain. The REC I domain is involved in crRNA binding, while REC II functions are unknown. The arginine-rich bridge domain, upon binding with the target

DNA, initiates the cleavage activity of Cas9, while the PI domain is used for PAM specificity. RuvC and HNH, which are the two nuclease domains, are used to cleave the foreign DNA (Anders et al., 2014; Jinek et al., 2012). The Cas9 system is used for efficient and specific research work in the arena of molecular biology, medicine, and biotechnology. In many species, it helps in gene-specific modification as guided by crRNA and, hence, is used for large-scale genome editing processes compared with traditional editing approaches. The Cas9 protein is also used for the regulation of the transcription process; therefore, scientists have applied the Cas9 system in diverse fields, like with microorganisms (Liu et al., 2020), plants (Zhang et al., 2018), animals (Lee et al., 2020), insects (Hillary and Ceasar, 2021), and humans (Garza-Elizondo et al., 2019).

1.4.3 Cas12 Protein

The Cas12 protein demonstrates significant versatility along with a variety of dynamic functions. The Cas12 protein has RNA-mediated DNA endonuclease activity, and it belongs to the type V system (Fonfara et al., 2016; Zetsche et al., 2015). Thus, during the genome editing process, it is used as an alternative for the Cas9 protein, but during pre-crRNA processing, the Cas12 protein does not require tra-crRNA or RNase-III, as required by Cas9 (Zetsche et al., 2015). The nuclease domains (RuvC and nuclease lobe) of the Cas12 protein degrade the target ssDNA or dsDNA, and during degradation, the Cas12 protein requires only crRNA, but not tra-crRNA (Fonfara et al., 2016). During the target degradation process, Cas12 mediates hybridization between crRNA and the target DNA and forms an R-loop-like structure (Anders et al., 2014). After R-loop formation, and with the help of the PAM sequence, the active RuvC domain degrades the non-target DNA strand (Fonfara et al., 2016). During pre-crRNA processing, the Cas12 protein does not require tra-crRNA or RNase-III, as required by Cas9 (Zetsche et al., 2015; Chen et al., 2018). Doudna and his co-workers used Cas12 protein and designed a new CRISPR-Cas-based tool called the DETECTR (DNA endonuclease–targeted CRISPR trans-reporter). The DETECTR tool is used for the detection of different viruses, like the human papillomavirus (Chen et al., 2023), African swine fever virus (He et al., 2020), and COVID-19 virus (Broughton et al., 2020).

1.4.4 Cas13 Protein

The Cas13 protein is a recently discovered Cas protein. In archaea and bacteria, Cas13 is used in adaptive immunity to defend against RNA viruses (Abudayyeh et al., 2017). Cas13 protein is classified into two subtypes: (1) *Lsh*Cas13a (C2c2) protein, which belongs to type VI and is isolated from the *Leptotrichia shahii* bacteria, and (ii) *Psp*Cas13b protein, which belongs to type III and is an isolated form of the *Prevotella* sp. This Cas13 protein cleaves the ssRNA only, but not the DNA (Liu et al., 2017).

Cas13 is like the Cas12 protein, as it is also activated by crRNA only. Cas13a contains one nuclease lobe domain and two HEPN domains to degrade their target ssRNA molecule (Gootenberg et al., 2018). In the Cas13a protein, the target sequence is at the 3' end, flanked with protospacer flanking sequence (PFS). In the Cas13b protein, the target sequence is at the 5' end, flanked with PFS and PAM at the 3' end (Konermann et al., 2018). The activated Cas13b protein degrades the non-specific ssRNA molecule (Smargon et al., 2017).

Cas13 has broad applications in the field of genome editing and diagnostics, and it is also used in the single-nucleotide deletion process (Zetsche et al., 2015). Like with DETECTR, Kellner and his colleagues used the Cas13 protein to develop a new tool, called SHERLOCK (specific high-sensitivity enzymatic reporter unlocking) (Kellner et al., 2019). This tool, along with qPCR, is used in the detection of RNA and DNA/RNA contamination (Kellner et al., 2019). SHERLOCK is also used in the differentiation of African/American Zika virus strains and Dengue virus serotyping.

1.4.5 Cas14 Protein

Doudna's group discovered the Cas14 protein, which is similar to the proteins Cas9, Cas12, and Cas13. Due to the small size (40–70 kDa) of Cas14, they demonstrated that without PAM, it can degrade ssDNA (Burstein et al., 2017). The Cas14 protein can develop immunity by degrading the ssDNA, but not dsDNA or ssRNA (Burstein et al., 2017). For the degradation process, Cas14 contains both crRNA and tra-crRNA, like that of the Cas9 protein. During the degradation process, the active Cas14 protein mediates base pairing between the crRNA and the target DNA and then cleaves the ssDNA without the presence of PAM (Harrington et al., 2018). Cas14 being superior can be combined with the DETECTR tool for high fidelity. The resulting technique is used for single-nucleotide polymorphism, virus phylogenetic association, epidemiology, and taxonomic studies (Harrington et al., 2018).

1.5 EFFECTOR COMPLEX TYPES AND THEIR FUNCTIONS

Within CRISPR-Cas systems, effector complexes play a crucial role in detecting and targeting foreign genetic entities, such as DNA and RNA, by employing crRNAs as their guiding components. The classification of these effector complexes into two categories is determined by the specific stage of target interference in the CRISPR-Cas mechanism, as detailed in the following sections.

1.5.1 Class 1 Effector Complexes

The class 1 effector complex is made up of a cascade of multiple Cas proteins and is used for the interference stage in the class 1 system. The subtype I-E effector complex of *E. coli* is well studied among all type I subtypes. The studied structure of subtype I-E consists of different Cas proteins with different functions and forms a complex with 450 KDa molecular weight: $(Cas5e)_1$-$(Cas6e)_1$-$(Cas7e)_6$-$(Cas8e)_1$-$(Cas11e)_2$. In this complex, the six Cas7e proteins form a helical backbone with the gRNA capped with the one Cas5e protein. After the formation of the backbone, the two smaller Cas11e proteins interact with Cas7e, followed by the larger Cas8e proteins interacting with the complex and forming the complex tail. In this complex, the Cas8e protein is responsible for PAM site recognition and the unwinding of the target nucleic acid (DNA). Finally, conformational change is induced in the Cas3 nuclease protein, resulting in its activation and the degradation of the target nucleic acid. The composition of cascade complexes involved in type I systems is similar in all, except in some subtypes (Makarova et al., 2015). The Csm and Cmr cascade effector complexes involved in subtypes III-A and III-B, respectively, are similar to the type I cascade complex, but they degrade both DNA and as well as RNA molecules (Makarova et al., 2017).

1.5.2 Class 2 Effector Complexes

A single Cas protein forms class 2 effector complex multiple domains and is used in the interference stage in class 2 systems. Among all, the Cas9 effector of the type II system is well characterized and functions via a single multifunctional Cas9 protein with two RNA molecules (crRNA and tra-crRNA). During the cleavage process, two different interactions occur among the effector complex and the target DNA. One interaction occurs between the spacer sequence of crRNA, which is complementary to the target nucleic acid. The PI domain facilitates additional interactions, complementing the region upstream of the PAM sequence. In the Cas9 effector complex, the two HNH and RuvC nuclease domains cleave the target and non-target strands of the foreign DNA, respectively, and hence make the dsDNA break (Anders et al., 2014). In types V and VI, Cas12 and Cas13 are used to cleave the target DNA and RNA, respectively, in a crRNA-dependent manner without a PAM site, but some subtypes contain PFS site (Zetsche et al., 2015; Abudayyeh et al., 2017).

1.6 HISTORICAL PERSPECTIVE OF CRISPR-CAS SYSTEMS

The CRISPR sequence was first determined by a group of researchers during the determination of the alkaline phosphate gene in *E. coli* (Ishino et al., 1987). Later on, in 1989, another sequence was discovered in the K-12 chromosome of other bacterial (*E. coli*) strains, like *Shigella dysenteriae* and *Salmonella enterica* (Nakata et al., 1989). But with subsequent research, the CRISPR sequence was also found in archaea, like *Haloferax mediterranei* and *Streptococcus thermopile*, but 90% in the bacterial genome and 40% in the archaeal genome (Mojica et al., 1993). The CRISPR-associated genes were discovered by Jansen and his co-workers in 2002 (Jansen et al., 2002). Later on, in 2005, the spacer sequences were found in genomes, and the CRISPR array contains unique spacer sequence regions (Haft et al., 2005). The CRISPR system's historical perspectives are divided into three phases.

1.6.1 Phase of Identification

Like many other novel discoveries, the CRISPR-Cas discovery was born out of an unexpected way. In 1987, Ishino and co-workers determined an unusual DNA sequence in the alkaline phosphatase coding gene in *E. coli* (Ishino et al., 1987). This DNA sequence consists of five highly homologous repeated sequences of size ~29 nts, and later known as direct repeat (DR). These DRs are separated by about 32 nt sequences and hence are called as spacer sequences (Ishino et al., 1987; Nakata et al.,

1989). The DR also contains a short sequence similar to the former and is called an inverted repeat (IR) (Mojica et al., 2000).

1.6.2 Phase of Functional and Structural Characterization

During this phase, the two independent research works linked the spacer sequence with bacteriophage sequences (Mojica et al., 2005). In 2002, scientists and their co-workers designated this highly homologous repeated DNA sequence as CRISPR, located in the DNA repair system (Jansen et al., 2002; Makarova et al., 2002). Hence, earlier, these genes were considered part of the DNA repair system, and the genes associated with CRISPR were called Cas genes. Later on, in 2006, scientists found the relationship between Cas genes and CRISPR functions and determined both of them as a single system, now called the CRISPR-Cas system. Researchers also found that this system shows similarity with the RNA interference system of the prokaryotic immune system (Makarova et al., 2002; Charpentier et al., 2015). During this period, CRISPR-Cas system reprogramming and transfer into other organisms were also done, like the transfer of the Cas9 system from *Streptococcus thermophilus* to *E. coli* to develop immunity against foreign particles, including viruses (Sapranauskas et al., 2011).

1.6.3 Phase of Applications

After the determination of similar sequences separated by repeated regular sequences by scientists in bacteria and archaea, they now focus on their biological functions. Thus, the CRISPR-Cas system continuously developed year after year across research labs worldwide. Therefore, the use of the CRISPR-Cas system in the eukaryotic gene-editing process to modify eukaryotes' genome leads to the development of various CRISPR-Cas applications in numerous biological fields, including agriculture, animal husbandry, medicine, and biotechnology. During this period, scientists also purified the Cas9 protein from bacteria (*Streptococcus thermophilus* and *Streptococcus pyogenes*) and degraded bacterial DNA in *in vitro* conditions (Deltcheva et al., 2011; Jinek et al., 2012).

1.7 CRISPR-CAS SYSTEM DEVELOPMENT AND DISCOVERY

The CRISPR-Cas array has been studied extensively throughout the world in the last three decades, until it received full research attention in different biological fields. The CRISPR-Cas-mediated gene editing is currently a hot research topic of molecular biological research scientists. In 2020, Emmanuelle and Jennifer used CRISPR-Cas methodology for the first time in the process of gene editing and won the Nobel prize for the same in 2020.

1.7.1 Early Detection of Repeated Intervals as a Special Sequence

Like many other novel discoveries, the CRISPR-Cas discovery was born out of an unexpected way. Briefly, Ishino and his co-workers determined an unusual DNA sequence at the 3' end of the alkaline phosphatase coding gene in *E. coli* (Ishino et al., 1987). This sequence consists of a highly repeated sequence, called a DR, separated by a sequence called the spacer sequence (Ishino et al., 1987; Nakata et al., 1989). In 2002, these repeated DNA sequences were termed CRISPRs (Jansen et al., 2002; Makarova et al., 2002), and the genes associated with CRISPR were referred to as Cas genes.

1.7.2 Use in Prokaryotes as a Defense System

Earlier, various studies showed that CRISPR-Cas array functions as a defense system in bacteria by providing viral resistance (Jinek et al., 2012; Bolotin et al., 2005). Later on, it was found that bacteria infected with viruses contain viral genome sequences integrated in their genome, which serve as spacer sequences in repeated infections (Barrangou et al., 2007). In 2005, researchers determined that the composition of spacer sequences is almost similar to exogenous foreign DNA sequences, with only a small difference. Scientists also found that new viruses that infect the host cells do not contain this homologous spacer sequence (Mojica et al., 2005). In its defense mechanism, the maturated crRNA recognizes the target DNA via a complementary base pairing (Brouns et al., 2008). After hybridization, the crRNA activates its associated Cas protein, which then degrades the target DNA sequence through its nuclease activity (Jinek et al., 2012).

1.7.3 Use in Eukaryotic Cells

Earlier, CRISPR-Cas system research was focused on prokaryotes, but then the system gained the interest of other fields, including human disease (Cong et al., 2013). Researchers tend to use this system in eukaryotic cells, like in the integration of tr-crRNA, pre-crRNA, host-specific RNase-III, and Cas9, along with their promoters in human 293T cell line from the *streptococcus pyogenes*, to target the empty spiracle homeobox 1 (EMX 1) of humans (Sapranauskas et al., 2011; Jinek et al., 2012). These studies suggest that the CRISPR system can work in eukaryotes, thereby providing more possibility for the use of the CRISPR-Cas system in eukaryotic cells, including humans.

1.7.4 First Use in Disease Treatment

After the discovery of the usage of the CRISPR system in human 293T cells, scientists have used this system for gene-editing processes in eukaryotic organisms, including both animals (mice, rat, fruit fly) and plants (rice and wheat) (Chen et al., 2019; Bassett and Liu, 2014; Wang et al., 2013; Ma et al., 2014; Jiang et al., 2013). In 2013, this system was also used to treat eukaryotic diseases, like the Cas9 system being used to treat cataract in mice caused by the deletion mutation (Wu et al., 2013; Zhao et al., 2010). During the study, they co-transferred the Cas9 mRNA along with sgRNA into mice-fertilized eggs that have cataract; from the results, they found that the cataract of some mice had been successfully cured (Wu et al., 2015; Fu et al., 2021). Cystic fibrosis (CF) caused by a mutation in the cystic fibrosis transmembrane conductor receptor (CFTR) has been treated using CRISPR-Cas9 system on a patient's intestinal stem cells (Maule et al., 2021).

1.7.5 First CRISPR-Cas Clinical Trial

In 2016, the first CRISPR-Cas9 clinical trial was conducted on humans in China. During this trial, scientists injected CRISPR-Cas9-edited T cells into selected patients to cause editing in the programmed cell death-1 (PD-1) gene (Lu et al., 2020). For this process, scientists first derived T cells from the patients and transferred the Cas9- and sgRNA-encoding plasmids into these cells through the electroporation process. From the results, they found that no adverse effect was developed in the patient, and there was a decrease in PD-1 gene expression in edited T cells compared to normal (Su et al., 2017).

1.8 CRISPR-CAS9 DELIVERY APPROACHES AND CHALLENGES

In CRISPR-Cas-based gene therapy, for precise and specific targeting, the main challenge is how to transfer CRISPR-Cas components into target cells, tissues, and organs efficiently. Among these, the delivery methods should also be non-toxic, highly efficient, low-cost, and biodegradable, with specific targeting properties (Lee and Kim, 2019). Till now, CRISPR components are delivered through physical, biological (viral), and chemical (non-viral) methods (Xu et al., 2019). Among the physical methods, microinjection and electroporation are the common approaches to transfer CRISPR-Cas elements in *in vitro* experiments because they are highly effective and are easy to use (Kotterman et al., 2015), whereas in viral vector methods, the lentivirus, adenovirus, and adeno-associated virus (AAVs) are commonly used and have great delivery efficiency, thus are used for both *in vitro/ ex vivo* and *in vivo* experiments. Nevertheless, there is a significant issue with the safety of viral transfer methods which must be resolved. As a result, studies are focusing on non-viral vector methods, like polymers, liposomes, and nanoparticles, for CRISPR-Cas system delivery (Mout et al., 2017); thus, it becomes a hotspot for the delivery of CRISPR components because of their high safety, accessibility, and affordability (Yin et al., 2016).

1.8.1 Delivery Approaches

For CRISPR-Cas delivery, plasmid vectors are the ideal ones because they cannot be easily degraded but can be easily modified and amplified (Lin et al., 2022). With the nuclear localization signal (NLS) present in the CRISPR plasmid, the plasmid can easily enter the target cell's nucleus and form the Cas9 mRNA and sgRNA (Liu et al., 2019). The various delivery vehicles used for the CRISPR system delivery are described in what follows.

1.8.1.1 Viral Vectors

Viral carrier approaches are the common systems used for gene transport in the biological system. Among viral vectors, the AAVs are mostly used because of their low host immunogenicity and because they can easily infect the target cell (Wang et al., 2020). As the CRISPR-Cas locus is more complex in nature as compared to other transferred genes, it exceeds the maximum AAV packaging capacity by about 4.7 kb (Sonntag et al., 2010). Therefore, to increase the AAV capacity of the CRISPR delivery system, some modifications need to be carried out in AAV vectors—for example, using smaller SaCas9 (smaller Cas9 ortholog) or using two vectors instead of one (Chew et al., 2016; Bohm et al., 2020). Himeda and colleagues developed the CRISPRi system, containing dead SaCas9, and successfully decreased DUX4 expression in *in vitro* conditions (Himeda et al., 2021). In the two-vector delivery method, two separately designed plasmids—one encoding Cas9, and the other for sgRNA—are used. In 2021, Nguyen and co-workers engineered the baculovirus (non-pathogenic insect virus) as a dCas9-VP64-p65-Rta (dCas9/VPR) delivery vector to induce the long non-coding RNA–mediated differentiation process in both bone marrow–derived mesenchymal and rat adipose–derived stem cells (Nguyen et al., 2021).

1.8.1.2 Lipid-Based Nanocarriers

As lipid-based carriers (liposomes) are used in the delivery of specific genes into the target cell, lipid nanoparticles (LNPs) can also be used for CRISPR-Cas delivery within the target cell or tissue (Behr et al., 2021). The LNPs consist of phospholipids, cholesterol, and polyethylene glycol but do not have a continuous bilayer of lipids and aqueous inner core as that of liposomes (Chen et al., 2020). Thus, due to their simple chemical composition and their presence in the blood, they can be used as gene carrier in *in vivo* conditions. But unfortunately, they are present in high concentration in the liver with respect to other organs and, therefore, do not show significant effects in other organs (Witzigmann et al., 2020). Hence, LNPs as a delivery system require some modification so that they can transfer the CRISPR system into extrahepatic organs.

1.8.1.3 Polymer-Based Nanoparticles

Polymer-based nanoparticles act as an efficient carrier system because of their good stability, good biocompatibility, low immunogenicity, and high modification potential (Mitchell et al., 2021). Liu and his co-workers designed a multistage delivery nanoparticle (MDNP) with a core-shell structure for CRISPR-Cas9 delivery (Liu et al., 2019). Like with the delivery of CRISPR-Cas9 into rat neurons to promote neurite growth, a nanofiber coated with self-assembled peptide (SAP) was used (Zhang et al., 2020). The polycaprolactone (PCL) carrier is also used for CRISPR-dCas9 transfer into the target cell (Tamay et al., 2019). The dual-locking nanoparticles (DLNP) are another polymer-based nanoparticle used for the transfer of the CRISPR-Cas13 system into the tumor cell while targeting the PD-L1 gene (Zhang et al., 2019).

1.8.1.4 Exosomes

The exosomes in living organisms act as intracellular carrier of cellular components, as they are membrane-bound vesicles originating from cell organelles (Li et al., 2018). The exosomes' chemical composition is similar to that of the plasma membrane, and therefore, they cannot be cleared by the immune system after transport. Thus, they can be used for the transfer of Cas9 components within the target cell. During the transport of Cas components, ribonucleoproteins (RNPs) are packed within the exosome vesicles through the electroporation method and form exosome-RNP vesicles. Wan and his co-workers, while working on liver disease, formed the sgRNA and Cas9-exosome vesicle to target different genes like cyclin E1, lysine acetyltransferase, and apoptosis modulators (Kwan et al., 2020). To increase the delivery efficiency of exosome-based methods in *in vivo* conditions, the genome editing with extracellular vesicle (GEDEX) method was formed for the delivery of dCas9/VPR (Lainscek et al., 2018).

1.8.1.5 Biomimetic Nanomaterials

Biomimetic nanomaterials are specifically crafted to emulate the structures, functions, or properties inherent in natural biological systems at the nanoscale. By harnessing the efficiency and complexity observed in nature, these materials apply the concepts of biomimicry to formulate advanced solutions for diverse applications, such as in medicine, energy, and materials science. The ability of nanomaterial-based delivery systems to accumulate at the target site, and their stability within the blood, can increase CRISPR-Cas genome editing efficiency (Luk and Zhang, 2015). But in *in vivo* conditions, the designated carrier molecules (organic and inorganic) can be easily cleared by the target's immune cells. Therefore, based on this, scientists focus on the material from the target cell itself to carry the genes and call it a biofilm (Bose et al., 2018).

1.8.2 Delivery Challenges

The CRISPR-Cas plasmid is transfected into the target cell in the same manner as the other gene-specific plasmids are transfected, using different transfecting reagents. But most of the CRISPR-Cas delivery methods are less suitable in the animal cell, including humans, because CRISPR-Cas delivery tools undergo a long delivery process and therefore face three major challenges in *in vivo* conditions, as discussed in the following:

1. The delivery system should remain stable within the target cell and without undergoing degradation or immune clearance.
2. The delivery system should accumulate at the specific target site.
3. The delivery system should perform specific genome editing or gene expression processes.

1.8.2.1 Immunogenicity Development

The immune response toward the Cas gene and the Cas protein is one of the most common obstacles in CRISPR-Cas delivery methods, because these genes are obtained from bacteria and can generate the host's immune response once transferred (Chew et al., 2016). It has been found that delivered Cas components produce immune response toward the Cas protein in *in vivo* conditions (Chew et al., 2016). Researchers also discovered anti-Cas9 T cells and antibodies in healthy humans, which suggests that human humoral and cellular immune responses already existed toward the Cas9 protein (Wagner et al., 2019). Therefore, for the clinical implementation of Cas systems, a decreased Cas protein immunogenicity should be figured out.

1.8.2.2 Cancer Development Risk

The lentiviruses' delivery methods integrate randomly into the host genome and may trigger immune responses and transformation processes (cancer development) into the host cell (Behr et al., 2021). This shows that during clinical implementation, CRISPR-Cas systems may give rise to the development of cancer, since genetically modified cells (gene editing) are likely to get converted into cancer-initiating cells.

1.8.2.3 Cell Growth and Cell Death Risk

The P53 gene is involved in cell growth and death (apoptosis). A study shows that the p53 pathway can be activated by CRISPR-Cas-driven double-strand DNA break (Haapaniemi et al., 2018). These studies give a warning that CRISPR-mediated genome editing may inhibit cell growth or initiate cell death. Thus, in this regard, more work is needed to be done on CRISPR-Cas systems.

1.8.2.4 Ethical Issues

CRISPR-mediated gene editing has gained a lot of interest because of its ability to alter any specific gene, including human germ cell line genes (Coller, 2019). However, germ cell gene editing may have unforeseen repercussions and raise ethical and safety issues, and it is prohibited in various nations (Gabel and Moreno, 2019). However, the ethical debate surrounding the editing of human embryos reached a peak when Chinese scientists published a work in 2015 regarding gene editing in human tri-pronuclear zygotes (Liang et al., 2015). Since then, morality and ethics have questioned genome editing, and the legalization of the practice has sparked contentious debates across the globe. A Chinese scientist in 2018 reported the birth of two gene-edited healthy babies, and this was the day before the second international human genome editing summit opened. These newborn babies are the first in history to have the CCR5 edited gene, which makes them immune to HIV from birth (Ormond et al., 2019). This announcement shocked global experts and surprised the general public, since it involved the modification of human embryonic gene for procreation, which raised ethical and security concerns and sparked some backlash from the scientific community. To guarantee that genome editing of human embryos proceeds in a safe and morally responsible manner, a specific global supervisory mechanism is also needed (Doxzen et al., 2020).

1.8.2.5 Virus Safety Issues

Viral vectors like the lentivirus, the adenovirus, and AAVs are also used for CRISPR-Cas system delivery because of their great delivery efficiency. Hence, they are frequently used to transport specific genes during gene therapy, but only a few of them have received approval for use in clinical trials (Liu et al., 2017), because there is still a major issue with viral pathogenicity (safety) in the host cell that needs to be resolved in the preclinical trials of CRISPR-Cas delivery.

1.9 MECHANISMS OF TARGET RECOGNITION AND DNA CLEAVAGE

1.9.1 Target Recognition

Prokaryotes are encountered by beneficial and non-beneficial nucleic acid (Thomas and Nielsen, 2005). To deal with these foreign nucleic acids, prokaryotic (bacterial) cells have developed various mechanisms, including the CRISPR-Cas system (Barrangou et al., 2007; Marraffini and Sontheimer, 2008). The Cas system can heritably alter the foreign nucleic acid by making crRNA, which specifically activates the Cas protein with nuclease activity (Jackson et al., 2017). In the CRISPR-Cas system, the mature crRNA defines the specificity, and its formation and maturation occur through different stages.

1.9.1.1 Recognition of Target Genome

In CRISPR-Cas systems, fragments of foreign DNA called spacers are incorporated into the target genome in between the CRISPR genes (Barrangou et al., 2007); the process is known as adaptation (Jackson et al., 2017). These small DNA fragments are then transcribed into individual crRNAs, followed by processing to form mature crRNA. But during this process, the Cas system can also acquire self-targeting spacers (Yosef et al., 2012; Levy et al., 2015), which develop autoimmunity (autoimmune diseases) and, finally, cell death (Bikard et al., 2012; Jiang et al., 2013). Thus, to avoid the self-destruction caused by CRISPR systems, different mechanisms are used to bias spacer acquisition with the target DNA, therefore differentiating the self from the nonself target molecules (Modell et al., 2017; Levy et al., 2015), like using the host's DNA repair system to produce the spacers—for example, gram-negative bacteria use the RecBCD repair system to generate the spacers (Levy et al., 2015). The RecBCD repair system binds to the free ends of target DNA and stimulates spacer acquisition from dsDNA breaks. The activity of the RecBCD system is slowed by *chi* sites (8 nts), and the host genome is rich in these sites with respect to the target genomes. Furthermore, bacterial DNA is mostly circular in nature, lacks free ends, and hence, cannot be used by the Cas protein, while the free phage dsDNA ends during the infection is exploited by the Cas system to form the spacers only from the phage DNA. The resulting spacer acquisition then recognizes and cleaves the foreign DNA only and develops an effective immunity (Modell et al., 2017). Moreover, spacer formation can also occur in the absence of the DNA repair pathway through an alternative pathway (Levy et al., 2015).

1.9.1.2 Site-Specific Integration of New Spacers

The integrase complex (Casl-Cas2) is used for the site-specific integration of new spacers into the leader sequence of the CRISPR array. During the integration process, Casl interacts with the Cas2 protein and forms a hetero-hexameric complex [$(Casl)_2$–$(Cas2)_2$], which acts as an integrase enzyme (Nunez et al., 2014, 2015; Wang et al., 2015; Xiao et al., 2017). The Casl-Cas2 complex contains two binding sites; one site binds with the CRISPR locus, and the other site binds with the new spacer/

protospacer. After new spacer loading, the hexameric complex catalyzes two cleavage-ligation reactions for spacer integration within the host genome. The first reaction occurs at the leader end of the first repeat of the CRISPR array, and the second reaction occurs at the spacer end of the repeat (Xiao et al., 2017; Wright and Doudna, 2016). During cleavage reaction, the 3'-OH group of the protospacer acts as a nucleophile and attacks the target DNA. After cleavage, the free 3'- ends of the protospacer DNA are ligated with the target DNA repeats, and the ssDNA gaps are filled by the DNA polymerase enzyme, which further causes duplication of DNA repeat (Xiao et al., 2017; Nunez et al., 2015; Yosef et al., 2012).

1.9.1.3 Functional Target Selection

After integration, only a small sequence in the target DNA can become a functional spacer element due to target specificity. The PAM site of types I and II CRISPR array in the target genome acts as a license for target specificity and prevents spacer degradation (Mojica et al., 2009). The Cas1-Cas2 complex of type I has direct and site-specific interactions with PAM (Wang et al., 2015), while in type II systems, the hexameric complex does not have direct PAM selection but causes PAM selection through the Cas9-PI domain (Xiao et al., 2017), although in the CRISPR-Cas type III system, flexibility of the flanking sequence and target transcription is required for target DNA specificity (Marraffini and Sontheimer, 2010). In other type I subtypes, the Cas4 protein is involved in biasing the spacer and the selection of functional targets (Shiimori et al., 2018). Thus, to obtain target specificity, the Cas system must select a spacer with a specific PAM site or a transcriptional active spacer.

1.9.2 DNA Cleavage

Prokaryotes (bacteria and archaea) are encountered by beneficial or non-beneficial exogenous nucleic acids (Thomas and Nielsen, 2005). To deal with these foreign nucleic acids (DNA), prokaryotic cells have developed various systems, including the CRISPR-Cas system (Barrangou et al., 2007; Marraffini and Sontheimer, 2008). In CRISPR-Cas systems, when bacteria are encountered with foreign DNA, it gets integrated into the spacer regions (Bolotin et al., 2005). The next time the bacteria are invaded by homologous DNA, the spacer regions activate the Cas system to produce crRNA. After maturation, the crRNA activates the Cas proteins to form single- or multiple-protein effector complexes, which degrade the invading nucleic acid (DNA). The multi-protein cascade effector complex is involved in class 1 systems, and single-protein effector is used in class 2 systems (Zetsche et al., 2015; Jinek et al., 2012; Marraffini and Sontheimer, 2010; Wang et al., 2011). The primary functions of both effector complexes are to recognize the target nucleic acid and then cleave it with the help of Cas3 (type I) or Cas9 (types II and V) (Makarova et al., 2020). But during this process, the CRISPR-Cas system can also acquire self-targeting spacers; to avoid self-destruction, the PAM site is used, which is present upstream (types I and V) or downstream (type II) of the specific spacer (Barrangou et al., 2007; Fonfara et al., 2016; Zetsche et al., 2015; Jiang et al., 2013). This PAM site is essential for Cas protein binding and followed by the specific target nucleic acid recognition and degradation. The PAM selection occurs through the hexameric complex in type I, but in type II systems, the PAM selection occurs through the Cas9-PI domain (Xiao et al., 2017).

In the CRISPR-Cas9 system, pre-crRNA is formed, and it contains the complementary sequence to the tra-crRNA (Liao and Beisel, 2021). The tra-crRNA is involved in the processing of pre-crRNAs into mature crRNA. During crRNA processing, the anti-repeat sequence of tra-crRNA hybridizes with the repeat sequence of pre-crRNA and forms a double-stranded RNA (tra-crRNA:crRNA). The formed double stranded RNA is then processed by the host's RNase-III and forms an intermediate crRNA, which further undergoes processing by an unknown mechanism and yields mature crRNA (Deltcheva et al., 2011). Moreover, some intermediate crRNAs can further undergo nucleotide (nt) trimming at their 5' or 3' end. The trimming of intermediate crRNAs is done by an unknown nuclease that yields the mature crRNA. The mature crRNA is composed of 5'-sequence (spacer sequence) and 3'-sequence (repeat sequence), which usually form a hairpin loop–like structure in most type I systems (Charpentier et al., 2015). After the maturation of crRNA, which is a primary process, the mature crRNA gains the ability to activate the specific Cas protein in a secondary process. The activated Cas protein, with its nuclease activity, then cleaves the repetitive and spacer sequence of the target nucleic acid (Brouns et al., 2008). During both the primary and secondary processes, the crRNA and the tra-crRNA remain bound together at a specific spacer region and form the sgRNA (Jiang and Doudna, 2017; Jinek et al., 2012). Normally, the Cas proteins remain inactive in the cell; the formation of sgRNA results in the conformational change in the Cas protein, which hence gets activated. Like during Cas9 protein activation, the designed sgRNA forms a T-like structure called a tetraloop or stem-loop (Jinek et al., 2012), and the 5' end of sgRNA is complementary to the target nucleic acid. The engineered sgRNA binds to the Cas9 and introduces conformational change, activating the Cas9 protein. The activated Cas protein then searches the target site by interacting with the sequence complementary to the PAM site (5'-NGG-3'). After binding to the target site, the active Cas9 protein uses its two HNH and RuvC nuclease domains to make a double-strand break in the target DNA about 3 bp upstream of the PAM site. In Cas9 protein, the HNH domain cleaves the target strand, which is complementary to the crRNA, while the RuvC domain cleaves the non-target strand, which is non-complementary to the crRNA (Anders et al., 2014; Jinek et al., 2012).

The most studied class 1 effector complex (subtype I-E) consists of different Cas proteins: $(Cas5e)_1$-$(Cas6e)_1$-$(Cas7e)_6$-$(Cas8e)_1$-$(Cas11e)_2$. In this complex, the Cas7e proteins form a helical backbone with the gRNA capped with Cas5e protein. After the formation of the backbone, the smaller Cas11e proteins interact with Cas7e, followed by the larger Cas8e proteins interacting with the complex. In this complex, Cas8e protein is responsible for PAM site recognition and the unwinding of the target nucleic acid (DNA). Finally, a conformational change is induced in the Cas3 nuclease protein, resulting in its activation and the degradation of the target nucleic acid. The Csm and Cmr cascade effector complexes involved in subtypes III-A and III-B, respectively, are similar to the type I cascade complex, but they degrade both DNA and as well as RNA molecules (Makarova et al., 2017).

1.10 HOW THE CRISPR-CAS9 SYSTEM REVOLUTIONIZED GENE EDITING

The CRISPR-Cas9 system mainly consists of a gRNA (20–30 nts), which hybridizes to the target nucleic acid (DNA), and Cas9 protein, which acts as a DNA endonuclease. The gRNA, also known as crRNA, guides the Cas9 protein to initiate the double-strand break in the foreign DNA. The broken DNA then initiates multiple changes at the cleavage site with the help of cellular repair systems, using either HDR or NHEJ pathways (Fu et al., 2021). The NHEJ repair pathway is used mostly, and it facilitates the repair, followed by the addition or deletion of some nts (gene editing). In the beginning, the gene-editing functions of the CRISPR system remain hidden, as various studies show that the Cas system is involved in bacterial immune system development against viruses (Bolotin et al., 2005). But after the discovery of the transfer of the CRISPR-Cas system into the eukaryotic cells, a revolution has been found in the field of molecular biology and biotechnology, and now it has become the leading technique in these fields (Rao et al., 2022). Because the CRISPR-Cas system is fast, cheap, and has extraordinary precision and versatility, it is a major breakthrough in the molecular science field because it performs better in terms of cost-efficiency and is relatively easy to use than previous genome editing methods, like ZFNs and TALENs. But the major difference between ZFN/TALEN and CRISPR-Cas system–mediated genome editing is that the Cas system mediates RNA-dependent target DNA recognition, followed by its degradation (Gaj et al., 2013). The CRISPR-Cas method also mediates genome editing in the eukaryotic cells, including both animals and plants. After the discovery of the CRISPR-Cas9 system, scientists also determined different Cas9 variants with various functions, including gene editing. For instigating gene editing through the Cas9 system, the system must be delivered safely into the target cell. Hence, researchers throughout the world developed various delivery vectors for the transport of the Cas9 system into the target cell. Like the CRISPR-LNPs (sgPLK1-cLNPs), carriers are used in the tumor cell to determine the gene-editing efficiency in *in vivo* conditions (Rosenblum et al., 2020). Furthermore, to facilitate the Cas9 genome editing process in eukaryotic cells, crRNA:tra-crRNA (sgRNA) and Cas9 transcripts are required. Scientists designed the transcripts of Cas9 system components to facilitate the genome editing process in the target cell (Su et al., 2017). Among these components, sgRNA construction is the key factor because it contains the specific target sequence required for the editing process. In 2016, the first CRISPR-Cas9 gene-edited T cells were injected into selected patients to cause editing in the PD-1 gene (Lu et al., 2020). During this process, scientists first derive T cells from the patients and transfer the Cas9 and the sgRNA encoding plasmids into these cells through the electroporation process. From the findings, no adverse effect was developed, and a decrease in PD-1 gene expression was reported (Su et al., 2017). After the successful construction of Cas9 components, the CRISPR-Cas9 genome editing technique holds great promise in the field of food, health care, agriculture, and medicine, for their improvement, like how researchers have achieved a great gene-editing yield in both animals (mice, fruit flies, rats) and plants (rice and wheat) while using the CRISPR-Cas9 method (Chen et al., 2019; Bassett and Liu, 2014; Wang et al., 2013; Jiang et al., 2013). Among all these, genetic disease treatment while using the CRISPR-Cas system was the greatest outcome. For example, in 2013 and 2015, the Cas9 system was used to treat cataract in mice caused by the deletion and spermatogonia stem cell gene editing in mice, respectively (Wu et al., 2013; Zhao et al., 2010; Wu et al., 2015). Research has also expanded this system to other animal models, where it has a wide range of therapeutic effects, such as the inhibition of viral infection; the reversal of crippling ailments, such as muscular dystrophy; and tumor removal.

1.11 ADVANTAGES OF CRISPR-CAS SYSTEMS

In comparison to traditional editing approaches, CRISPR-Cas systems offer numerous advantages, which include the following:

1. **Efficiency.** With its high efficiency in producing genetic edits, CRISPR-Cas is effective for knockouts, insertions, and point mutations. The system's customizable nature allows for the targeting of multiple genes at the same time and can be directly applied to embryos.
2. **Simplicity.** CRISPR distinguishes itself from older editing techniques, such as TALENs and ZFNs, by employing a single protein, like Cas9, in conjunction with a guide RNA, thereby facilitating a more straightforward design and implementation process. The design of a guide RNA for a specific target sequence is uncomplicated and can be customized with bioinformatics tools.
3. **Precision.** CRISPR-Cas systems are highly accurate tools for DNA modification, capable of specifically targeting individual genes without inducing other unintended effects.
4. **Cost-effective development.** The creation and application of CRISPR tools are significantly less expensive than earlier gene-editing approaches.
5. **Versatility.** The CRISPR mechanism can be utilized to edit, activate, or repress genes, as well as to modify DNA or RNA in a diverse array of organisms.
6. **Flexibility in delivery.** Available components include DNA, RNA, or ribonucleoprotein complexes, all of which are compatible with various delivery systems, whether viral or non-viral.
7. **Transient expression.** The transient delivery of Cas proteins and gRNAs minimizes off-target effects and mitigates immune responses.

1.12 CONCLUSION

The CRISPR-Cas system signifies a revolutionary advancement in the fields of molecular biology and biotechnology, functioning as exact "molecular scissors" that have transformed the landscape of gene editing. Initially developed as a defense mechanism in bacteria, this system has been repurposed to allow for targeted alterations in various genomes with unmatched precision. Its uses span from fundamental research to prospective treatments for genetic diseases, improvements in agriculture, and cutting-edge therapies. There is a pressing need to implement the CRISPR-Cas technology across a range of platforms, involving specialists from the fields of ethics, society, religion, law, and technology, to create a permanent policy addressing the advantages and drawbacks of the technology. Notwithstanding the many social and legal obstacles that CRISPR genome editing must overcome, its potential uses are too great to be overlooked. CRISPR-Cas offers new avenues for treatment and other scientific applications, and it can greatly improve our understanding of illnesses at the genetic level. Numerous ethical and societal concerns are raised by CRISPR-Cas9, affecting the environment as well as the people. It is also necessary to carry out risk assessments for ecological and environmental issues.

REFERENCES

Abudayyeh, Omar O., Jonathan S. Gootenberg, Patrick Essletzbichler, Shuo Han, Julia Joung, Joseph J. Belanto, Vanessa Verdine et al. "RNA targeting with CRISPR–Cas13." *Nature* 550, no. 7675 (2017): 280–284.

Alkhnbashi, Omer S., Shiraz A. Shah, Roger A. Garrett, Sita J. Saunders, Fabrizio Costa, and Rolf Backofen. "Characterizing leader sequences of CRISPR loci." *Bioinformatics* 32, no. 17 (2016): i576–i585.

Anders, Carolin, Ole Niewoehner, Alessia Duerst, and Martin Jinek. "Structural basis of PAM-dependent target DNA recognition by the Cas9 endonuclease." *Nature* 513, no. 7519 (2014): 569–573.

Babu, Mohan, Natalia Beloglazova, Robert Flick, Chris Graham, Tatiana Skarina, Boguslaw Nocek, Alla Gagarinova et al. "A dual function of the CRISPR–Cas system in bacterial antivirus immunity and DNA repair." *Molecular Microbiology* 79, no. 2 (2011): 484–502.

Barrangou, Rodolphe, Christophe Fremaux, Hélène Deveau, Melissa Richards, Patrick Boyaval, Sylvain Moineau, Dennis A. Romero, and Philippe Horvath. "CRISPR provides acquired resistance against viruses in prokaryotes." *Science* 315, no. 5819 (2007): 1709–1712.

Bassett, Andrew R., and Ji-Long Liu. "CRISPR/Cas9 and genome editing in Drosophila." *Journal of Genetics and Genomics* 41, no. 1 (2014): 7–19.

Behr, Matthew, Jing Zhou, Bing Xu, and Hongwei Zhang. "In vivo delivery of CRISPR-Cas9 therapeutics: Progress and challenges." *Acta Pharmaceutica Sinica B* 11, no. 8 (2021): 2150–2171.

Bikard, David, Asma Hatoum-Aslan, Daniel Mucida, and Luciano A. Marraffini. "CRISPR interference can prevent natural transformation and virulence acquisition during in vivo bacterial infection." *Cell Host & Microbe* 12, no. 2 (2012): 177–186.

Bohm, Sybille, Victoria Splith, Lisa Maria Riedmayr, René Dominik Rötzer, Gilles Gasparoni, Karl J. V. Nordström, Johanna Elisabeth Wagner et al. "A gene therapy for inherited blindness using dCas9-VPR–mediated transcriptional activation." *Science Advances* 6, no. 34 (2020): eaba5614.

Bolotin, Alexander, Benoit Quinquis, Alexei Sorokin, and S. Dusko Ehrlich. "Clustered regularly interspaced short palindrome repeats (CRISPRs) have spacers of extrachromosomal origin." *Microbiology* 151, no. 8 (2005): 2551–2561.

Bose, Rajendran J. C., Ramasamy Paulmurugan, James Moon, Soo-Hong Lee, and Hansoo Park. "Cell membrane-coated nanocarriers: The emerging targeted delivery system for cancer theranostics." *Drug Discovery Today* 23, no. 4 (2018): 891–899.

Broughton, James P., Xianding Deng, Guixia Yu, Clare L. Fasching, Venice Servellita, Jasmeet Singh, Xin Miao et al. "CRISPR–Cas12-based detection of SARS-CoV−2." *Nature Biotechnology* 38, no. 7 (2020): 870–874.

Brouns, Stan J. J., Matthijs M. Jore, Magnus Lundgren, Edze R. Westra, Rik J. H. Slijkhuis, Ambrosius P. L. Snijders, Mark J. Dickman, Kira S. Makarova, Eugene V. Koonin, and John Van Der Oost. "Small CRISPR RNAs guide antiviral defense in prokaryotes." *Science* 321, no. 5891 (2008): 960–964.

Burstein, David, Lucas B. Harrington, Steven C. Strutt, Alexander J. Probst, Karthik Anantharaman, Brian C. Thomas, Jennifer A. Doudna, and Jillian F. Banfield. "New CRISPR–Cas systems from uncultivated microbes." *Nature* 542, no. 7640 (2017): 237–241.

Carte, Jason, Ruiying Wang, Hong Li, Rebecca M. Terns, and Michael P. Terns. "Cas6 is an endoribonuclease that generates guide RNAs for invader defense in prokaryotes." *Genes & Development* 22, no. 24 (2008): 3489–3496.

Charpentier, Emmanuelle, and Luciano A. Marraffini. "Harnessing CRISPR-Cas9 immunity for genetic engineering." *Current Opinion in Microbiology* 19 (2014): 114–119.

Charpentier, Emmanuelle, Hagen Richter, John van der Oost, and Malcolm F. White. "Biogenesis pathways of RNA guides in archaeal and bacterial CRISPR-Cas adaptive immunity." *FEMS Microbiology Reviews* 39, no. 3 (2015): 428–441.

Chen, Janice S., Enbo Ma, Lucas B. Harrington, Maria Da Costa, Xinran Tian, Joel M. Palefsky, and Jennifer A. Doudna. "CRISPR-Cas12a target binding unleashes indiscriminate single-stranded DNase activity." *Science* 360, no. 6387 (2018): 436–439.

Chen, Ye, Lirong Meng, Bin Lang, Lifang Li, Jiashuo Liu, Yuyan Wang, Zheying Huang, Xun Tian, Rui Tian, and Zheng Hu. "A Cas12a ortholog with distinct TTNA PAM enables sensitive detection of HPV16/18." *Cell Reports Methods* 3, no. 4 (2023): 1–9.

Chen, Fengqian, Martin Alphonse, and Qi Liu. "Strategies for nonviral nanoparticle-based delivery of CRISPR/Cas9 therapeutics." *Wiley Interdisciplinary Reviews: Nanomedicine and Nanobiotechnology* 12, no. 3 (2020): e1609.

Chen, Kunling, Yanpeng Wang, Rui Zhang, Huawei Zhang, and Caixia Gao. "CRISPR/Cas genome editing and precision plant breeding in agriculture." *Annual Review of Plant Biology* 70, no. 1 (2019): 667–697.

Chew, Wei Leong, Mohammadsharif Tabebordbar, Jason K. W. Cheng, Prashant Mali, Elizabeth Y. Wu, Alex H. M. Ng, Kexian Zhu, Amy J. Wagers, and George M. Church. "A multifunctional AAV–CRISPR–Cas9 and its host response." *Nature Methods* 13, no. 10 (2016): 868–874.

Coller, Barry S. "Ethics of human genome editing." *Annual Review of Medicine* 70, no. 1 (2019): 289–305.

Cong, Le, F. Ann Ran, David Cox, Shuailiang Lin, Robert Barretto, Naomi Habib, Patrick D. Hsu et al. "Multiplex genome engineering using CRISPR/Cas systems." *Science* 339, no. 6121 (2013): 819–823.

Cox, David Benjamin Turitz, Randall Jeffrey Platt, and Feng Zhang. "Therapeutic genome editing: Prospects and challenges." *Nature Medicine* 21, no. 2 (2015): 121–131.

Deltcheva, Elitza, Krzysztof Chylinski, Cynthia M. Sharma, Karine Gonzales, Yanjie Chao, Zaid A. Pirzada, Maria R. Eckert, Jörg Vogel, and Emmanuelle Charpentier. "CRISPR RNA maturation by trans-encoded small RNA and host factor RNase III." *Nature* 471, no. 7340 (2011): 602–607.

Doxzen, Kevin, and Jodi Halpern. "Focusing on human rights: A framework for CRISPR germline genome editing ethics and regulation." *Perspectives in Biology and Medicine* 63, no. 1 (2020): 44–53.

Fonfara, Ines, Hagen Richter, Majda Bratovič, Anaïs Le Rhun, and Emmanuelle Charpentier. "The CRISPR-associated DNA-cleaving enzyme Cpf1 also processes precursor CRISPR RNA." *Nature* 532, no. 7600 (2016): 517–521.

Fu, Ya-Wen, Xin-Yue Dai, Wen-Tian Wang, Zhi-Xue Yang, Juan-Juan Zhao, Jian-Ping Zhang, Wei Wen et al. "Dynamics and competition of CRISPR–Cas9 ribonucleoproteins and AAV donor-mediated NHEJ, MMEJ and HDR editing." *Nucleic Acids Research* 49, no. 2 (2021): 969–985.

Gabel, Isabel, and Jonathan Moreno. "Genome editing, ethics, and politics." *AMA Journal of Ethics* 21, no. 12 (2019): E1105.

Gaj, Thomas, Charles A. Gersbach, and Carlos F. Barbas. "ZFN, TALEN, and CRISPR/Cas-based methods for genome engineering." *Trends in Biotechnology* 31, no. 7 (2013): 397–405.

Garza-Elizondo, Mario Alberto, D. R. Rodríguez-Rodríguez, R. Ramírez-Solís, M. D. L. Garza-Rodríguez, and H. A. Barrera-Saldaña. "Genome editing: A perspective on the application of CRISPR/Cas9 to study human diseases." *International Journal of Molecular Medicine* 43 (2019): 1559–1574.

Gootenberg, Jonathan S., Omar O. Abudayyeh, Max J. Kellner, Julia Joung, James J. Collins, and Feng Zhang. "Multiplexed and portable nucleic acid detection platform with Cas13, Cas12a, and Csm6." *Science* 360, no. 6387 (2018): 439–444.

Haapaniemi, Emma, Sandeep Botla, Jenna Persson, Bernhard Schmierer, and Jussi Taipale. "CRISPR–Cas9 genome editing induces a p53-mediated DNA damage response." *Nature Medicine* 24, no. 7 (2018): 927–930.

Haft, Daniel H., Jeremy Selengut, Emmanuel F. Mongodin, and Karen E. Nelson. "A guild of 45 CRISPR-associated (Cas) protein families and multiple CRISPR/Cas subtypes exist in prokaryotic genomes." *PLoS Computational Biology* 1, no. 6 (2005): e60.

Harrington, Lucas B., David Burstein, Janice S. Chen, David Paez-Espino, Enbo Ma, Isaac P. Witte, Joshua C. Cofsky, Nikos C. Kyrpides, Jillian F. Banfield, and Jennifer A. Doudna. "Programmed DNA destruction by miniature CRISPR-Cas14 enzymes." *Science* 362, no. 6416 (2018): 839–842.

He, Qian, Dongmei Yu, Mengdi Bao, Grant Korensky, Juhong Chen, Mingyeong Shin, Juwon Kim, Myeongkee Park, Peiwu Qin, and Ke Du. "High-throughput and all-solution phase African Swine Fever Virus (ASFV) detection using CRISPR-Cas12a and fluorescence based point-of-care system." *Biosensors and Bioelectronics* 154 (2020): 112068.

Hillary, V. Edwin, and S. Antony Ceasar. "Genome engineering in insects for the control of vector borne diseases." *Progress in Molecular Biology and Translational Science* 179 (2021): 197–223.

Himeda, Charis L., Takako I. Jones, and Peter L. Jones. "Targeted epigenetic repression by CRISPR/dSaCas9 suppresses pathogenic DUX4-fl expression in FSHD." *Molecular Therapy-Methods & Clinical Development* 20 (2021): 298–311.

Ishino, Yoshizumi, Hideo Shinagawa, Kozo Makino, Mitsuko Amemura, and Atsuo Nakata. "Nucleotide sequence of the iap gene, responsible for alkaline phosphatase isozyme conversion in Escherichia coli, and identification of the gene product." *Journal of Bacteriology* 169, no. 12 (1987): 5429–5433.

Jackson, Simon A., Rebecca E. McKenzie, Robert D. Fagerlund, Sebastian N. Kieper, Peter C. Fineran, and Stan J. J. Brouns. "CRISPR-Cas: Adapting to change." *Science* 356, no. 6333 (2017): eaal5056.

Jansen, Ruud, Jan D. A. van Embden, Wim Gaastra, and Leo M. Schouls. "Identification of genes that are associated with DNA repeats in prokaryotes." *Molecular Microbiology* 43, no. 6 (2002): 1565–1575.

Jiang, Fuguo, and Jennifer A. Doudna. "CRISPR–Cas9 structures and mechanisms." *Annual Review of Biophysics* 46, no. 1 (2017): 505–529.

Jiang, Wenyan, David Bikard, David Cox, Feng Zhang, and Luciano A. Marraffini. "CRISPR-assisted editing of bacterial genomes." *Nature Biotechnology* 31, no. 3 (2013): 233.

Jinek, Martin, Krzysztof Chylinski, Ines Fonfara, Michael Hauer, Jennifer A. Doudna, and Emmanuelle Charpentier. "A programmable dual-RNA–guided DNA endonuclease in adaptive bacterial immunity." *Science* 337, no. 6096 (2012): 816–821.

Karginov, Fedor V., and Gregory J. Hannon. "The CRISPR system: Small RNA-guided defense in bacteria and archaea." *Molecular Cell* 37, no. 1 (2010): 7–19.

Kellner, Max J., Jeremy G. Koob, Jonathan S. Gootenberg, Omar O. Abudayyeh, and Feng Zhang. "SHERLOCK: Nucleic acid detection with CRISPR nucleases." *Nature Protocols* 14, no. 10 (2019): 2986–3012.

Konermann, Silvana, Peter Lotfy, Nicholas J. Brideau, Jennifer Oki, Maxim N. Shokhirev, and Patrick D. Hsu. "Transcriptome engineering with RNA-targeting type VI-D CRISPR effectors." *Cell* 173, no. 3 (2018): 665–676.

Koonin, Eugene V., and Kira S. Makarova. "Origins and evolution of CRISPR-Cas systems." *Philosophical Transactions of the Royal Society B* 374, no. 1772 (2019): 20180087.

Kotterman, Melissa A., Thomas W. Chalberg, and David V. Schaffer. "Viral vectors for gene therapy: Translational and clinical outlook." *Annual Review of Biomedical Engineering* 17, no. 1 (2015): 63–89.

Kwan, Suet-Yan, Ankur Sheel, Chun-Qing Song, Xiao-Ou Zhang, Tingting Jiang, Hien Dang, Yueying Cao et al. "Depletion of TRRAP induces p53-independent senescence in liver cancer by down-regulating mitotic genes." *Hepatology* 71, no. 1 (2020): 275–290.

Lainscek, Duško, Lucija Kadunc, Mateja Mancek Keber, Iva Hafner Bratkovic, Rok Romih, and Roman Jerala. "Delivery of an artificial transcription regulator dCas9-VPR by extracellular vesicles for therapeutic gene activation." *ACS Synthetic Biology* 7, no. 12 (2018): 2715–2725.

Lee, Hyunji, Da Eun Yoon, and Kyoungmi Kim. "Genome editing methods in animal models." *Animal Cells and Systems* 24, no. 1 (2020): 8–16.

Lee, Minyoung, and Hyongbum Kim. "Therapeutic application of the CRISPR system: Current issues and new prospects." *Human Genetics* 138, no. 6 (2019): 563–590.

Levy, Asaf, Moran G. Goren, Ido Yosef, Oren Auster, Miriam Manor, Gil Amitai, Rotem Edgar, Udi Qimron, and Rotem Sorek. "CRISPR adaptation biases explain preference for acquisition of foreign DNA." *Nature* 520, no. 7548 (2015): 505–510.

Li, Ming, Rui Wang, Dahe Zhao, and Hua Xiang. "Adaptation of the Haloarcula hispanica CRISPR-Cas system to a purified virus strictly requires a priming process." *Nucleic Acids Research* 42, no. 4 (2014): 2483–2492.

Li, Song-pei, Zhong-xiao Lin, Xue-yan Jiang, and Xi-yong Yu. "Exosomal cargo-loading and synthetic exosome-mimics as potential therapeutic tools." *Acta Pharmacologica Sinica* 39, no. 4 (2018): 542–551.

Li, Tianxiang, Yanyan Yang, Hongzhao Qi, Weigang Cui, Lin Zhang, Xiuxiu Fu, Xiangqin He, Meixin Liu, Pei-feng Li, and Tao Yu. "CRISPR/Cas9 therapeutics: Progress and prospects." *Signal Transduction and Targeted Therapy* 8, no. 1 (2023): 36.

Liang, Puping, Yanwen Xu, Xiya Zhang, Chenhui Ding, Rui Huang, Zhen Zhang, Jie Lv et al. "CRISPR/Cas9-mediated gene editing in human tripronuclear zygotes." *Protein & Cell* 6, no. 5 (2015): 363–372.

Liao, Chunyu, and Chase L. Beisel. "The tracrRNA in CRISPR biology and technologies." *Annual Review of Genetics* 55, no. 1 (2021): 161–181.

Lin, Yi, Ernst Wagner, and Ulrich Lächelt. "Non-viral delivery of the CRISPR/Cas system: DNA versus RNA versus RNP." *Biomaterials Science* 10, no. 5 (2022): 1166–1192.

Liu, Chang, Li Zhang, Hao Liu, and Kun Cheng. "Delivery strategies of the CRISPR-Cas9 gene-editing system for therapeutic applications." *Journal of Controlled Release* 266 (2017): 17–26.

Liu, Qi, Kai Zhao, Chun Wang, Zhanzhan Zhang, Chunxiong Zheng, Yu Zhao, Yadan Zheng et al. "Multistage delivery nanoparticle facilitates efficient CRISPR/dCas9 activation and tumor growth suppression in vivo." *Advanced Science* 6, no. 1 (2019): 1801423.

Liu, Wenyi, Luoxi Li, Jianxin Jiang, Min Wu, and Ping Lin. "Applications and challenges of CRISPR-Cas gene-editing to disease treatment in clinics." *Precision Clinical Medicine* 4, no. 3 (2021): 179–191.

Liu, Zhenquan, Huina Dong, Yali Cui, Lina Cong, and Dawei Zhang. "Application of different types of CRISPR/Cas-based systems in bacteria." *Microbial Cell Factories* 19 (2020): 1–14.

Lu, You, Jianxin Xue, Tao Deng, Xiaojuan Zhou, Kun Yu, Lei Deng, Meijuan Huang et al. "Safety and feasibility of CRISPR-edited T cells in patients with refractory non-small-cell lung cancer." *Nature Medicine* 26, no. 5 (2020): 732–740.

Luk, Brian T., and Liangfang Zhang. "Cell membrane-camouflaged nanoparticles for drug delivery." *Journal of Controlled Release* 220 (2015): 600–607.

Ma, Yuanwu, Xu Zhang, Bin Shen, Yingdong Lu, Wei Chen, Jing Ma, Lin Bai, Xingxu Huang, and Lianfeng Zhang. "Generating rats with conditional alleles using CRISPR/Cas9." *Cell Research* 24, no. 1 (2014): 122–125.

Makarova, Kira S., L. Aravind, Nick V. Grishin, Igor B. Rogozin, and Eugene V. Koonin. "A DNA repair system specific for thermophilic Archaea and bacteria predicted by genomic context analysis." *Nucleic Acids Research* 30, no. 2 (2002): 482–496.

Makarova, Kira S., Feng Zhang, and Eugene V. Koonin. "SnapShot: Class 2 CRISPR-Cas systems." *Cell* 168, no. 1 (2017): 328–328.

Makarova, Kira S., Yuri I. Wolf, Omer S. Alkhnbashi, Fabrizio Costa, Shiraz A. Shah, Sita J. Saunders, Rodolphe Barrangou et al. "An updated evolutionary classification of CRISPR–Cas systems." *Nature Reviews Microbiology* 13, no. 11 (2015): 722–736.

Makarova, Kira S., Yuri I. Wolf, Jaime Iranzo, Sergey A. Shmakov, Omer S. Alkhnbashi, Stan J. J. Brouns, Emmanuelle Charpentier et al. "Evolutionary classification of CRISPR–Cas systems: A burst of class 2 and derived variants." *Nature Reviews Microbiology* 18, no. 2 (2020): 67–83.

Marraffini, Luciano A., and Erik J. Sontheimer. "CRISPR interference limits horizontal gene transfer in staphylococci by targeting DNA." *Science* 322, no. 5909 (2008): 1843–1845.

Marraffini, Luciano A., and Erik J. Sontheimer. "CRISPR interference: RNA-directed adaptive immunity in bacteria and archaea." *Nature Reviews Genetics* 11, no. 3 (2010): 181–190.

Maule, Giulia, Marjolein Ensinck, Mattijs Bulcaen, and Marianne S. Carlon. "Rewriting CFTR to cure cystic fibrosis." *Progress in Molecular Biology and Translational Science* 182 (2021): 185–224.

Mitchell, Michael J., Margaret M. Billingsley, Rebecca M. Haley, Marissa E. Wechsler, Nicholas A. Peppas, and Robert Langer. "Engineering precision nanoparticles for drug delivery." *Nature Reviews Drug Discovery* 20, no. 2 (2021): 101–124.

Modell, Joshua W., Wenyan Jiang, and Luciano A. Marraffini. "CRISPR–Cas systems exploit viral DNA injection to establish and maintain adaptive immunity." *Nature* 544, no. 7648 (2017): 101–104.

Mojica, Francisco J. M., Cesar Díez-Villaseñor, Elena Soria, and Guadalupe Juez. "Biological significance of a family of regularly spaced repeats in the genomes of Archaea, Bacteria and mitochondria." *Molecular Microbiology* 36, no. 1 (2000): 244–246.

Mojica, Francisco J. M., César Díez-Villaseñor, Jesús García-Martínez, and Cristóbal Almendros. "Short motif sequences determine the targets of the prokaryotic CRISPR defence system." *Microbiology* 155, no. 3 (2009): 733–740.

Mojica, Francisco J. M., César Díez-Villaseñor, Jesús García-Martínez, and Elena Soria. "Intervening sequences of regularly spaced prokaryotic repeats derive from foreign genetic elements." *Journal of Molecular Evolution* 60 (2005): 174–182.

Mojica, Francisco J. M., Guadalupe Juez, and Francisco Rodriguez-Valera. "Transcription at different salinities of Haloferax mediterranei sequences adjacent to partially modified PstI sites." *Molecular Microbiology* 9, no. 3 (1993): 613–621.

Mout, Rubul, Moumita Ray, Gulen Yesilbag Tonga, Yi-Wei Lee, Tristan Tay, Kanae Sasaki, and Vincent M. Rotello. "Direct cytosolic delivery of CRISPR/Cas9-ribonucleoprotein for efficient gene editing." *ACS Nano* 11, no. 3 (2017): 2452–2458.

Nakata, A. T. S. U. O., M. I. T. S. U. K. O. Amemura, and K. O. Z. O. Makino. "Unusual nucleotide arrangement with repeated sequences in the Escherichia coli K-12 chromosome." *Journal of Bacteriology* 171, no. 6 (1989): 3553–3556.

Nam, Ki Hyun, Charles Haitjema, Xueqi Liu, Fran Ding, Hongwei Wang, Matthew P. DeLisa, and Ailong Ke. "Cas5d protein processes pre-crRNA and assembles into a cascade-like interference complex in subtype IC/Dvulg CRISPR-Cas system." *Structure* 20, no. 9 (2012): 1574–1584.

Nguyen, Nuong Thi Kieu, Yu-Han Chang, Vu Anh Truong, Mu-Nung Hsu, Nam Ngoc Pham, Chin-Wei Chang, Yi-Hsiu Wu, Yi-Hao Chang, Hung Li, and Yu-Chen Hu. "CRISPR activation of long non-coding RNA DANCR promotes bone regeneration." *Biomaterials* 275 (2021): 120965.

Nunez, James K., Lucas B. Harrington, Philip J. Kranzusch, Alan N. Engelman, and Jennifer A. Doudna. "Foreign DNA capture during CRISPR–Cas adaptive immunity." *Nature* 527, no. 7579 (2015): 535–538.

Nunez, James K., Philip J. Kranzusch, Jonas Noeske, Addison V. Wright, Christopher W. Davies, and Jennifer A. Doudna. "Cas1–Cas2 complex formation mediates spacer acquisition during CRISPR–Cas adaptive immunity." *Nature Structural & Molecular Biology* 21, no. 6 (2014): 528–534.

Ormond, Kelly E., Yvonne Bombard, Vence L. Bonham, Lily Hoffman-Andrews, Heidi C. Howard, Rosario Isasi, Kiran Musunuru, Kirsten A. Riggan, Marsha Michie, and Megan Allyse. "The clinical application of gene editing: Ethical and social issues." *Personalized Medicine* 16, no. 4 (2019): 337–350.

Rao, Yuchun, Xi Yang, Chenyang Pan, Chun Wang, and Kejian Wang. "Advance of clustered regularly interspaced short palindromic repeats-Cas9 system and its application in crop improvement." *Frontiers in Plant Science* 13 (2022): 839001.

Rosenblum, Daniel, Anna Gutkin, Ranit Kedmi, Srinivas Ramishetti, Nuphar Veiga, Ashley M. Jacobi, Mollie S. Schubert et al. "CRISPR-Cas9 genome editing using targeted lipid nanoparticles for cancer therapy." *Science Advances* 6, no. 47 (2020): eabc9450.

Sapranauskas, Rimantas, Giedrius Gasiunas, Christophe Fremaux, Rodolphe Barrangou, Philippe Horvath, and Virginijus Siksnys. "The Streptococcus thermophilus CRISPR/Cas system provides immunity in Escherichia coli." *Nucleic Acids Research* 39, no. 21 (2011): 9275–9282.

Sashital, Dipali G., Martin Jinek, and Jennifer A. Doudna. "An RNA-induced conformational change required for CRISPR RNA cleavage by the endoribonuclease Cse3." *Nature Structural & Molecular Biology* 18, no. 6 (2011): 680–687.

Shiimori, Masami, Sandra C. Garrett, Brenton R. Graveley, and Michael P. Terns. "Cas4 nucleases define the PAM, length, and orientation of DNA fragments integrated at CRISPR loci." *Molecular Cell* 70, no. 5 (2018): 814–824.

Smargon, Aaron A., David B. T. Cox, Neena K. Pyzocha, Kaijie Zheng, Ian M. Slaymaker, Jonathan S. Gootenberg, Omar A. Abudayyeh et al. "Cas13b is a type VI-B CRISPR-associated RNA-guided RNase differentially regulated by accessory proteins Csx27 and Csx28." *Molecular Cell* 65, no. 4 (2017): 618–630.

Sonntag, Florian, Kristin Schmidt, and Jürgen A. Kleinschmidt. "A viral assembly factor promotes AAV2 capsid formation in the nucleolus." *Proceedings of the National Academy of Sciences* 107, no. 22 (2010): 10220–10225.

Su, Shu, Zhengyun Zou, Fangjun Chen, Naiqing Ding, Juan Du, Jie Shao, Lin Li et al. "CRISPR-Cas9-mediated disruption of PD-1 on human T cells for adoptive cellular therapies of EBV positive gastric cancer." *Oncoimmunology* 6, no. 1 (2017): e1249558.

Tamay, Dilara Goksu, Tugba Dursun Usal, Ayse Selcen Alagoz, Deniz Yucel, Nesrin Hasirci, and Vasif Hasirci. "3D and 4D printing of polymers for tissue engineering applications." *Frontiers in Bioengineering and Biotechnology* 7 (2019): 164.

Thomas, Christopher M., and Kaare M. Nielsen. "Mechanisms of, and barriers to, horizontal gene transfer between bacteria." *Nature Reviews Microbiology* 3, no. 9 (2005): 711–721.

Vorontsova, Daria, Kirill A. Datsenko, Sofia Medvedeva, Joseph Bondy-Denomy, Ekaterina E. Savitskaya, Ksenia Pougach, Maria Logacheva et al. "Foreign DNA acquisition by the IF CRISPR–Cas system requires all components of the interference machinery." *Nucleic Acids Research* 43, no. 22 (2015): 10848–10860.

Wagner, Dimitrios L., Leila Amini, Desiree J. Wendering, Lisa-Marie Burkhardt, Levent Akyüz, Petra Reinke, Hans-Dieter Volk, and Michael Schmueck-Henneresse. "High prevalence of Streptococcus pyogenes Cas9-reactive T cells within the adult human population." *Nature Medicine* 25, no. 2 (2019): 242–248.

Wang, Dan, Feng Zhang, and Guangping Gao. "CRISPR-based therapeutic genome editing: Strategies and in vivo delivery by AAV vectors." *Cell* 181, no. 1 (2020): 136–150.

Wang, Haoyi, Hui Yang, Chikdu S. Shivalila, Meelad M. Dawlaty, Albert W. Cheng, Feng Zhang, and Rudolf Jaenisch. "One-step generation of mice carrying mutations in multiple genes by CRISPR/Cas-mediated genome engineering." *Cell* 153, no. 4 (2013): 910–918.

Wang, Jiuyu, Jiazhi Li, Hongtu Zhao, Gang Sheng, Min Wang, Maolu Yin, and Yanli Wang. "Structural and mechanistic basis of PAM-dependent spacer acquisition in CRISPR-Cas systems." *Cell* 163, no. 4 (2015): 840–853.

Wang, Ruiying, Gan Preamplume, Michael P. Terns, Rebecca M. Terns, and Hong Li. "Interaction of the Cas6 riboendonuclease with CRISPR RNAs: Recognition and cleavage." *Structure* 19, no. 2 (2011): 257–264.

Wei, Yunzhou, Megan T. Chesne, Rebecca M. Terns, and Michael P. Terns. "Sequences spanning the leader-repeat junction mediate CRISPR adaptation to phage in Streptococcus thermophilus." *Nucleic Acids Research* 43, no. 3 (2015a): 1749–1758.

Wei, Yunzhou, Rebecca M. Terns, and Michael P. Terns. "Cas9 function and host genome sampling in Type II-A CRISPR–Cas adaptation." *Genes & Development* 29, no. 4 (2015b): 356–361.

Wiedenheft, Blake, Kaihong Zhou, Martin Jinek, Scott M. Coyle, Wendy Ma, and Jennifer A. Doudna. "Structural basis for DNase activity of a conserved protein implicated in CRISPR-mediated genome defense." *Structure* 17, no. 6 (2009): 904–912.

Witzigmann, Dominik, Jayesh A. Kulkarni, Jerry Leung, Sam Chen, Pieter R. Cullis, and Roy van der Meel. "Lipid nanoparticle technology for therapeutic gene regulation in the liver." *Advanced Drug Delivery Reviews* 159 (2020): 344–363.

Wright, Addison V., and Jennifer A. Doudna. "Protecting genome integrity during CRISPR immune adaptation." *Nature Structural & Molecular Biology* 23, no. 10 (2016): 876–883.

Wu, Yuxuan, Dan Liang, Yinghua Wang, Meizhu Bai, Wei Tang, Shiming Bao, Zhiqiang Yan, Dangsheng Li, and Jinsong Li. "Correction of a genetic disease in mouse via use of CRISPR-Cas9." *Cell Stem Cell* 13, no. 6 (2013): 659–662.

Wu, Yuxuan, Hai Zhou, Xiaoying Fan, Ying Zhang, Man Zhang, Yinghua Wang, Zhenfei Xie et al. "Correction of a genetic disease by CRISPR-Cas9-mediated gene editing in mouse spermatogonial stem cells." *Cell Research* 25, no. 1 (2015): 67–79.

Xiao, Yibei, Sherwin Ng, Ki Hyun Nam, and Ailong Ke. "How type II CRISPR–Cas establish immunity through Cas1–Cas2-mediated spacer integration." *Nature* 550, no. 7674 (2017): 137–141.

Xu, Xiaojie, Tao Wan, Huhu Xin, Da Li, Hongming Pan, Jun Wu, and Yuan Ping. "Delivery of CRISPR/Cas9 for therapeutic genome editing." *The Journal of Gene Medicine* 21, no. 7 (2019): e3107.

Yin, Hao, Chun-Qing Song, Joseph R. Dorkin, Lihua J. Zhu, Yingxiang Li, Qiongqiong Wu, Angela Park et al. "Therapeutic genome editing by combined viral and non-viral delivery of CRISPR system components in vivo." *Nature Biotechnology* 34, no. 3 (2016): 328–333.

Yosef, Ido, Moran G. Goren, and Udi Qimron. "Proteins and DNA elements essential for the CRISPR adaptation process in Escherichia coli." *Nucleic Acids Research* 40, no. 12 (2012): 5569–5576.

Zetsche, Bernd, Jonathan S. Gootenberg, Omar O. Abudayyeh, Ian M. Slaymaker, Kira S. Makarova, Patrick Essletzbichler, Sara E. Volz et al. "Cpf1 is a single RNA-guided endonuclease of a class 2 CRISPR-Cas system." *Cell* 163, no. 3 (2015): 759–771.

Zhang, Kunyu, Wai Hon Chooi, Shuang Liu, Jiah Shin Chin, Aoife Murray, Dean Nizetic, Du Cheng, and Sing Yian Chew. "Localized delivery of CRISPR/dCas9 via layer-by-layer self-assembling peptide coating on nanofibers for neural tissue engineering." *Biomaterials* 256 (2020): 120225.

Zhang, Yi, Karen Massel, Ian D. Godwin, and Caixia Gao. "Applications and potential of genome editing in crop improvement." *Genome Biology* 19, no. 1 (2018): 210.

Zhang, Zhanzhan, Qixue Wang, Qi Liu, Yadan Zheng, Chunxiong Zheng, Kaikai Yi, Yu Zhao et al. "Dual-Locking nanoparticles disrupt the PD-1/PD-L1 pathway for efficient cancer immunotherapy." *Advanced Materials* 31, no. 51 (2019): 1905751.

Zhao, Liya, Kai Li, Shimin Bao, Yuxun Zhou, Yinming Liang, Guoji Zhao, Ye Chen, and Junhua Xiao. "A 1-bp deletion in the γC-crystallin leads to dominant cataracts in mice." *Mammalian Genome* 21 (2010): 361–369.

2 Beyond the CRISPR-Cas9 System

Other Gene-Editing Technologies

Aarif Ali, Mohammad Saleem Bhat, Javeed Iqbal Ahmad Bhat, Sadhna Aggarwal, and Sheikh Bilal Ahmad

2.1 OVERVIEW OF GENE-EDITING TECHNOLOGIES

Over the years, the gradual rise of genome editing approaches has empowered researchers to proficiently and cost-effectively make targeted changes to the genetic makeup of various organisms and cell types. In the 1970s, genetic engineering marked a groundbreaking advancement in genome editing by manipulating DNA or RNA. The rapid evolution of gene-editing methods over the last ten years can be attributed to advancements in engineered and bacterial nucleases. These technologies have shown remarkable applicability in diverse sectors, ranging from essential research to the fields of biomedical and biotechnology investigation (Cornu et al., 2017). Genome editing is a method of genetic engineering that involves the precise insertion, deletion, modification, or replacement of DNA within an organism's genome. Unlike older techniques that relied on random insertion of genetic material, current methods allow for specific targeting of desired locations. This targeted approach minimizes off-target effects and facilitates the control of precise gene sequences within the organism genome. The editing of genomes can be carried out either *in vitro* or *in vivo* by introducing directly the editing machinery, allowing for precise modifications to genes and other targeted genomic changes (Ghosh et al., 2019; Gaj et al., 2013). The alterations in the DNA begin with the formation of double-strand breaks (DSBs) made by nucleases that, in turn, activate effective recombination mechanisms in the mammalian cells (Kosicki et al., 2018). The DSBs induced can be repaired through homology-directed repair (HDR) or non-homologous end joining (NHEJ), thereby leading to gene disruptions, respectively (O'Driscoll and Jeggo, 2006).

Historically, homologous recombination (HR) has primarily been the method for achieving the targeted gene editing, inactivation, or replacement by utilizing templates like undamaged homologous fragments of DNA. However, its efficiency model organisms and the mammalian cells are heavily limited (Kaniecki et al., 2018). The insight that DSBs can significantly raise the rate of HDR by quite a few

DOI: 10.1201/9781003537151-2

orders of magnitude has prompted the identification of targeted nucleases as an alternate strategy to improve the efficacy of HDR-facilitated genetic alterations. When a specific DSB is established, HDR at the damaged site can utilize an external DNA template that closely resembles the sequence to repair the cleaved DNA. In contrast, NHEJ-mediated repair is often error-prone, leading to the generation of gene insertions or deletions (indels) of various lengths at the DSB site, which may result in gene inactivation. The occurrence of insertions or deletions in the coding regions can lead to mutations, particularly frameshift, which may cause degradation of mRNA or the formation of non-functional truncated proteins through the process of nonsense-mediated decay. This technique and its applicability are viewed as less complex than those involving HR, as it eliminates the necessity for a repair template and is less affected by the cell type, since NHEJ can function throughout the cell cycle. Therefore, similar to RNA interference (RNAi), NHEJ can be employed in immortalized cell lines to induce the deactivation of one or more genes, ultimately causing permanent gene inactivation through the production of loss of functional mutations (Kim and Kim, 2014).

2.2 GENOME EDITING APPROACHES

The primary technologies currently utilized for genome editing, as depicted in **Figure 2.1**, encompass zinc finger nucleases (ZFNs), transcription activator–like effector nucleases (TALENs), clustered regularly interspaced short palindromic repeats (CRISPR) and CRISPR–associated protein 9 (Cas9), and homing endonucleases (meganucleases).

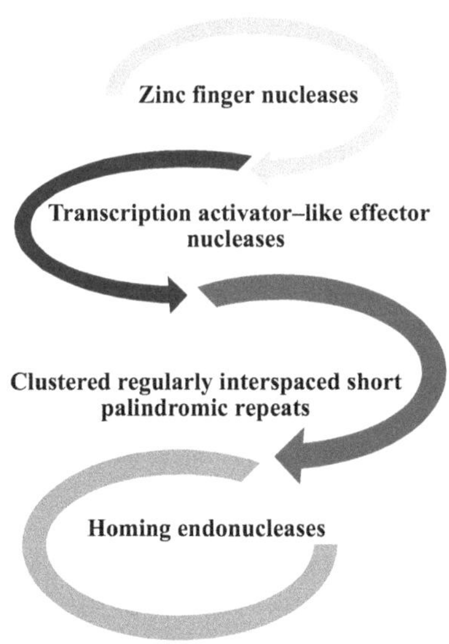

FIGURE 2.1 Primary technologies currently utilized for genome editing.

ZFNs and TALENs are generally recognized amongst the custom hybrid nucleases. For a long time, these artificially created restriction enzymes have been used successfully as genome editing tools. The primary characteristics of editing a genome are the ability of an endonuclease to identify certain lengthy target sequences and its capacity to retarget to sequences designated by users. A new paradigm shift has been sparked by the potential ability of TALENs and CRISPR-Cas9 in targeting and editing the new sequences in genome, resulting in intensified technological advances and discoveries in areas of synthetic biology, illness modelling, development of drugs, gene therapy, neuroscience, and agriculture.

2.2.1 Zinc Finger Nucleases

ZFNs represent the most prevalent and basic form of endonucleases, specifically engineered to act as synthetic restriction enzymes (Yang et al., 2017). In eukaryotes, the most widely recognized DNA-binding motif is the Cys2-His2 zinc finger domain. In the human genome, this is the second most abundant protein domain. The eukaryotic transcription factors serve as a source of origin to the zinc finger proteins (ZFPs) and serve as DNA-binding domain for the ZFNs (De Souza, 2012). From *Flavobacterium okeanokoites*, the nucleotide chopper domain Folk 1 originates (De Souza, 2012). Made up of 30 amino acids, ZFPs are characterized by two antiparallel beta sheets that lie opposite to the alpha helix (De Souza, 2012). A cleavage domain is connected to three to six ZFPs that are uniquely adapted for a designated target site. ZFPs possess a high degree of accuracy in target-specific genes as they have 18–base pair specificity in comparison to the DNA-binding domains (9–base pair specificity) (Sander et al., 2011a). In diverse prokaryotic and eukaryotic systems, ZFNs have shown a significant effectiveness in facilitating HDR gene knock-in (KI) and NHEJ knockout (KO) techniques (Carroll, 2011). Despite the favorable outcome, ZFNs are currently not being contemplated for industrial applications due to limitations such as the inability to target all possible nucleotide genomic sequences and the potential disruption of sequence specificity by neighboring protein domains (Sander et al., 2011a). Additionally, the complex and costly design of folks and ZFPs makes them less preferred.

2.2.2 Transcription Activator–Like Effector Nucleases

TALENs are moderately innovative, with TAL proteins playing a vital part. The 34 amino acid repeat units found in these proteins are secreted by the Xanthomonas bacteria and constructed well for individual-specific recognition and binding efficiency (Li et al., 2012). TALENs are capable of recognizing single nucleotides with high precision, unaffected by nearby domains. It is often used in KI and KO procedures because it works very well. For the cleavage process in TALENs, there are two protein site domains present, which also possess TAL repeats to recognize a precise cut (Christian et al., 2010). In DNA, sticky ends are induced by TALENs, leading to decreased off-target outcomes (Aryan et al., 2013). TALE specificity relies on two hypervariable amino acids called repeat-variable diresidues (RVDs) (Deng et al., 2012). The continuous DNA sequences are identified by the engineered modular TALE repeats similar to the zinc fingers. The efficiency of targeted KO with TALENs falls within the range of 30 to 100%, while KI typically ranges from 1 to 10%. TALENs have been utilized to genetically modify numerous animals, including rats, chickens, mammalian cells, frogs, and zebrafish (Sander et al., 2011b; Aryan et al., 2013). However, unlike ZFPs, the connection between repeats in TALEs does not require modification to form long arrays that can target specific sites in the organism genome. Effector domains including site-specific recombinases, transcriptional activators, and nucleases have been fused to TALE repeats for precise genetic modifications following years of research on ZFPs. In comparison to the triplet constrained ZFPs, a great flexibility is provided by the TALE-DNA-binding repeats. A great technical challenge is caused by the cloning of repeat TALE arrays, owing to the existence of identical long repetitive sequences.

2.2.3 Clustered Regularly Interspaced Short Palindromic Repeat (CRISPR)–Associated 9 (Cas9)

The unearthing of CRISPRs in *Escherichia coli (E. coli)* and other bacteria in early 1987 marked a significant advancement in understanding bacterial and archaeal adaptive immune defense mechanisms in contradiction to the foreign DNA (Ishino et al., 1987). Initially, the role of these short repetitive sequences was not known, until studies revealed their similarity to phage DNA and their involvement in RNA-guided DNA cleavage (Jinek et al., 2012).

The CRISPR-Cas system is a natural prokaryotic immunity system that involves capturing and inserting small plasmid or viral DNA fragments into the bacterium's genome, providing resistance against future invasions. Depending upon the structural variation and organization style of Cas genes, these are subdivided into two categories, namely, class 1 and class 2 systems (Jinek et al., 2012). Class 1 systems are composed of complexes that include multiple effector proteins, while class 2 systems are represented by a lone effector protein. Presently, the CRISPR-Cas system is categorized into six types and more than 29 subtypes, with ongoing research actively contributing to the evolution of this classification (Makarova et al., 2015, 2011). In particular, class 1 systems are characterized by the presence of multi-protein effector complexes, whereas class 2 systems only consist of a single effector protein. Researchers have developed a CRISPR-Cas9 method that uses guide RNA and the Cas9 endonuclease. While it is commonly referred to as CRISPR, it is important to note that other endonucleases, such as Cpf1, have also been used. In the context of DNA editing in eukaryotic cells, *Streptococcus pyogenes* Cas9 (SpCas9) is the most widely used endonuclease. This enzyme, which is derived from *Streptococcus pyogenes*, has a bilobed structure that contains domains dedicated to the recognition lobe (REC) and nuclease activity lobe (NUC). The NUC lobe houses the catalytic domains RuvC and HNH, as well as the PAM interaction domain (PI) (Jinek et al., 2012).

The Cas9 enzyme requires activation and targeting to its specific target to function. In bacteria's natural immunity process, this is achieved with the help of two RNA molecules: crRNA and tracrRNA. However, for laboratory use, researchers have developed a single guide RNA (sgRNA or gRNA) by combining crRNA and tracrRNA to simplify the process (Jinek et al., 2012). The CRISPR tool consists of three primary components: a nuclease, which is generally wild-type Cas9 derived from *Streptococcus pyogenes*; a molecule of sgRNA; and the target DNA, respectively. This configuration enhances the efficiency of the experimental process. A specific 20–base pair sequence is designed meticulously within the sgRNA that corresponds to the target DNA. This sequence is succeeded by a brief DNA segment that is crucial for ensuring compatibility with the Cas9 protein, referred to as "protospacer adjacent motif" (PAM), which can be either "NGG" or "NAG" (Sternberg et al., 2014). The sgRNA associates via the Watson–Crick base pairing with the target sequence that permits Cas9 to effectively cleave the DNA and produce a DSB (Gasiunas et al., 2012). Following the existence of a DSB in the DNA molecule, the repair mechanisms are engaged to begin the process of genomic restoration. The CRISPR-Cas9 system allows for targeted modifications in the genome through pathways such as HDR or NHEJ, thus enabling the introduction of minor insertions and deletions (indels) (Ran et al., 2013a).

2.2.4 Homing Endonucleases

Meganucleases, or *homing nucleases*, are endoribonucleases that recognize and cleave target DNA sequence at particular position to promote the displacement of elements in the genome to the allelic sites, respectively. Within the unoccupied target sequence, this process generates DSBs. With the discovery of numerous meganucleases, the likelihood of finding an enzyme targeting a desired locus is minimal, leading to induced DSBs and reliance on NHEJ-prone DNA repair mechanism, resulting in the random insertion of DNA templates in unintended locations. Although attempts have been made to enhance this mechanism, specificity remains notably low (Rosen et al., 2006). These proteins have the capacity to bind to 14–40 bp long target DNA sites and are encoded by introns, freestanding intronic genes, and inteins (Chevalier et al., 2001). The advantages of meganucleases include high specificity, small size for easy delivery, and the generation of a 3′ overhang after DNA cleavage, enhancing recombinogenicity for HDR. These proteins have been considered as useful tools for studying DSB repair, manipulation/analysis of genome, cloning, therapeutic gene editing, gene replacements, mapping, and gene knockouts.

Table 2.1 shows the characteristics of modern gene-editing approaches, along with their advantages and limitations (Lee et al., 2015; Ain et al., 2015).

2.3 ALTERNATIVE GENE-EDITING TOOLS

In addition to the main editing approaches, some of the alternate gene-editing tools that are being used are described in what follows.

2.3.1 Base Editors

Base editors represent advanced molecular instruments utilized in the fields of genetic engineering and genome modification. They are intended to allow exact modifications to individual DNA bases without causing DSBs in the DNA molecule.

TABLE 2.1
Characteristic Features of Recent Gene-Editing Approaches

	Characteristics					
Gene-Editing Tool	**Target Site**	**Binding Specificity**	**Rate of CpG Methylation**	**Success Rate of Nuclease Design**	**Advantages**	**Limitations**
ZFNs	18–36 bp	3 nucleotides	Low	Not known	Variable target efficiency. The delivery to the desired target gene is very easy.	The off-target frequency is high. There is no high-throughput targeting.
TALENs	30–40 bp	1 nucleotide	High	Sensitive	It is easy to design with high specificity, and target efficiency is high.	The off-target effect is low, and there is low/limited high-throughput targeting. Moreover, unintended cuts may be caused by the repetitive sequences to the DNA.
CRISPR-Cas9	20–22 bp	1:1 nucleotide	High	No effect	The target efficiency and specificity are high. The multiplexed gene editing is very easy.	There are no limitations in high-throughput targeting. However, there are some/variable off-target effects.

This feature distinguishes them from standard genome editing approaches, like CRISPR-Cas9, which normally rely on causing DSBs to activate cellular repair systems. Base editors may perform exact base substitutions without the need for DSBs or donor DNA templates, and they are independent of HDR, making them a promising therapeutic option for human genetic illnesses characterized by single-nucleotide mutations (Eichler, 2019). Presently, base editors are developed by integrating DNA deaminase enzymes with catalytically deficient Cas nucleases, enabling precise modification of a single nucleotide within a designated sequence (Anzalone et al., 2020). The process of base editing merges the gene targeting abilities of a Cas protein with the DNA base modifying activities of a deaminase. Occasionally, to ensure the desired efficacy, necessary additional components are required. Base editors have been primarily grouped into two types: cytosine base editors (CBEs) and adenine base editors (ABEs), respectively.

2.3.1.1 Cytosine Base Editors

CBEs are a type of gene-editing enzymes that were developed in 2016 and are considered as the first base editors. In the living cell's genomic DNA, these editors use a cytidine deaminase enzyme that changes C-G base pairs to T-A base pairs (Komor et al., 2016; Gaudelli et al., 2017). The transformation of this chemical compound occurs via hydrolytic enzymatic deamination that converts cytosine into uracil, which is subsequently recognized by DNA polymerases as thymine (Komor et al., 2016). As of now, CBEs usually possess four key components, namely, a naturally occurring cytidine deaminase (AID or CDA, APOBEC), altered Cas9, and one or more uracil glycosylase inhibitor (UGI) peptides (Anzalone et al., 2020; Komor et al., 2017). Typically, these components are covalently bonded, but they can also be assembled through noncovalent means (Collantes et al., 2021). CBEs have been broadly applied in gene reversion and cellular engineering, holding promise for therapeutic interventions for individuals affected by debilitating genetic conditions or malignancies (Rees et al., 2021). The present CBEs tools can achieve a high on-target gene-editing efficiency (Anzalone et al., 2019).

2.3.1.2 Adenine Base Editors

ABEs are specialized gene-editing enzymes designed to introduce A-T to G-C point mutations at specific locations (Komor et al., 2016; Gaudelli et al., 2017). This method is facilitated by a laboratory-evolved TadA deaminase, which effectively transforms adenine into inosine (Gaudelli et al., 2017). Within the active site of the enzyme DNA polymerase, the inosine base pairs with cytosine, thereby leading to a mutation (inosine to guanine) following DNA replication. The limited effectiveness of initial ABEs significantly restricted their use until the discovery of various high-activity protein variants, such as ABE8e, ABE8s, and more recently, ABEmax-KR (Gaudelli et al., 2020).

2.3.2 Prime Editing

CBEs and ABEs have been known for a long period to possess the capacity to edit DNA, but until recently, they were limited in their precision to produce base modifications that go further than the four transition mutations. Anzalone et al. (2019) have recently described a method to overcome these limitations, known as prime editing (PE). It is a cutting-edge genome editing technology that enables precise base conversion, small deletions, and insertions of various types. In comparison to CRISPR-based base editing, PE operates without the need for DSBs.

In this method, Cas9 nickase attached to prime editing guide RNA (pegRNA) is linked to an especially designed reverse transcriptase (RT) that plays a vital part in the system's function as it contains both the complementary sequences to the target sites guiding Cas9 to its target sequence and other additional sequence implied with the desired changes (Anzalone et al., 2019). The pegRNA is significantly different from regular small guide RNAs (sgRNAs) and is essential to the system's function. The non-complementary strand is revealed by the primer binding site (PBS) present on the DNA that attaches to the pegRNA's 5' end. A primer for RT linked to Cas9 is created when the unbound DNA of the PAM-containing strand is nicked by

Cas9. The nicked PAM strand is then extended by the RT, with pegRNA internal sequences serving as a template, thereby facilitating programmable alterations to the target region. As a consequence, two redundant PAM DNA flaps are produced via this step, namely, the original unedited 5' flap and the edited 3' flap. From the pegRNA, the edited 3' flap is reverse-transcribed. The fortitude of which flap binds with the non-PAM DNA strand is an equilibria process wherein the 5' complementary flap is perfectly likely to be favored thermodynamically. During the process of DNA synthesis of lagging strand, cellular endonucleases are usually active and preferentially influence the 5' flap degradation (Hosfield et al., 1998). Ultimately, the heteroduplex formed by the edited 3' flap and the unedited strand is resolved and, finally, via cellular replication and repair mechanisms, is integrated into host genome. PE presents unique benefits compared to earlier CRISPR-based base editing techniques, including more flexible requirements of PAM owing to the varying lengths of the RT template and the elimination of "bystander" editing. Nonetheless, PE is still in its nascent phase, necessitating further investigation into its specificity and likely for its off-target effects. The latest iterations of base editors have been more extensively characterized, especially *in vivo*, showcasing improved efficiency and reduced indel formation. Consequently, these tools are currently preferred over prime editors. However, the PE system marks a noteworthy progress in the quest for a universal genome editing technique, with its potential clinical applications for correcting known pathogenic mutations likely to have a profound impact.

2.3.3 CRISPR-Associated Natural Transposases

CRISPR-associated transposons (CAST) are genetic mobile elements capable of inserting enormous DNA cargo via an RNA-guided method (Klompe et al., 2022). Comprising multiple conserved proteins, this system includes a CRISPR effector (Cas12k or Cascade), an AAA+ regulator (TnsC), a transposase (TnsA–TnsB), and a target site–associated factor (TniQ). These proteins function in unison to integrate DNA by forming a multisubunit transposition integration complex, commonly known as a transpososome. A potential approach is presented by the combination of CRISPR effectors in CAST systems and transposase subunits, allowing RNA-guided transposition of mobile genetic elements for the targeted and accurate insertion of large segments of DNA (Klompe et al., 2022). Most CASTs identified are linked to Tn7-like transposons, which preserve the key genes necessary for transposition but do not contain genes for target selection. CAST systems utilize CRISPR-Cas proteins that are deficient in nuclease activity to initiate RNA-guided transposition. Experimental investigations have characterized two unique subtypes of these systems: V-K CAST and the I-F CAST systems. Usually, the V-K CAST system comprises of Cas 12k effectors with inactive domains of nuclease, whereas the I-F CAST systems consist of cascade complexes without Cas3 nuclease (Strecker et al., 2019). In comparison to the other identified CAST systems, type I-B CAST systems are comparatively rare (Faure et al., 2019). In *E. coli*, transposition driven by the prototypical Tn7 transposon is contingent upon the core transposase proteins TnsA, TnsB, and TnsC, respectively. These proteins perform their roles in two distinct manners: through homing, which involves interaction with the site-specific DNA-binding protein TnsD for insertion into the attachment site of chromosomal glmS, and through mobile element transposition, where they work alongside the protein TnsE to engage with mobile elements like conjugative plasmids (Waddell and Craig, 1988). The effective and safe transposon travel from one host to another is facilitated by two pathways. In the TnsD pathway, across the hosts, a conserved site is targeted afterward, passing over a mobile element to integrate into the bacterial chromosomal safely.

2.3.4 Programmable and RNA-Guided Nuclease Transposon

Transposons are abundant in both prokaryotic and eukaryotic genomes, serving as a diverse class of mobile genetic elements. In prokaryotes, RNA-programmable DNA nucleases have multiple functions, including defense against and propagation of mobile elements. Programmable RNA-guided DNA nucleases are involved in various functions within prokaryotes, but their presence in non-prokaryotic organisms remains uncertain. In prokaryotes, RNA-guided endonucleases are crucial components of CRISPR-Cas adaptive immune systems and are utilized in a range of editing tools, including Cas9 or Cas12a and reprogrammable guide RNAs. The IS200/605 transposon family, characterized by its high abundance, contains two genes: TnpA, which is responsible for the mechanism of single-strand "peel and paste" transposition, and TnpB, whose specific role in the transposon life cycle has yet to be determined (Kapitonov et al., 2016; He et al., 2013). TnpB features a RuvC-like nuclease domain that bears resemblance to the nuclease domain of the type V CRISPR effector Cas12, suggesting a plausible direct evolutionary trajectory from TnpB to Cas12 (Karvelis et al., 2021; Bao and Jurka, 2013; Altae-Tran et al., 2021). An analysis of the phylogenetic relationships among the RuvC-like domains substantiates this association, indicating that the different type V subtypes of Cas12s have emerged independently from distinct TnpB groups (Altae-Tran et al., 2021; Shmakov et al., 2017). Within the framework of obligate mobile element–guided activity (OMEGA) systems, TnpBs are key components that encode the guide ωRNA, which is situated next to the nuclease gene. Biochemical and cellular studies have confirmed that the ωRNA–TnpB complex acts as an RNA-guided, programmable-DNA endonuclease (Karvelis et al., 2021; Altae-Tran et al., 2021). RuvC domain–containing proteins are not exclusive to prokaryotic organisms; a category of TnpB homologs known as Fanzors is also present in eukaryotic cells (Bao and Jurka, 2013). Fanzors have been recognized in multiple eukaryotic lineages, such as metazoans, fungi, algae, amorphea, and certain large dsDNA viruses, paralleling the diversity of TnpBs observed in bacteria and archaea. The identified Fanzors can be classified into two principal categories: Fanzor1 proteins, which are associated with eukaryotic transposons and are predominantly found in a variety of eukaryotic species, and Fanzor2 proteins, which are linked to IS607-like transposons and are present in the genomes of dsDNA viruses. Despite the resemblances between TnpB and Fanzors, a detailed investigation of Fanzors across the full range of eukaryotic diversity has not yet been performed, and they have not been subjected to experimental characterization.

2.3.5 Replication Interrupted Template-Driven DNA Modification (RITDM)

It is a genetic engineering strategy that uses cellular DNA replication machinery to carry out precise modifications in the genome without the need for DSBs or donor DNA templates, as is the case with traditional methods, such as HDR. RITDM capitalizes on the natural DNA replication process to introduce targeted modifications into the genome (Rickman and Smogorzewska, 2019). By manipulating the cellular machinery during DNA replication, it is possible to halt or slow down the process at specific locations, thereby allowing for the desired changes to be implemented (Cimprich and Cortez, 2008). Through the use of engineered elements, a replication stalling at a selected genomic site can be induced. This stalling can be achieved using various techniques, such as employing binding proteins to physically impede the replication fork or utilizing small molecules to hinder replication progression. After replication stalling, targeted modifications can be implemented at the stalled replication fork. These modifications may involve incorporating modified nucleotides, utilizing base editing enzymes, or employing other molecular tools to enable precise alterations to the DNA sequence. Following the introduction of modifications, replication can proceed as usual, integrating the changes into both daughter strands of the DNA molecule.

2.4 POTENTIAL APPLICATIONS OF GENE-EDITING APPROACHES

Prior to the development of engineered nucleases, altering mammalian cell lines genetically was a time-consuming, expensive process that was typically restricted to specialized laboratories. Nevertheless, thanks to the introduction of affordable and easy-to-use gene-editing tools, personalized cell lines with a widespread array of genetic modifications can now be created in just a few weeks. Routine outcomes resulting from the use of targeted nucleases include knockout, inversion, addition, deletion, correction in genes, along with chromosomal translocations (Mali et al., 2013; Ran et al., 2013b; Hockemeyer et al., 2011). Utilizing engineered or bacterial nucleases, these editing approaches have made it probable to modify and target sequences of interest in the genome of eukaryotic cells. This development has provided researchers with the tools to delve into the genetic causes of diseases, facilitating the establishment of precise cellular and animal models for the analysis of pathological processes. The revolutionary influence of genome editing extends across multiple fields, encompassing both basic research and its practical uses in biotechnology and biomedical sciences. Additionally, the advent of targeted nucleases has significantly accelerated the development of genetically developed organisms, all of which play a crucial part in enhancing human disease modeling and identification of novel therapeutics (Sander et al., 2011b; Hwang et al., 2013; Wang et al., 2013). Chimeric editing tools offer a robust way for targeted gene alteration, allowing the precise manipulation of cellular behavior and function, as well as the assessment of gene function. These tools have given researchers the ability to employ genetically engineered animals to achieve a more profound comprehension of the origins of diverse ailments and to elucidate the molecular mechanistics for the development of more effective treatment approaches.

2.4.1 Cancer

The promising prospects of editing methods are mainly relevant in the study of mutant tumor suppressor genes and oncogenes (Vogelstein et al., 2013). The advancements in genome editing approaches have successfully enabled gene editing, with its applicability in the modification of the targeted tumor T cells and hematopoietic stem cells (HSCs). This method has presented novel concepts for genetic modifications and has made strides into the realm of cancer research. ZFN-mediated targeting has established itself as a fundamental approach for programmable DNA cleavage, permitting the alteration of multiple genes in diverse organisms. This has significantly impacted the field of genome editing technologies. In 1994, a key milestone was reached with the success of ZFN-driven gene disruption, which involved engineering a three-finger protein to silence the BCR-ABL human oncogene in a mouse cell line (Choo et al., 1994). Targeted modulation was effectively accomplished through engineered ZFNs, disrupting endogenous T cell receptor (TCR) β- and α-chain genes. Treated lymphocytes exhibited a lack of surface expression of cluster of differentiation 3 (CD3) TCR and underwent expansion with elevated levels of interleukin-7 (IL-7) and IL-15 (Provasi et al., 2012). An innovative therapeutic ZFN has been carried out in an *in vivo* model of adult T cell leukemia (ATL) that primarily focuses on the long terminal repeat (LTR) of human T cell leukemia virus type 1 (HTLV-1) and has proven successful in destroying the infected cells (Tanaka et al., 2013). This illustrates the capability of tailored nucleases to facilitate precise genomic alterations in pertinent cell types (Piganeau et al., 2013). Additionally, the deployment of a cell-penetrating peptide (CPP) linked to a mammalian target of rapamycin (mTOR)–specific ZFN aimed to target human epidermal growth factor receptor 2 (HER2)–positive cells that effectively inhibited critical cancer signaling pathways and rendered the mTOR locus inactive. This finding indicates the potential for innovative molecular targeted therapies for breast cancer and various other malignancies. The introduction of alterations to a coding sequence through NHEJ or HDR mechanisms to effectively impede indigenous genes neurotrophin 3 (NTF3) and C-C chemokine receptor type 5 (CCR5) in leukemia cells of humans marked a turning point for TALENs and showed that they may well be engineered for particular endogenous gene cleavage (Miller et al., 2011). It is interesting to note that TALENs significantly reduced cytotoxicity as compared to ZFNs when they were compared in two human loci. In addition, relative to the ZFN approach, the CCR5-specific TALEN showed the proficiency to identify the CCR5 target locus while distinguishing it from a highly analogous region in CCR2 (Mussolino et al., 2011). Recent findings have demonstrated that TALEN is a valuable tool for knocking out genes in cancer cells, such as those found in breast cancer, prostate cancer, and hepatocellular carcinoma (HCC), allowing for in-depth exploration of genetic mutations at the molecular level (Cai et al., 2016; Xiao et al., 2018). Using the CRISPR-Cas method, organoid tumor models have been developed as well (Sanchez-Rivera and Jacks, 2015). Anticipating the immense potential of the gene-editing approaches, particularly CRISPR-Cas9, it could be concluded that accurate cancer models could be created that may help in analyzing functional cancer genomics, progression, and development of advanced treatment strategies.

2.4.2 Viral Diseases

The beginning of gene-editing approaches paved a novel pathway for the development of antiviral therapies aimed at combating infectious diseases. These technologies are capable of targeting either the host genes that the virus relies on or the viral genes that are critical for its replication, as evidenced by successful instances (Cox et al., 2015). A furthermost prominent success in this field was achieved through disruption of HIV coreceptor CCR5 mediated by ZFN, thus enabling the introduction of resistant HIV into CD4+ T cells (Perez et al., 2008) as well as CD34+ hematopoietic stem/progenitor cells (HSPCs), respectively (Holt et al., 2010). In individuals with HIV/AIDS, the infusion of gene-modified T cells in phase I of clinical trials has demonstrated safety and tolerability (Tebas et al., 2014). According to Gaj et al. (2016), specific nucleases can be effectively utilized alongside viral vectors, such as adeno-associated viruses (AAV), for in situ genome modification. A strong correlation of cervical cancer has been found to be associated with persistent expression of E6 and E7 oncogenes of human papillomavirus (HPV) (Moody and Laimins, 2010). Utilizing targeted mutagenesis on these high-risk genes with the editing gene approaches may pave the way for innovative genetic therapies, potentially reversing cervical cancer in its original location. Ding et al. (2014) successfully developed a ZFN that can precisely recognize and cut the E7 DNA of HPV16/18. Their research revealed that the ZFN-mediated disruption of HPV16/18 E7 DNA led to a direct reduction in E7 expression in cervical cancer cells, which in turn caused effective specific apoptosis and growth inhibition in cancerous cells that are positive for HPV16/18, as shown *in vitro* study (Ding et al., 2014). By employing TALEN technology to focus on HPV oncogenes E6 and E7, researchers were able to reactivate the tumor suppressor genes p53 and RB1 in the host DNA. This innovative method not only diminished HPV-positive cell lines tumorigenicity but also yielded improved results in cervical malignancy within transgenic mouse models (Hu et al., 2015). Furthermore, the amalgamation of CRISPR-Cas9 with HPV16 E6/E7 enhanced the vulnerability of cervical cancer cells to cisplatin, indicating a promising avenue for cervical cancer treatment (Zhen et al., 2016).

2.4.3 Cardiovascular Disorders

The prevalence of cardiovascular disease (CVD) presents a grave risk to human well-being and ranks as the primary contributor to mortality in several advanced economies. It is often observed that various types of CVDs are linked to specific genetic mutations or a combination of rare inherited heterozygous mutations (Li et al., 2015). In the discipline of cardiovascular medicine, these methodologies are regarded as promising prospects for the clinical integration of germline genome editing techniques. A congenital form of heart diseases is long QT syndrome (LQTS) that is autosomal-dominant, with hybrid mutations in various genes that can lead to the onset of LQTS. For example, mutations in the hERG gene that encodes potassium voltage-gated channel's pore-forming subunit alpha protein can lead to fatal ventricular arrhythmia (Wallace et al., 2019). The hERG gene is primarily expressed in myocardial and smooth muscle cells. Using the CRISPR approach to repair gene mutations in hERG cardiomyocytes may offer a promising treatment approach for LQTS (Wallace et al., 2019). Marfan syndrome (MFS), which arises from a heterozygous mutation in the fibrillin-1 (FBN1) gene, is characterized by a range of cardiovascular issues and skeletal deformities.

2.4.4 Metabolic Disorders

Metabolic disorders consist of a collection of syndromes resulting from genetic and environmental influences. Screening of functional gens, gene therapy, and the modeling of metabolic disorders are all applications of gene-editing methods.

Hypercholesterolemia is a condition marked by increased levels of cholesterol in the blood and is often the result of mutations in proprotein convertase subtilisin/kexin type 9 (PCSK9) or low-density lipoprotein receptor (LDLR) (Musunuru et al., 2021). The role of PCSK9 is to facilitate LDLR degradation, whereas LDL in the bloodstream is removed by LDLR, respectively (Musunuru et al., 2021). Two related studies have illustrated the application of lipid nanoparticles (LNPs) for the delivery of ABE mRNA and sgRNA, aimed at inhibiting the Pcsk9 expression in both cynomolgus monkeys and mice through the excision of canonical splice site (Musunuru et al., 2021; Rothgangl et al., 2021). In their research, Musunuru et al. (2021) employed ABE8.8, resulting in a 66% editing efficiency in the liver of monkeys, which corresponded to reduction in LDL cholesterol (60%) and PCSK9 protein levels (90%) in the bloodstream. In monkeys, an editing rate of 34% was observed, along with 32% reduction in PCSK9 levels, and decreased in levels of LDL by 14% by applying ABEmax (Rothgangl et al., 2021). The findings of these studies revealed that off-target DNA editing and bystander occurred at minimal levels, respectively. Additionally, the off-target RNA editing levels noted were pretty low, which disappeared 17 days following injection due to the transient nature of LNPs (Rothgangl et al., 2021). These findings imply that ABE mRNA delivery via LNPs is precise and effective enough to warrant further optimization for human use. An innovative CRISPR interference technique known as evolved engineered transcriptional repressor (EvoETR) has been established to facilitate long-lasting epigenetic silencing following the temporary introduction of epigenome editors in vivo in mice (Cappelluti et al., 2024). To activate the epigenome editors of mRNA, LNPs were utilized for delivery of ZFPs or dCas9, thereby effectively silencing PCSK9 expression in the murine model (Cappelluti et al., 2024). ZFP administration caused 75% decrease in blood levels of PCSK9, while dCas9 led to a 70% decrease, sustained until day 43.

A severe and progressive condition is transthyretin amyloidosis (ATTR amyloidosis), marked by the accumulation of misfolded transthyretin (TTR) protein within various tissues (Finn et al., 2018). A preclinical investigation in rat and mouse models that utilized LNPs demonstrated the effective delivery of SpCas9 nuclease along with a chemically altered Ttr-specific sgRNA, thereby achieving a 70% editing efficiency in the liver and a 90% reduction in serum TTR levels, while minimal editing was noted in the spleen and the kidney (Finn et al., 2018).

Alpha-1 antitrypsin deficiency (AATD) is a hereditary condition that occurs in individuals with alterations in the SERPINA1 gene. This disorder impacts the liver's ability to produce alpha-1 antitrypsin (AAT). A study utilized Cas9 nuclease, along with a specific sgRNA and an HDR template, where researchers administered AAV8 and AAV9 via intraperitoneal injection in a mouse model harboring the E342K mutation in SERPINA1. The findings revealed a 2% precise correction and 22% on-target insertions or deletions, contributing to a partial recovery of AAT levels in the serum of the mice, with no observed off-target effects (Song et al., 2018).

2.4.5 Neurological Disorders

The exploration of therapeutic gene editing as a means to combat neurological disorders, such as amyotrophic lateral sclerosis (ALS), Alzheimer's disease (AD), Huntington's disease (HD), and Niemann-Pick disease, is currently underway. Researchers are focused on decelerating or stopping the progression of these diseases by specifically targeting and modifying the genes linked to these ailments. ALS, a terminal neurodegenerative condition, predominantly affects adults and is marked by the gradual degeneration of neurons, particularly motor, in the brain and the spinal cord, thereby leading to muscle wasting, paralysis, and ultimately, death (Rowland and Shneider, 2001). A recent study explored the SOD1G93A gene in an ALS mouse model that involved the systemic intake of AAV9 vector that encodes SaCas9 nuclease and a specific sgRNA during the neonatal period, causing a frameshift decline in hSOD1G93A transgene expression (Rosen et al., 1993). The study also revealed a 0.4% editing rate, along with reduction in SOD1 mutant by threefold, along with enhanced motor neuron survival, diminished atrophy in muscle, and a 25% increase in survival rate (Rosen et al., 1993). In a preclinical study addressing AD, researchers employed two AAV9 vectors to deliver SpCas9 nuclease and a mutant allele-specific sgRNA. This strategy aimed to generate indels and silence the APPswe allele expression in the hippocampus of adult APPswe transgenic mice (Gyorgy et al., 2018). The findings indicated an editing efficiency of 2%, along with 100 copies of APPswe gene and lower levels of pathogenic Aβ secretion (Gyorgy et al., 2018). Utilizing AAV2 vectors, a team of researchers established an allele-specific gene-editing strategy that involves SpCas9 nuclease along with a pair of sgRNAs (Monteys et al., 2017). This technique was implemented to induce targeted deletions that silence the expression of the huntingtin gene (HTT) in a BacHD mouse model that carries an altered form of the mutant huntingtin (mHTT) gene featuring 97 CAG repeats (Monteys et al., 2017). The study's outcomes demonstrated a 40% reduction in the production of the toxic mHTT protein while preserving the integrity of the wild-type allele (Monteys et al., 2017).

2.4.6 Muscular Disorders

A significant potential has been shown by gene-editing processes in treating various genetic disorders of muscles. A degenerative type of muscle illness is the genetically inherited muscular dystrophy that predominantly causes muscle loss. In the developed mouse models of Duchenne muscular dystrophy (DMD), enhancement of muscle function, along with the restoration of expression of the dystrophin gene, was achieved via the *in vivo* gene therapy (Nelson et al., 2016). Initial experimental investigations utilized Cas9 nuclease along with sgRNA Cas to rectify or eradicate the mutant exon 23, thereby reframing DMD in the X-linked muscular dystrophy mouse model (mdx) (Nelson et al., 2016). Additionally, Larcher and colleagues successfully created Dmd^{mdx} rats by employing TALENs to target exon 23 of DMD (Larcher et al., 2014). The observed rats exhibited a notable decline in muscle strength, along with a reduction in spontaneous motor activity (Larcher et al., 2014). The findings of targeting efficiency revealed that CRISPR-Cas9 may cause mosaic mutations in 87% of the dystrophin alleles within the monkey muscle (Chen et al., 2015a, 2015b). In an AAV vector, the systemic administration of gRNA and *Staphylococcus aureus* Cas9 restored the reading frame in DMD, which led to the expression in about 40% of the cardiac muscle fibers (El Refaey et al., 2017). As a result, improvements in cardiac myofiber architecture, papillary muscle contractibility, and cardiac fibrosis, were reported (El Refaey et al., 2017). A similar study carried in a DMD dog model revealed enhanced dystrophin expression in the skeletal and cardiac muscle after treatment with intravenous AAV9 carrying Cas9 and gRNA (Amoasii et al., 2018).

2.5 FUTURE PROSPECTS

The current era presents an exciting period for genetics, thanks in large part to progress in genetic analysis and manipulation. The influence of developments in high-throughput DNA sequencing and genome editing has been widespread. Recent advancements in gene modifying approaches have made it feasible to accomplish specific genetic modifications across virtually all cell types and organisms (Carrol, 2014; Gaj et al., 2013). A crucial factor in enlightening the efficacy of editing approaches is the capacity to form a targeted DSB at the specific chromosomal location of interest. The realization that inducing such a break could facilitate gene targeting and localized mutagenesis emerged from extensive research into DNA damage and repair processes. Currently, there are three potent classes of nucleases at our disposal, which can be instructed to create DSBs at virtually any specified target (Carrol, 2014). While the CRISPR-Cas platform at this time is the most commonly used approach in research facilities globally, the other two are still utilized for research studies as well as in other allied fields (medical and agricultural). The use of editing methods is projected to persist as a valuable tool in research, commercial applications, and medical fields. The question remains whether CRISPR-Cas is the ultimate programmable nuclease or if there are superior alternatives on the horizon. While improvements may be made, they are likely to be variations on existing themes rather than entirely new concepts. Various approaches, from triples-forming oligonucleotides to peptide nucleic acids to polyamines, have been explored without achieving the desired cleavage efficiency and recognition range (Kim et al., 2006; Doss et al., 2006).

Artificial intelligence (AI) significantly enhances the accuracy, efficiency, and cost-effectiveness of genome editing for a range of diseases, including sickle cell anemia and thalassemia. AI-driven models are employed to create guide RNAs (gRNAs) for CRISPR-Cas systems, utilizing tools such as DeepCRISPR, CRISTA, and DeepHF to predict the most effective gRNAs for specific target sequences. The utilization of artificial intelligence in genetic editing and diagnostics is crucial and has the potential to reform the health-care area. CRISPR-Cas9 facilitates accurate and targeted alterations to the organism's genomic makeup, marking a significant advancement in the field of biotechnology (Tyagi et al., 2020). The combination of artificial intelligence with CRISPR technology improves the genetic editing and diagnostics processes, providing novel insights, enhanced capabilities, and new opportunities for the manipulation and comprehension of genetic information. Recent progress in multiomics technologies, which generate extensive data from diverse sources, including modifications in DNA, genes, RNA, and proteins, has rendered AI indispensable for the analysis of medical data (Hamet and Tremblay, 2017). The integration of deep learning (DL) and machine learning (ML) models has been pivotal in the analysis and comprehension of extensive and intricate genomic datasets. These research efforts hold the promise of discovering more relevant features for AI models, which could enhance their predictive accuracy concerning editing outcomes, including off-target edits. In the field of oncology, AI-based approaches can analyze genomic datasets to differentiate cancer subtypes, while CRISPR-based editing plays a crucial role in creating immune cells that can specifically target these subtypes or interfere with oncogenes (Katti et al., 2022). Leenay and his team (2019) introduced an ML algorithm named SPROUT which reliably predicts the outcome in edited primary T cells. Consequently, the importance of AI methodologies is set to rise. The field of genome editing is experiencing rapid development, with numerous potential applications and seemingly endless possibilities. In light of this progress, it is crucial to carefully evaluate the best uses of the technology, ensure its safety and effectiveness, and consider how to make its benefits accessible to those who need it most.

2.6 CONCLUSION

Genome editing tools are becoming increasingly important in genetic engineering, despite their pros and cons. These tools are feasible, effective, and now more budget-friendly than they were ten years ago. While TALEN is emerging as a more effective option compared to ZFN, Crispr-Cas9 possesses distinct qualities that set it apart. Drawing direct comparisons between these methods is complex due to their specific strengths and weaknesses. Therefore, the selection of a technique should be guided by the desired impact, target sequence, and cost considerations. Moreover, the CRISPR-Cas9 technology presents a valuable tool for manipulating noncoding regions of the cancer genome, expediting research into poorly understood areas. The substantial strides made in the development of engineered nucleases have enabled the translation of genome editing from theory to clinical application. With increased accessibility, patients are expected to experience significant advantages from this innovative technology. A multitude of human diseases are currently being researched worldwide, and these technologies are expected to undergo further refinement, potentially becoming mainstream therapeutics for numerous diseases in the future. Maintaining awareness of the possible cytotoxic effects of the materials involved in editing is of utmost importance. Moreover, it is very important to continuously conduct research and assess risks to advance the field responsibly and sustainably.

REFERENCES

Ain, Qurrat Ul, Jee Young Chung, and Yong-Hee Kim. "Current and future delivery systems for engineered nucleases: ZFN, TALEN and RGEN." *Journal of Controlled Release* 205 (2015): 120–127.

Altae-Tran, Han, Soumya Kannan, F. Esra Demircioglu, Rachel Oshiro, Suchita P. Nety, Luke J. McKay, Mensur Dlakić et al. "The widespread IS200/IS605 transposon family encodes diverse programmable RNA-guided endonucleases." *Science* 374, no. 6563 (2021): 57–65.

Amoasii, Leonela, John C. W. Hildyard, Hui Li, Efrain Sanchez-Ortiz, Alex Mireault, Daniel Caballero, Rachel Harron et al. "Gene editing restores dystrophin expression in a canine model of Duchenne muscular dystrophy." *Science* 362, no. 6410 (2018): 86–91.

Anzalone, Andrew V., Luke W. Koblan, and David R. Liu. "Genome editing with CRISPR–Cas nucleases, base editors, transposases and prime editors." *Nature Biotechnology* 38, no. 7 (2020): 824–844.

Anzalone, Andrew V., Peyton B. Randolph, Jessie R. Davis, Alexander A. Sousa, Luke W. Koblan, Jonathan M. Levy, Peter J. Chen et al. "Search-and-replace genome editing without double-strand breaks or donor DNA." *Nature* 576, no. 7785 (2019): 149–157.

Aryan, Azadeh, Michelle A. E. Anderson, Kevin M. Myles, and Zach N. Adelman. "TALEN-based gene disruption in the dengue vector Aedes aegypti." *PLoS One* 8, no. 3 (2013): e60082.

Bao, Weidong, and Jerzy Jurka. "Homologues of bacterial TnpB_ IS605 are widespread in diverse eukaryotic transposable elements." *Mobile DNA* 4 (2013): 1–16.

Cai, Yanyan, Jonathan Crowther, Tibor Pastor, Layka Abbasi Asbagh, Maria Francesca Baietti, Magdalena De Troyer, Iria Vazquez et al. "Loss of chromosome 8p governs tumor progression and drug response by altering lipid metabolism." *Cancer Cell* 29, no. 5 (2016): 751–766.

Cappelluti, Martino Alfredo, Valeria Mollica Poeta, Sara Valsoni, Piergiuseppe Quarato, Simone Merlin, Ivan Merelli, and Angelo Lombardo. "Durable and efficient gene silencing in vivo by hit-and-run epigenome editing." *Nature* 627, no. 8003 (2024): 416–423.

Carroll, Dana. "Genome engineering with zinc-finger nucleases." *Genetics* 188, no. 4 (2011): 773–782.

Carroll, Dana. "Genome engineering with targetable nucleases." *Annual Review of Biochemistry* 83, no. 1 (2014): 409–439.

Chen, Yongchang, Yuyu Niu, Yanjiao Li, Zongyong Ai, Yu Kang, Hong Shi, Zheng Xiang et al. "Generation of cynomolgus monkey chimeric fetuses using embryonic stem cells." *Cell Stem Cell* 17, no. 1 (2015a): 116–124.

Chen, Yongchang, Yinghui Zheng, Yu Kang, Weili Yang, Yuyu Niu, Xiangyu Guo, Zhuchi Tu et al. "Functional disruption of the dystrophin gene in rhesus monkey using CRISPR/Cas9." *Human Molecular Genetics* 24, no. 13 (2015b): 3764–3774.

Chevalier, Brett S., and Barry L. Stoddard. "Homing endonucleases: Structural and functional insight into the catalysts of intron/intein mobility." *Nucleic Acids Research* 29, no. 18 (2001): 3757–3774.

Choo, Yen, Isidro Sanchez-Garcia, and Aaron Klug. "In vivo repression by a site-specific DNA-binding protein designed against an oncogenic sequence." *Nature* 372, no. 6507 (1994): 642–645.

Christian, Michelle, Tomas Cermak, Erin L. Doyle, Clarice Schmidt, Feng Zhang, Aaron Hummel, Adam J. Bogdanove, and Daniel F. Voytas. "Targeting DNA double-strand breaks with TAL effector nucleases." *Genetics* 186, no. 2 (2010): 757–761.

Cimprich, Karlene A., and David Cortez. "ATR: An essential regulator of genome integrity." *Nature Reviews Molecular Cell Biology* 9, no. 8 (2008): 616–627.

Collantes, Juan Carlos, Victor M. Tan, Huiting Xu, Melany Ruiz-Urigüen, Amer Alasadi, Jingjing Guo, Hanlin Tao et al. "Development and characterization of a modular CRISPR and RNA aptamer mediated base editing system." *The CRISPR Journal* 4, no. 1 (2021): 58–68.

Cornu, Tatjana I., Claudio Mussolino, and Toni Cathomen. "Refining strategies to translate genome editing to the clinic." *Nature Medicine* 23, no. 4 (2017): 415–423.

Cox, David Benjamin Turitz, Randall Jeffrey Platt, and Feng Zhang. "Therapeutic genome editing: Prospects and challenges." *Nature Medicine* 21, no. 2 (2015): 121–131.

De Souza, Natalie. "Primer: Genome editing with engineered nucleases." *Nature Methods* 9, no. 1 (2012): 27–27.

Deng, Dong, Chuangye Yan, Xiaojing Pan, Magdy Mahfouz, Jiawei Wang, Jian-Kang Zhu, Yigong Shi, and Nieng Yan. "Structural basis for sequence-specific recognition of DNA by TAL effectors." *Science* 335, no. 6069 (2012): 720–723.

Ding, Wencheng, Zheng Hu, Da Zhu, Xiaohui Jiang, Lan Yu, Xiaoli Wang, Changlin Zhang et al. "Zinc finger nucleases targeting the human papillomavirus E7 oncogene induce E7 disruption and a transformed phenotype in HPV16/18-positive cervical cancer cells." *Clinical Cancer Research* 20, no. 24 (2014): 6495–6503.

Doss, Raymond M., Michael A. Marques, Shane Foister, David M. Chenoweth, and Peter B. Dervan. "Programmable oligomers for minor groove DNA recognition." *Journal of the American Chemical Society* 128, no. 28 (2006): 9074–9079.

Eichler, Evan E. "Genetic variation, comparative genomics, and the diagnosis of disease." *New England Journal of Medicine* 381, no. 1 (2019): 64–74.

El Refaey, Mona, Li Xu, Yandi Gao, Benjamin D. Canan, T. M. Ayodele Adesanya, Sarah C. Warner, Keiko Akagi et al. "In vivo genome editing restores dystrophin expression and cardiac function in dystrophic mice." *Circulation Research* 121, no. 8 (2017): 923–929.

Faure, Guilhem, Kira S. Makarova, and Eugene V. Koonin. "CRISPR–Cas: Complex functional networks and multiple roles beyond adaptive immunity." *Journal of Molecular Biology* 431, no. 1 (2019): 3–20.

Finn, Jonathan D., Amy Rhoden Smith, Mihir C. Patel, Lucinda Shaw, Madeleine R. Youniss, Jane van Heteren, Tanner Dirstine et al. "A single administration of CRISPR/Cas9 lipid nanoparticles achieves robust and persistent in vivo genome editing." *Cell Reports* 22, no. 9 (2018): 2227–2235.

Gaj, Thomas, Benjamin E. Epstein, and David V. Schaffer. "Genome engineering using adeno-associated virus: Basic and clinical research applications." *Molecular Therapy* 24, no. 3 (2016): 458–464.

Gaj, Thomas, Charles A. Gersbach, and Carlos F. Barbas. "ZFN, TALEN, and CRISPR/Cas-based methods for genome engineering." *Trends in Biotechnology* 31, no. 7 (2013): 397–405.

Gasiunas, Giedrius, Rodolphe Barrangou, Philippe Horvath, and Virginijus Siksnys. "Cas9–crRNA ribonucleoprotein complex mediates specific DNA cleavage for adaptive immunity in bacteria." *Proceedings of the National Academy of Sciences* 109, no. 39 (2012): E2579–E2586.

Gaudelli, Nicole M., Alexis C. Komor, Holly A. Rees, Michael S. Packer, Ahmed H. Badran, David I. Bryson, and David R. Liu. "Programmable base editing of A• T to G• C in genomic DNA without DNA cleavage." *Nature* 551, no. 7681 (2017): 464–471.

Gaudelli, Nicole M., Dieter K. Lam, Holly A. Rees, Noris M. Solá-Esteves, Luis A. Barrera, David A. Born, Aaron Edwards et al. "Directed evolution of adenine base editors with increased activity and therapeutic application." *Nature Biotechnology* 38, no. 7 (2020): 892–900.

Ghosh, Debarati, Prabhadevi Venkataramani, Saikat Nandi, and Sonali Bhattacharjee. "CRISPR–Cas9 a boon or bane: The bumpy road ahead to cancer therapeutics." *Cancer Cell International* 19 (2019): 1–10.

Gyorgy, Bence, Camilla Lööv, Mikołaj P. Zaborowski, Shuko Takeda, Benjamin P. Kleinstiver, Caitlin Commins, Ksenia Kastanenka et al. "CRISPR/Cas9 mediated disruption of the Swedish APP allele as a therapeutic approach for early-onset Alzheimer's disease." *Molecular Therapy-Nucleic Acids* 11 (2018): 429–440.

Hamet, Pavel, and Johanne Tremblay. "Artificial intelligence in medicine." *Metabolism* 69 (2017): S36–S40.

He, Susu, Catherine Guynet, Patricia Siguier, Alison B. Hickman, Fred Dyda, Mick Chandler, and Bao Ton-Hoang. "IS 200/IS 605 family single-strand transposition: Mechanism of IS 608 strand transfer." *Nucleic Acids Research* 41, no. 5 (2013): 3302–3313.

Hockemeyer, Dirk, Haoyi Wang, Samira Kiani, Christine S. Lai, Qing Gao, John P. Cassady, Gregory J. Cost et al. "Genetic engineering of human pluripotent cells using TALE nucleases." *Nature Biotechnology* 29, no. 8 (2011): 731–734.

Holt, Nathalia, Jianbin Wang, Kenneth Kim, Geoffrey Friedman, Xingchao Wang, Vanessa Taupin, Gay M. Crooks et al. "Human hematopoietic stem/progenitor cells modified by zinc-finger nucleases targeted to CCR5 control HIV-1 in vivo." *Nature Biotechnology* 28, no. 8 (2010): 839–847.

Hosfield, David J., Clifford D. Mol, Binghui Shen, and John A. Tainer. "Structure of the DNA repair and replication endonuclease and exonuclease FEN-1: Coupling DNA and PCNA binding to FEN-1 activity." *Cell* 95, no. 1 (1998): 135–146.

Hu, Zheng, Wencheng Ding, Da Zhu, Lan Yu, Xiaohui Jiang, Xiaoli Wang, Changlin Zhang et al. "TALEN-mediated targeting of HPV oncogenes ameliorates HPV-related cervical malignancy." *The Journal of Clinical Investigation* 125, no. 1 (2015): 425–436.

Hwang, Woong Y., Yanfang Fu, Deepak Reyon, Morgan L. Maeder, Shengdar Q. Tsai, Jeffry D. Sander, Randall T. Peterson, J. R. Joanna Yeh, and J. Keith Joung. "Efficient genome editing in zebrafish using a CRISPR-Cas system." *Nature Biotechnology* 31, no. 3 (2013): 227–229.

Ishino, Yoshizumi, Hideo Shinagawa, Kozo Makino, Mitsuko Amemura, and Atsuo Nakata. "Nucleotide sequence of the iap gene, responsible for alkaline phosphatase isozyme conversion in Escherichia coli, and identification of the gene product." *Journal of Bacteriology* 169, no. 12 (1987): 5429–5433.

Jinek, Martin, Krzysztof Chylinski, Ines Fonfara, Michael Hauer, Jennifer A. Doudna, and Emmanuelle Charpentier. "A programmable dual-RNA–guided DNA endonuclease in adaptive bacterial immunity." *Science* 337, no. 6096 (2012): 816–821.

Kaniecki, Kyle, Luisina De Tullio, and Eric C. Greene. "A change of view: Homologous recombination at single-molecule resolution." *Nature Reviews Genetics* 19, no. 4 (2018): 191–207.

Kapitonov, Vladimir V., Kira S. Makarova, and Eugene V. Koonin. "ISC, a novel group of bacterial and archaeal DNA transposons that encode Cas9 homologs." *Journal of Bacteriology* 198, no. 5 (2016): 797–807.

Karvelis, Tautvydas, Gytis Druteika, Greta Bigelyte, Karolina Budre, Rimante Zedaveinyte, Arunas Silanskas, Darius Kazlauskas, Česlovas Venclovas, and Virginijus Siksnys. "Transposon-associated TnpB is a programmable RNA-guided DNA endonuclease." *Nature* 599, no. 7886 (2021): 692–696.

Katti, Alyna, Bianca J. Diaz, Christina M. Caragine, Neville E. Sanjana, and Lukas E. Dow. "CRISPR in cancer biology and therapy." *Nature Reviews Cancer* 22, no. 5 (2022): 259–279.

Kim, Hyongbum, and Jin-Soo Kim. "A guide to genome engineering with programmable nucleases." *Nature Reviews Genetics* 15, no. 5 (2014): 321–334.

Kim, Ki-Hyun, Peter E. Nielsen, and Peter M. Glazer. "Site-specific gene modification by PNAs conjugated to psoralen." *Biochemistry* 45, no. 1 (2006): 314–323.

Klompe, Sanne E., Nora Jaber, Leslie Y. Beh, Jason T. Mohabir, Aude Bernheim, and Samuel H. Sternberg. "Evolutionary and mechanistic diversity of Type IF CRISPR-associated transposons." *Molecular Cell* 82, no. 3 (2022): 616–628.

Komor, Alexis C., Yongjoo B. Kim, Michael S. Packer, John A. Zuris, and David R. Liu. "Programmable editing of a target base in genomic DNA without double-stranded DNA cleavage." *Nature* 533, no. 7603 (2016): 420–424.

Komor, Alexis C., Kevin T. Zhao, Michael S. Packer, Nicole M. Gaudelli, Amanda L. Waterbury, Luke W. Koblan, Y. Bill Kim, Ahmed H. Badran, and David R. Liu. "Improved base excision repair inhibition and bacteriophage Mu Gam protein yields C: G-to-T: A base editors with higher efficiency and product purity." *Science Advances* 3, no. 8 (2017): eaao4774.

Kosicki, Michael, Kärt Tomberg, and Allan Bradley. "Repair of double-strand breaks induced by CRISPR–Cas9 leads to large deletions and complex rearrangements." *Nature Biotechnology* 36, no. 8 (2018): 765–771.

Larcher, Thibaut, Aude Lafoux, Laurent Tesson, Severine Remy, Virginie Thepenier, Virginie François, Caroline Le Guiner et al. "Characterization of dystrophin deficient rats: A new model for Duchenne muscular dystrophy." *PLoS One* 9, no. 10 (2014): e110371.

Lee, Jae Seong, Thomas Beuchert Kallehauge, Lasse Ebdrup Pedersen, and Helene Faustrup Kildegaard. "Site-specific integration in CHO cells mediated by CRISPR/Cas9 and homology-directed DNA repair pathway." *Scientific Reports* 5, no. 1 (2015): 8572.

Leenay, Ryan T., Amirali Aghazadeh, Joseph Hiatt, David Tse, Theodore L. Roth, Ryan Apathy, Eric Shifrut et al. "Large dataset enables prediction of repair after CRISPR–Cas9 editing in primary T cells." *Nature Biotechnology* 37, no. 9 (2019): 1034–1037.

Li, Alexander H., Alanna C. Morrison, Christie Kovar, L. Adrienne Cupples, Jennifer A. Brody, Linda M. Polfus, Bing Yu et al. "Analysis of loss-of-function variants and 20 risk factor phenotypes in 8,554 individuals identifies loci influencing chronic disease." *Nature Genetics* 47, no. 6 (2015): 640–642.

Li, Ting, Bo Liu, Martin H. Spalding, Donald P. Weeks, and Bing Yang. "High-efficiency TALEN-based gene editing produces disease-resistant rice." *Nature Biotechnology* 30, no. 5 (2012): 390–392.

Makarova, Kira S., Daniel H. Haft, Rodolphe Barrangou, Stan J. J. Brouns, Emmanuelle Charpentier, Philippe Horvath, Sylvain Moineau et al. "Evolution and classification of the CRISPR–Cas systems." *Nature Reviews Microbiology* 9, no. 6 (2011): 467–477.

Makarova, Kira S., Yuri I. Wolf, Omer S. Alkhnbashi, Fabrizio Costa, Shiraz A. Shah, Sita J. Saunders, Rodolphe Barrangou et al. "An updated evolutionary classification of CRISPR–Cas systems." *Nature Reviews Microbiology* 13, no. 11 (2015): 722–736.

Mali, Prashant, Luhan Yang, Kevin M. Esvelt, John Aach, Marc Guell, James E. DiCarlo, Julie E. Norville, and George M. Church. "RNA-guided human genome engineering via Cas9." *Science* 339, no. 6121 (2013): 823–826.

Miller, Jeffrey C., Siyuan Tan, Guijuan Qiao, Kyle A. Barlow, Jianbin Wang, Danny F. Xia, Xiangdong Meng et al. "A TALE nuclease architecture for efficient genome editing." *Nature Biotechnology* 29, no. 2 (2011): 143–148.

Monteys, Alex Mas, Shauna A. Ebanks, Megan S. Keiser, and Beverly L. Davidson. "CRISPR/Cas9 editing of the mutant huntingtin allele in vitro and in vivo." *Molecular Therapy* 25, no. 1 (2017): 12–23.

Moody, Cary A., and Laimonis A. Laimins. "Human papillomavirus oncoproteins: Pathways to transformation." *Nature Reviews Cancer* 10, no. 8 (2010): 550–560.

Mussolino, Claudio, Robert Morbitzer, Fabienne Lütge, Nadine Dannemann, Thomas Lahaye, and Toni Cathomen. "A novel TALE nuclease scaffold enables high genome editing activity in combination with low toxicity." *Nucleic Acids Research* 39, no. 21 (2011): 9283–9293.

Musunuru, Kiran, Alexandra C. Chadwick, Taiji Mizoguchi, Sara P. Garcia, Jamie E. DeNizio, Caroline W. Reiss, Kui Wang et al. "In vivo CRISPR base editing of PCSK9 durably lowers cholesterol in primates." *Nature* 593, no. 7859 (2021): 429–434.

Nelson, Christopher E., Chady H. Hakim, David G. Ousterout, Pratiksha I. Thakore, Eirik A. Moreb, Ruth M. Castellanos Rivera, Sarina Madhavan et al. "In vivo genome editing improves muscle function in a mouse model of Duchenne muscular dystrophy." *Science* 351, no. 6271 (2016): 403–407.

O'Driscoll, Mark, and Penny A. Jeggo. "The role of double-strand break repair—insights from human genetics." *Nature Reviews Genetics* 7, no. 1 (2006): 45–54.

Perez, Elena E., Jianbin Wang, Jeffrey C. Miller, Yann Jouvenot, Kenneth A. Kim, Olga Liu, Nathaniel Wang et al. "Establishment of HIV-1 resistance in CD4+ T cells by genome editing using zinc-finger nucleases." *Nature Biotechnology* 26, no. 7 (2008): 808–816.

Piganeau, Marion, Hind Ghezraoui, Anne De Cian, Lionel Guittat, Mark Tomishima, Loic Perrouault, Oliver René et al. "Cancer translocations in human cells induced by zinc finger and TALE nucleases." *Genome Research* 23, no. 7 (2013): 1182–1193.

Provasi, Elena, Pietro Genovese, Angelo Lombardo, Zulma Magnani, Pei-Qi Liu, Andreas Reik, Victoria Chu et al. "Editing T cell specificity towards leukemia by zinc finger nucleases and lentiviral gene transfer." *Nature Medicine* 18, no. 5 (2012): 807–815.

Ran, F. A., Patrick D. Hsu, Jason Wright, Vineeta Agarwala, David A. Scott, and Feng Zhang. "Genome engineering using the CRISPR-Cas9 system." *Nature Protocols* 8, no. 11 (2013a): 2281–2308.

Ran, F. Ann, Patrick D. Hsu, Chie-Yu Lin, Jonathan S. Gootenberg, Silvana Konermann, Alexandro E. Trevino, David A. Scott et al. "Double nicking by RNA-guided CRISPR Cas9 for enhanced genome editing specificity." *Cell* 154, no. 6 (2013b): 1380–1389.

Rees, Holly A., Alex C. Minella, Cameron A. Burnett, Alexis C. Komor, and Nicole M. Gaudelli. "CRISPR-derived genome editing therapies: Progress from bench to bedside." *Molecular Therapy* 29, no. 11 (2021): 3125–3139.

Rickman, Kimberly, and Agata Smogorzewska. "Advances in understanding DNA processing and protection at stalled replication forks." *Journal of Cell Biology* 218, no. 4 (2019): 1096–1107.

Rosen, Daniel R., Teepu Siddique, David Patterson, Denise A. Figlewicz, Peter Sapp, Afif Hentati, Deirdre Donaldson et al. "Mutations in Cu/Zn superoxide dismutase gene are associated with familial amyotrophic lateral sclerosis." *Nature* 362, no. 6415 (1993): 59–62.

Rosen, Laura E., Holly A. Morrison, Selma Masri, Michael J. Brown, Brendan Springstubb, Django Sussman, Barry L. Stoddard, and Lenny M. Seligman. "Homing endonuclease I-CreI derivatives with novel DNA target specificities." *Nucleic Acids Research* 34, no. 17 (2006): 4791–4800.

Rothgangl, Tanja, Melissa K. Dennis, Paulo J. C. Lin, Rurika Oka, Dominik Witzigmann, Lukas Villiger, Weihong Qi et al. "In vivo adenine base editing of PCSK9 in macaques reduces LDL cholesterol levels." *Nature Biotechnology* 39, no. 8 (2021): 949–957.

Rowland, Lewis P., and Neil A. Shneider. "Amyotrophic lateral sclerosis." *New England Journal of Medicine* 344, no. 22 (2001): 1688–1700.

Sanchez-Rivera, Francisco J., and Tyler Jacks. "Applications of the CRISPR–Cas9 system in cancer biology." *Nature Reviews Cancer* 15, no. 7 (2015): 387–393.

Sander, Jeffry D., Lindsay Cade, Cyd Khayter, Deepak Reyon, Randall T. Peterson, J. Keith Joung, and Jing-Ruey J. Yeh. "Targeted gene disruption in somatic zebrafish cells using engineered TALENs." *Nature Biotechnology* 29, no. 8 (2011a): 697–698.

Sander, Jeffry D., Elizabeth J. Dahlborg, Mathew J. Goodwin, Lindsay Cade, Feng Zhang, Daniel Cifuentes, Shaun J. Curtin et al. "Selection-free zinc-finger-nuclease engineering by context-dependent assembly (CoDA)." *Nature Methods* 8, no. 1 (2011b): 67–69.

Shmakov, Sergey, Aaron Smargon, David Scott, David Cox, Neena Pyzocha, Winston Yan, Omar O. Abudayyeh et al. "Diversity and evolution of class 2 CRISPR–Cas systems." *Nature Reviews Microbiology* 15, no. 3 (2017): 169–182.

Song, Chun-Qing, Dan Wang, Tingting Jiang, Kevin O'Connor, Qiushi Tang, Lingling Cai, Xiangrui Li et al. "In vivo genome editing partially restores alpha1-antitrypsin in a murine model of AAT deficiency." *Human Gene Therapy* 29, no. 8 (2018): 853–860.

Sternberg, Samuel H., Sy Redding, Martin Jinek, Eric C. Greene, and Jennifer A. Doudna. "DNA interrogation by the CRISPR RNA-guided endonuclease Cas9." *Biophysical Journal* 106, no. 2 (2014): 695a.

Strecker, Jonathan, Alim Ladha, Zachary Gardner, Jonathan L. Schmid-Burgk, Kira S. Makarova, Eugene V. Koonin, and Feng Zhang. "RNA-guided DNA insertion with CRISPR-associated transposases." *Science* 365, no. 6448 (2019): 48–53.

Tanaka, A., S. Takeda, R. Kariya, K. Matsuda, E. Urano, S. Okada, and J. Komano. "A novel therapeutic molecule against HTLV-1 infection targeting provirus." *Leukemia* 27, no. 8 (2013): 1621–1627.

Tebas, Pablo, David Stein, Winson W. Tang, Ian Frank, Shelley Q. Wang, Gary Lee, S. Kaye Spratt et al. "Gene editing of CCR5 in autologous CD4 T cells of persons infected with HIV." *New England Journal of Medicine* 370, no. 10 (2014): 901–910.

Tyagi, Swati, Robin Kumar, Aparup Das, So Youn Won, and Pratyoosh Shukla. "CRISPR-Cas9 system: A genome-editing tool with endless possibilities." *Journal of Biotechnology* 319 (2020): 36–53.

Vogelstein, Bert, Nickolas Papadopoulos, Victor E. Velculescu, Shibin Zhou, Luis A. Diaz Jr, and Kenneth W. Kinzler. "Cancer genome landscapes." *Science* 339, no. 6127 (2013): 1546–1558.

Waddell, Candace S., and Nancy L. Craig. "Tn7 transposition: Two transposition pathways directed by five Tn7-encoded genes." *Genes & Development* 2, no. 2 (1988): 137–149.

Wallace, Eimear, Linda Howard, Min Liu, Timothy O'Brien, Deirdre Ward, Sanbing Shen, and Terence Prendiville. "Long QT syndrome: Genetics and future perspective." *Pediatric Cardiology* 40 (2019): 1419–1430.

Wang, Haoyi, Yueh-Chiang Hu, Styliani Markoulaki, G. Grant Welstead, Albert W. Cheng, Chikdu S. Shivalila, Tatyana Pyntikova et al. "TALEN-mediated editing of the mouse Y chromosome." *Nature Biotechnology* 31, no. 6 (2013): 530–532.

Xiao, Lijia, Yuliang Wang, Weicheng Liang, Liping Liu, Nannan Pan, Huimin Deng, Luqian Li, Chang Zou, Franky Leung Chan, and Yiwen Zhou. "LRH-1 drives hepatocellular carcinoma partially through induction of c-myc and cyclin E1, and suppression of p21." *Cancer Management and Research* (2018): 2389–2400.

Yang, Sheng, Shufang Ding, Qianhua Xu, Xiong Li, and Qiong Xiong. "Genetic manipulation by zinc-finger nucleases in rat-induced pluripotent stem cells." *Cellular Reprogramming* 19, no. 3 (2017): 180–188.

Zhen, Shuai, Jiao-Jiao Lu, Li-Jie Wang, Xiao-Min Sun, Jia-Qi Zhang, Xu Li, Wen-Juan Luo, and Le Zhao. "In vitro and in vivo synergistic therapeutic effect of cisplatin with human papillomavirus16 E6/E7 CRISPR/Cas9 on cervical cancer cell line." *Translational Oncology* 9, no. 6 (2016): 498–504.

3 CRISPR-Cas Systems in Genetic Diseases

Zainab Manzoor, Umar Rasool, Arshid Reshi, and Aarif Ali

3.1 OVERVIEW OF GENETIC DISEASES

Subtle variations in the genetic code can lead to a spectrum of serious and debilitating conditions. The hereditary basis of human traits has been recognized since classical eras. In the history of modern medicine, the first documented genetic disorder, alkaptonuria, was noted at the turn of the twentieth century, which initiated the exploration of inherited metabolic anomalies (Speicher et al., 2010). The hereditary material that defines an organism's structure is composed of DNA or RNA. In adult humans, this genetic material can be segmented into compartments that differ in size, heritability, and diversity. The mitochondrial genome, although the smallest at 16.5 kb, is the most abundant and is inherited from the mother, exhibiting little variation among individuals. The nuclear genome, which is significantly larger, contains mutations that are responsible for the majority of genetic diseases. Both the nuclear and mitochondrial genomes are established at conception, but somatic mutations can result in mosaic disorders, in cancer, and may even influence the aging process. Genetic disorders result from defects in a person's genetic makeup. These abnormalities range from modest mutations in a single DNA base to major chromosomal alterations that result in the acquisition or loss of entire chromosomes or chromosome sets. Such illnesses might be inherited from parents or caused by post-natal alterations. *Epigenetics* is the study of how environmental variables can influence genetic mutations, which can occur spontaneously (Nussbaum et al., 2015). Moreover, the manifestation and severity of genetic disorders are often modulated by environmental factors, including diet, exposure to toxins, prenatal conditions, or the cellular environment of the gene. Disorders with both genetic and environmental influences are referred to as multifactorial conditions (Bangham et al., 2023). Almost 6,000 genetic illnesses have been recognized, and new ones are being documented on a regular basis (Kumar et al., 2001). Approximately 1 in 50 persons has a recognized single-gene abnormality, while 1 in 263 has a chromosomal disorder (Kumar et al., 2001). Around 65% of people suffer from health problems caused by congenital genetic mutations (Kumar et al., 2001). Given the huge number of genetic abnormalities, a rare genetic disorder affects approximately 1 in every 21 people. The majority of genetic abnormalities occur infrequently. Our genetic makeup significantly influences various disease processes, including prevalent disorders, due to the numerous variations present in our DNA. These genetic differences can, either individually or in combination,

DOI: 10.1201/9781003537151-3

increase an individual's vulnerability to certain conditions, such as specific cancers, while simultaneously decreasing susceptibility to unrelated diseases, like diabetes. Environmental factors, including lifestyle choices, also play a crucial role in many health issues, as seen with the impact of diet and exercise on diabetes. However, our cellular and physiological reactions to these environmental influences can vary based on our genetic background. The diversity in immune system genetics across the population shapes our responses to infections caused by pathogens. Additionally, most cancers arise from a buildup of genetic alterations that accumulate over a person's lifetime, often influenced by environmental factors. Therefore, a comprehensive understanding of genetics and genomic variation within the human population is essential for grasping disease mechanisms, which in turn lays the groundwork for effective therapies, beneficial treatments, and preventive strategies.

3.2 CLASSES OF GENETIC DISORDERS

Genetic disorders are classified into four classes based on the extent to which they are inherited: disorders caused by mutations in single gene, called single-gene disorder; disorders caused by chromosomal mutations; mitochondrial diseases; and multifactorial diseases.

3.2.1 Chromosomal Disorders

Chromosomal abnormalities, also known as chromosomal aberrations, are physical or numerical anomalies in chromosomes that affect either autosomes, sex chromosomes, or both of them. Gene deficiencies on whole chromosomes or

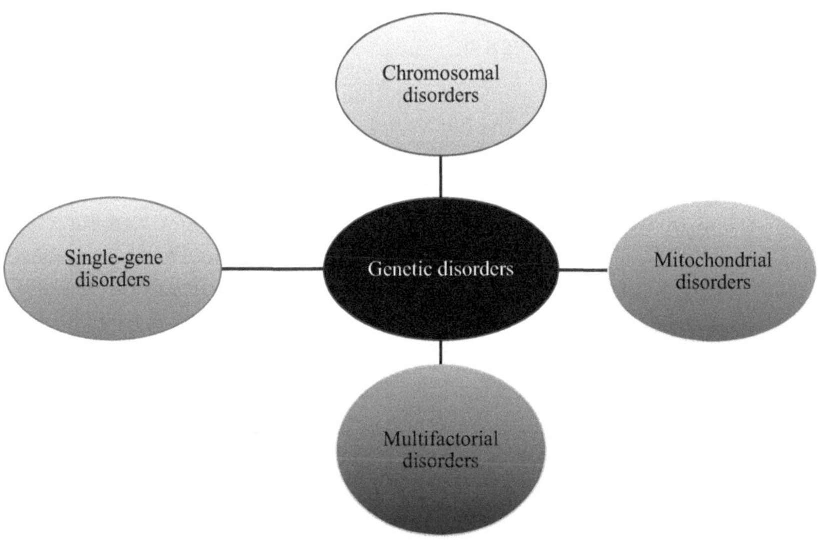

FIGURE 3.1 Overview of different classes of genetic disorders.

specific chromosomal regions may be the cause of certain medical disorders. For instance, an extra copy of chromosome 21, a common chromosomal aneuploidy, causes Down syndrome. Complex birth abnormalities like DiGeorge syndrome can result from duplications or deletions of smaller chromosomal segments, which can range in size from a single gene to a few percent of a chromosome's length. The normal human karyotype is made up of about 2 meters of DNA organized into a total of 46 chromosomes, which consist of 01 pair sex chromosomes (XY belongs to males, and XX belongs to females) and 22 pairs of homologous chromosomes. Defects in cell division, which can happen during the early pregnancy, after pregnancy, or before the implantation of embryo into the uterus, called pre-implantation phases, are the main cause of anomalies in chromosomes (Nussbaum et al., 2015).

3.2.2 Single-Gene Disorders

These disorders are caused by abnormalities in a specific gene's DNA sequence that modify or eliminate the protein it encodes. The nature of these illnesses is determined by the affected gene and the protein's function (McInnes et al., 1991). Mutations can develop on either one or both chromosomes in a pair. These illnesses typically have traditional patterns of inheritance; they can show X-linked inheritance or autosomal dominant/recessive patterns of inheritance. Various examples of single-gene disorders include Tay–Sachs disease, hemochromatosis (a single-gene disorder that causes iron overload), sickle cell anemia, cystic fibrosis (CF) (a single-gene disorder that leads to salt imbalance and dehydration), polycystic kidney diseases, Duchenne and Becker muscular dystrophy, hemophilia, Huntington's disease (HD), and BRCA1/BRCA2-related malignancies (McInnes et al., 1991).

3.2.3 Mitochondrial Disorders

The mitochondrial genome, which is required for energy synthesis in cells, is substantially conserved in humans but varies greatly in terms of size and genetic content among eukaryotes. Mitochondrial genomic sequences are maintained through purifying selection due to their high replication error rate, which is around 100 times that of nuclear DNA. Because of the genome's tight structure, major changes, like insertions and deletions, are usually impossible. During oogenesis, a significant purifying selection occurs, with one out of every five children carrying entirely novel mitochondrial genetic alterations, the most of which are homologous (Zaidi et al., 2019). In somatic cells, mitochondria are constantly purified by fission and fusion cycles. This continuous division results in genetic changes, resulting in obvious somatic mosaicism in adulthood. Older people's mitochondria display numerous additional genetic changes compared to their inherited genome (Li et al., 2015, 2019).

Given the importance of the function of the mitochondria, alterations to mitochondrial DNA can have a significant impact on health. While many mitochondrial defects result in ineffective oocytes that are destroyed right before ovulation, illnesses, including myopathies and neuropathies, are detected in around one out of every 5,000 live births (Schaefer et al., 2004). Clinical illnesses are associated with germline mutations in all the mitochondrial genome's protein-coding genes. Furthermore, the buildup of somatic mitochondrial mutations is likely to contribute to a variety of ageing-related illnesses (Larsson, 2010).

3.2.4 Multifactorial Disorders

Multifactorial diseases are conditions influenced by both genetic and environmental factors. Examples include cardiovascular diseases (CVDs), asthma, autoimmune disorders, and various neurodegenerative diseases, such as Alzheimer's and psychiatric conditions (McInnes et al., 1991). Advances in technology have significantly improved the accuracy of genetic variation measurements, leading to the identification of hundreds of predisposing genes. In many of these diseases, there isn't a single genetic error; instead, they result from the combined effects of variants in numerous genes. Each variety, frequently in conjunction with or caused by environmental variables, has the potential to spread, prevent, or incline an individual to the disorder (McInnes et al., 1991). Advances in technology have improved the detection of genetic variations, identifying hundreds of predisposing genes. These diseases are often the result of combined effects from variants in multiple genes, which may interact with or be triggered by environmental factors (McInnes et al., 1991).

Genetics plays a role in all diseases, varying in significance. DNA variations and their functional implications, whether alone or in combination, along with environmental influences (including lifestyle), contribute to disease processes. Approximately 8,000 monogenic diseases have been identified (Rodwell et al., 2013). Despite advances in genome editing tools like transcription activator–like effector nucleases (TALENs), zinc finger nucleases (ZFNs), clustered regularly interspaced short palindromic repeats and CRISPR-associated protein 9 (CRISPR-Cas9), across 75,000 disease-associated genetic mutations have been recognized, posing substantial obstacles for correction across multiple cell types (Landrum et al., 2016). Diagnosing and treating these disorders remain a major problem. CRISPR-Cas9, in specific, has transformed the discipline by providing unprecedented precision, performance, and versatility in targeted genomic changes. However, its clinical use in health care is still in its early phases, and there are still issues to be resolved about delivery methods, effects of off-targets, and ethical considerations (Zhang et al., 2016; Mojica et al., 2005). Thus, while genome editing offers great promise, achieving widespread clinical application for diagnosing and curing genetic disorders remains a complex challenge.

3.3 CURRENT DIAGNOSTIC METHODS

Accurate diagnosis is crucial for effective treatment and in improving patient and family well-being. Rare genetic disorders (RDs) impose a significant societal burden, impacting patients, their families, and the health-care system. Many rare genetic disorders are life-limiting, with 57–67% leading to reduced life expectancy (Retterer et al., 2016). Diagnostic techniques for rare genetic diseases include thus:

1. **Genetic counseling.** The role of genetic counseling in addressing genetic diseases is to provide individuals and families with essential information and support regarding these conditions. It enhances understanding of the genetic factors involved, evaluates inheritance risks, and aids in making informed choices about health, family planning, and medical interventions. Genetic counseling is usually conducted by trained genetic counselors, who have specialized knowledge in genetics, counseling practices, and ethical issues in health care.
2. **Genetic testing.** Genetic testing identifies genetic abnormalities early through the use of different invasive and non-invasive techniques, as well as whole genome sequencing.
3. **Prenatal screening.** Prenatal screening for genetic diseases encompasses tests conducted during pregnancy to evaluate the potential risk of a fetus developing a genetic disorder or congenital anomaly. While these screenings do not yield a conclusive diagnosis, they offer an estimation of the probability that a newborn may be impacted by specific conditions. Based on the findings, additional diagnostic testing may be suggested.

3.4 CRISPR-CAS: A NEW RAY OF HOPE IN GENETIC DISORDERS

The landscape of genetic disorder treatment has undergone a dramatic transformation with the advent of genome editing technologies. Previously, treatments were mainly focused on alleviating symptoms or slowing the progression of these diseases. However, recent developments have shifted toward addressing the root genetic causes. Over the last few years, a novel approach has been developed that allows for direct manipulation of nearly any gene within various cell types and organisms. This groundbreaking approach, known as "genome editing," makes use of synthetic and engineered nucleases. These nuclease enzymes are hybrid structures with a domain that binds DNA that targets specific sequences and a DNA-cleaving component. This enables precise genetic alterations by generating directed double-strand breaks (DSBs) in DNA sequences, which activate the cell's natural repairing steps, including an almost error-free, homology-directed repair system (HDR) and error-prone, non-homology-directed end joining (NHDR) (Carroll et al., 2011; Wyman et al., 2006). Numerous methods for changing genomes have been developed, including as ZFNs, TALENs, CRISPR-Cas systems, etc. Among these modification techniques, CRISPR-Cas9's unmatched precision and potency have made it a revolutionary instrument. Because ZFNs and TALENs can be modified, they are highly flexible. They can now identify almost any sequence of DNA. ZFNs and TALENs are very adaptable because of their ability to be customized. This allows them to recognize practically any sequence of DNA. These effector domains, which regulate the genome's makeup and functioning, can be connected to these DNA-binding domains. Enzymes such as nucleases, transcription enhancers and inhibitors, transposases, recombinases, and histone methyltransferases/acetyltransferases are examples of these effector units. The degree to which these genetic alterations are efficacious depends on the sensitivity and suitability of the changed zinc finger and TALE proteins for DNA binding. These effector domains, which regulate the genome's makeup and functioning, can be connected to these DNA-binding domains. Enzymes such as nucleases, transcription enhancers and inhibitors, transposases, recombinases, and histone methyltransferases/acetyltransferases are examples of these effector units. The degree to which these genetic alterations are efficacious depends on the sensitivity and suitability of the changed zinc finger and TALE proteins for DNA binding. These effector domains, which regulate the genome's makeup and functioning, can be connected to these DNA-binding domains. Enzymes such as nucleases, transcription enhancers and inhibitors, transposases, recombinases, and histone methyltransferases/acetyltransferases are examples of these effector units. The degree to which these genetic alterations are efficacious depends on the sensitivity and suitability of the changed zinc finger and TALE Proteins for DNA binding (Gaj et al., 2013).

In comparison to ZFNs and TALENs genetic modification techniques, the CRISPR-Cas genetic modification tool offers several advantages. It is simpler to use, more cost-effective, and capable of editing multiple sites within the genome simultaneously (Cong et al., 2013). Furthermore, because it can edit multiple loci at once, CRISPR-Cas9 is more straightforward, efficient, and expandable than other genome editing techniques. For these reasons, it has become an essential tool in the research of biology. Investigators may now use genetically modified animals to look into the fundamental causes of a variety of ailments and find biological processes that could be exploited to better medical treatments thanks to the development of innovative genome modification technology (Cong et al., 2013).

Working in concert with the CRISPR-associated Cas protein, the CRISPR system may be a simpler and more effective alternative to ZFNs and TALENs for inducing specific genetic alterations (Gupta et al., 2019). This is not to be confused with the previously discussed site-specific nucleases. First identified in bacteria, the CRISPR technology provides acquired protection against foreign DNA by RNA-guided DNA breakdown. With its many uses in human genomics, the CRISPR-Cas9 system has evolved into a flexible gene-editing instrument that allows for precise gene regulation or repair (Gupta et al., 2019). CRISPR-Cas9 can be employed for a variety of directed genome editing applications due to its cost and scalability (Chakraborty et al., 2017; Peng et al., 2016). Since CRISPR's initial discovery in 1987, the technology has evolved incrementally. Today, CRISPR-Cas9-based genetic experiments are being conducted in various models, including plants, yeast, worms, fruit flies (drosophila), zebrafish, mice, and even humans (Bortesi and Fischer, 2015; Ryan et al., 2016; Liu et al., 2017a; Friedland et al., 2013; Gratz et al., 2015; Zhu et al., 2017).

3.5 COMPOSITION OF CRISPR-CAS

A guide RNA (gRNA) and an enzyme known as Cas9 endonuclease make up the CRISPR-Cas9 system. Together, tracrRNA, called the trans-activating CRISPR-RNA, and this gRNA forms an intricate structure that functions as an active unit. By maximizing these two short RNAs, a single guide RNA, known as single guide RNA (sgRNA), which is more widely employed, can be produced. Cas9 is an RNA-guided enzyme that can cut double-stranded DNA at certain DNA sites because of a 20-nucleotide sequence in the guide RNA that binds to complementary DNA sequences through base pairing. When the Cas9 endonuclease recognizes the target DNA site—which is guided by the crRNA sequence—it causes DSB or DNA cleavage (Rodríguez-Rodríguez et al., 2019).

The CRISPR array, which consists of switching "repeat interval repeat" sequences, was first discovered in the genome of bacteria. The repeat sequences are unique to bacteria and archaea, but not other organisms, like eukaryotes and viruses. They are composed of identical 29-nucleotide sequences that are separated by 32-nucleotide spacers called nucleotide spacers that are derived from conjugative plasmids or attacking bacteriophages (Jansen et al., 2002; Ishino et al., 1987). When these archaea and bacteria come into contact with the same phages or plasmids again, they are not reinfected because the CRISPR array directs the Cas endonuclease enzyme to cut the viral DNA sequences complementary to the gRNA by introducing DSBs (Barrangou et al., 2007). The presence of a protospacer adjacent motif (PAM), a brief guanine-rich nucleotide stretch in the non-target DNA strand, is necessary for the recognition and fragmentation of double-stranded DNA. The process also depends on the complementarity of the target DNA strand to the RNA guide template (Hsu et al., 2014). Cellular DNA repair techniques might utilize HDR, non-homologous end joining (NHEJ) repair systems, or customized DNA sequences, depending on the methodology employed. NHEJ errors can cause minor mutations at the target sites that can be insertions or deletions (indels), which can impair or completely abolish the function of the digested genes (Barrangou et al., 2007; Hsu et al., 2014). On the other hand, high-fidelity genome correction, HDR, may be able to rectify genetic flaws that cause hereditary diseases by using a homologous repair template (Barrangou et al., 2007; Hsu et al., 2014). Beyond depending solely on disease models that duplicate particular conditions, the CRISPR-Cas system was first identified as a useful study tool for comprehending the causative functions of unique genetic variations (Zhan et al., 2019).

CRISPR-Cas9 offers the ability to precisely target almost any desired genomic region by modifying the nucleotide arrangement of sequences of a brief guide RNA segment. This provides the possibility of correcting mutations that cause diseases or silencing genes linked to the beginning of the disease (Hsu et al., 2014). Nonetheless, certain extensively chromatinized areas of the genome might continue to be inaccessible to the CRISPR-Cas system. The potential uses of CRISPR-Cas9 technology are manifold. Among these is the possibility of using genome editing techniques to treat a variety of common human ailments, including sickle cell anemia, cancer, cardiovascular problems, and neurological diseases (Zhan et al., 2019; Vermersch et al., 2020; Newby et al., 2021). There is increasing evidence that the CRISPR-Cas9 technology may effectively edit the genome of mammalian cells (Mali et al., 2013), and the technology is currently being evaluated in clinical trials across the United States, backed up by the National Institute of Health, NIH Recombinant DNA Advisory Committee, which has authorized the use of CRISPR in cancer therapies that involve alterations of the T cell populations of patients at the NIH (Baylis et al., 2017).

3.6 APPLICATIONS OF CRISPR-CAS IN GENETIC DISEASES

3.6.1 CRISPR-Cas9-Based Gene Therapy and Disease Remodeling

CRISPR-Cas technology is instrumental in correcting genetic mutations in model organisms, which demonstrates its potential for developing therapeutic approaches for human diseases. These genetic modifications are achieved through either gene knock-in or knockout strategies. It is now feasible to quickly create new *in vivo* and *in vitro* disease models thanks to CRISPR-based genome editing. Here are some key applications:

3.6.1.1 CRISPR-Cas in *In Vivo* Models

Using this method, single-cell embryos are administered directly with mRNA of Cas endonuclease, along with sgRNA enabling precise modifications to the genome. This method has been frequently applied to the creation of models for rats (Shao et al., 2014), mice (Shen et al., 2013), and monkeys (Niu et al., 2014). The ability to swiftly and effectively create genetically modified animals with one or more simultaneous genetic alterations has been demonstrated by the CRISPR-Cas9 technology.

3.6.1.2 Use of CRISPR-Cas Technology for *In Vivo* and *In Vitro* Gene Knock-In and Knockout Models

Delivering the CRISPR-Cas system directly into certain cells within their native tissue locations is known as "*in vivo* gene editing." Many delivery systems, including polymers, liposomes, peptides, or viral vectors, like adeno-associated viruses (AAVs), can be used to administer the CRISPR-Cas9 components. Because of their small size and non-pathogenic nature, AAVs are particularly useful for the precise editing of genes. For instance, the use of AAV to deliver the CRISPR-Cas system has shown potential in treating diseases such as atherosclerosis in mutant mice. Numerous other illnesses have also been treated with this method, such as sickle cell anemia (Yin et al., 2020), neurological problems (Park et al., 2021), muscular dystrophy (Fuentes et al., 2018), and metabolic disorders associated with the hepatic system (Nasrallah et al., 2022).

3.6.1.3 CRISPR-Cas in *In Vitro* Models

CRISPR gene editing is used in conjunction with *in vitro* studies employing induced pluripotent stem cells (iPSCs) from humans to explore complicated genetic disorders. iPSCs are capable of transforming into any cell type present in the body of an organism, allowing for the creation of disease models. These cells can be used to evaluate possible treatments and mimic the mechanisms underlying diseases. CRISPR can reverse specific mutations in iPSCs derived from patients, providing a foundation for developing gene therapies (Wong et al., 2016).

3.6.1.4 CRISPR-Based Screening Techniques

Using the CRISPR-Cas9 technology, collections of gene variants linked to particular traits or functions have been produced. These libraries are utilized in genome-wide screening efforts, where genes are analyzed through functional assays and sequencing to identify their roles. CRISPR-based screening is particularly useful in drug discovery and understanding disease mechanisms. It can also reveal how gene expression changes in response to cancer drugs, identifying new targets for precision medicine and offering insights into disease progression (Wong et al., 2016; Jason et al., 2019; Luo, 2016).

3.6.1.5 CRISPR-Cas for RNA Targeting

Initially recognized for its ability to cleave double-stranded DNA, CRISPR-Cas9 is directed by sgRNA to specific sites, creating breaks that are repaired by cellular mechanisms. A variant of Cas9, known as Cas9 nickase (Cas9n), creates single-strand breaks instead, which reduces off-target effects and enhances editing precision. Additionally, the discovery of Cas12a (previously Cpf1), a CRISPR-associated nuclease, has expanded the CRISPR toolkit. Cas12a introduces sticky-end cuts, which may facilitate more precise gene insertions and repairs (Mao et al., 2022; Stella et al., 2017; Paul and Montoya, 2020).

3.6.1.6 Base Editing for Altering Point Mutations

Base editing, a method developed to enable precise nucleotide changes without causing DSBs, was pioneered by David R. Liu's team. By combining a cytidine deaminase enzyme with Cas9 nickase, they produced cytosine base editors (CBEs). Cytidine (C) can be changed by CBEs into uridine (U), which can subsequently be changed into thymine (T). This technique is particularly promising for correcting single-nucleotide mutations, such as those causing sickle cell anemia, where a single change in the HBB gene leads to altered hemoglobin structure and function (Komor et al., 2016; Rees and Liu, 2018).

3.6.1.7 Prime Editing Based on CRISPR-Cas

Prime editing is an advanced genome editing technique introduced in 2019 by the Liu group. This technique uses a primary editing guide RNA (pegRNA) in place of the conventional gRNA and combines a reverse transcriptase with the Cas9 nickase. The pegRNA directs the editing apparatus to the target location, allowing for precise alterations to the DNA sequence, and this is achieved without the need of a DNA sequence that can act as a donor template DNA sequence. This technique provides enhanced flexibility and accuracy in genome editing, making it applicable to a broad spectrum of genetic modifications (Anzalone et al., 2019).

3.7 METHOD FOR CRISPR DELIVERY

Delivery methods for the CRISPR-Cas9 system have been developed into three primary tactics, which are as follows.

3.7.1 Plasma-Mediated Delivery of CRISPR-Cas

This approach utilizes a plasmid to encode both the Cas9 protein and sgRNA. Plasmid DNA (pDNA) is stable, can be produced in large amounts, and is easily modified, making it an optimal vector for CRISPR delivery. This method provides long-lasting expression of Cas9 and sgRNA, minimizing the need for repeated transfections (Cong et al., 2013; Mali et al., 2013; Doudna et al., 2014).

3.7.2 Cas9 and sgRNA Intracellular Introduction

This technique involves directly introducing Cas9 mRNA and sgRNA into cells. While it is fast-acting, the main limitation is the instability of mRNA, which results in short-term gene-editing effects (Liu et al., 2017b).

3.7.3 Direct Delivery of sgRNA and Cas9 Protein

Delivering the protein Cas and sgRNA directly into cells offers several benefits, including quick action, stability, and low antigenic properties, making it a highly convenient method for genome modification (Biagioni et al., 2018).

3.8 DISEASE CORRECTION AND TREATMENT USING CRISPR-CAS

The technology of CRISPR-Cas has immense promise for directly altering the genetic code associated with genetic disorders. By targeting and editing specific mutations responsible for these genetic diseases, CRISPR-Cas offers the possibility of correcting the underlying defect, potentially leading to more effective treatments or even cures. The modification system has great promise for directly modifying the genetic code linked to hereditary illnesses. CRISPR-Cas offers the option of addressing the underlying deficiency by targeting and editing particular mutations that cause various genetic disorders. This could result in more effective treatments or possibly a cure.

The rapid advancement and broad application of the CRISPR-Cas genome modification technology to fix genetic abnormalities linked to a range of human diseases are demonstrated by the following notable examples.

3.8.1 Cystic Fibrosis

The gene called cystic fibrosis transmembrane conductance regulator (CFTR) gene is mutated in cystic fibrosis (CF), a hereditary autosomal disorder. This CFTR gene codes for a chloride channel that is essential to ion transport and is regulated by cAMP molecule. It is found in the apical membrane of epithelial cells. Dysregulation of this channel impairs salt and fluid balance, leading to multi-organ dysfunction and, ultimately, respiratory failure. For more than 70 years, the main focus of CF therapies was on managing the symptoms rather than the underlying genetic abnormality. Many mutations have been found in the CFTR gene, which are dispersed throughout its intronic and exonic regions (Anderson et al., 1991). About 90% of CF patients benefit from the very effective CFTR modulators that have been developed as a result of recent research developments. These modulators don't treat the whole spectrum of CFTR mutations, though, and they are not therapeutic. Since CF is a genetic disorder, the ideal treatment would involve directly repairing the genetic defects. Intestinal stem cells from CF patients have been used to modify the CFTR gene using the CRISPR-Cas system via HDR (Schwank et al., 2013). Analysis of the phenylalanine 508 deletion (CFTR F508del) in exon 11 has revealed early CFTR protein breakdown, folding errors, and endoplasmic ER persistence. The CFTR gene's mutation site is where the Cas9 enzyme breaks DNA, influenced by a particular RNA sequence. The alteration in exon 11 of the CFTR gene can then be corrected by the cell's repair mechanisms via HDR (Cheng et al., 1990). CRISPR-Cas9 has been utilized in recent research to modify the CFTR locus in ovine fetal fibroblasts, resulting in the production of CFTR–/– and CFTR+/+ lambs. The genetically altered lambs had liver damage, pancreatic fibrosis, and intestinal blockage, all of which are hallmarks of human CF pathology (Fan et al., 2018). Furthermore, employing a piggyBac transposase as an appropriate indicator, CRISPR has been used to precisely correct the CFTR gene with a homozygous F508 deletion in induced pluripotent stem cells (iPSCs) from CF patients (Firth et al., 2015), achieving up to 90% correction effectiveness. By electroporating CRISPR-Cas RNP into patient-derived iPSC cell lines with the F508del mutation, a repair rate of above 20% was obtained (Ruan et al., 2019). In airway epithelial cells or proximal lung organoids produced from iPSCs, the CFTR gene has been restored through the use of CRISPR for genetic repair.

3.8.2 Atherosclerosis

The term "cardiovascular diseases" (CVDs) refers to a group of conditions that affect the cardiovascular system and blood vessel arteries, such as peripheral vascular disease, rheumatic heart disease, congenital heart disease, hypertension, coronary heart disease, heart failure, and cardiomyopathies (Gidding and Allen, 2019). Long-term exposure to high low-density lipoprotein (LDL) cholesterol causes fibro-fatty lesions in the arterial walls, which are a hallmark of atherosclerosis, a primary cause of CVD (Gidding and Allen, 2019). Atherosclerosis has a protracted preclinical phase; therefore, early identification can help identify those who are at risk and provide possibilities for prevention. Both subclinical and clinical atherosclerosis are recognized to have genetic components, and genetic research has revealed new treatment targets with potential causation. For example, a multi-phenotype, genome-wide association study (GWAS) has identified serpin peptidase inhibitor, clade A member 1 (SERPINA1) and aquaporin 9 (AQP9) as putative gene candidates (Inouye et al., 2012). Alpha 1-antitrypsin (A1AT), a protease inhibitor that facilitates LDL breakdown, is encoded by SERPINA1 (Mashiba et al., 2001). In the liver, AQP9 controls the transport of glycerol and aids in the decrease of hepatocyte lipid buildup (Rodríguez et al., 2011). Furthermore, among people exposed to traffic air pollution, unique genetic variations in the Fc alpha and mu receptor (FCAMR), and polymeric immunoglobulin receptor (PIGR) locus have been linked to coronary atherosclerosis, according to a genome-wide interaction study (GWIS) (Ward-Caviness et al., 2017). By causing mRNA cleavage, RNA-targeted therapies, such as antisense oligonucleotides (ASOs) and small interfering RNAs (siRNAs), have demonstrated promise in suppressing proteins implicated in lipid homeostasis, including apolipoprotein B (APOB), proprotein convertase subtilisin/kexin type 9 (PCSK9), angiopoietin-like protein 3 (ANGPTL3), and apolipoprotein C3 (APOC3) (Nair et al., 2014). Although methods based on nucleic acids have demonstrated promise,

CRISPR-Cas9 modification of genes presents a unique treatment approach for atherosclerosis. The athero-protective effects associated with inactivating gene targets, such as PCSK9 (Cohen et al., 2005), APOC3 (Jørgensen et al., 2014), and ANGPTL3 (Musunuru et al., 2010), have been demonstrated. For example, mice's serum PCSK9 levels and plasma cholesterol were reduced by 30–40% as a result of CRISPR-mediated suppression of PCSK9 (Ding et al., 2014). With only minor consequences of off-target specific base engineering with base editor 3 (BE3), a CRISPR-Cas9 coupled with a cytosine deaminase domain, led to a 50% drop in plasma PCSK9 protein levels and a 30-percentage point decrease in levels of cholesterol (Chadwick et al., 2017). Opportunities have been demonstrated by ANGPTL3, another intriguing target that affects human lipoprotein metabolism by blocking endothelial lipase and lipoprotein lipase (LPL). Mice with ANGPTL3 gene silencing had lower plasma cholesterol levels (Xu et al., 2018), and CRISPR-Cas9 base editing of ANGPTL3 resulted in lower cholesterol and triglyceride levels, demonstrating the prospective use of CRISPR-Cas9 technologies in therapeutic gene therapy (Chadwick et al., 2018).

3.8.3 Leber Congenital Amaurosis (LCA)

One of the main causes of childhood blindness is Leber's congenital amaurosis (LCA), a serious hereditary retinal dystrophy marked by nystagmus, amaurotic pupils, and early-onset impairment of vision (Perrault et al., 1999). Approximately 5% of inherited retinal dystrophies (IRDs) are caused by LCA, which was first reported by Dr. Theodore Leber in 1869. Its prevalence ranges from 1/81,000 to 1/30,000 (Koenekoop, 2004). Patients with LCA may experience a spectrum of impairment that ranges from light perception solely to functional visual perception, with some cases exhibiting brief oscillations in vision ahead to a steady deterioration (Koenekoop, 2004). Guanylate cyclase 2D, retinal (GUCY2D), retinal pigment epithelium-specific (RPE65), crumbs homolog 1 (CRB1), centrosomal protein 290 (CEP290), and retinol Dehydrogenase 12 (RDH12) are among the genes often mutated (Den Hollander et al., 2008; Sweeney et al., 2007). The frequency of these mutations varies according to ethnic background: CRB1 is more common in Chinese populations (Wang et al., 2015), whereas GUCY2D, RPE65, and CEP290 are more common in Caucasians (Sweeney et al., 2007; Coppieters et al., 2010). Treating LCA with gene enhancement therapy is a crucial strategy. Using lentiviral or AAV vectors, studies have mostly concentrated on genes such as GUCY2D, RPE65, aryl-hydrocarbon interacting protein-like 1 (AIPL1), leber congenital amaurosis 5 (LCA5), CEP290, and RDH12 (Kumaran et al., 2017). There are currently clinical trials underway for genes such CEP290, RPE65, and GUCY2D (Varela et al., 2022). Treatment for RPE65 mutations has advanced significantly (Weleber et al., 2016; Russell et al., 2017). Research on gene functions and disorders has increased since the introduction of CRISPR-Cas9 (Zhong et al., 2015). The most prevalent LCA mutation, CEP290, has a size that is too big for AAV delivery. Dual recombinant AAV vectors were used by Ruan et al. (2017) to overcome this issue by causing the CRISPR-Cas system–facilitated deletion of a particular intronic region in the mouse photoreceptor CEP290 gene (Ruan et al., 2017). Furthermore, a smaller *Staphylococcus aureus*

CRISPR-Cas9 system was used to introduce the Cas gene and two gRNAs for dual-cut excision of the CEP290 alteration in primary fibroblasts from Leber congenital amaurosis 10 (LCA10) patients using a single AAV vector (Cho et al., 2017). Maeder et al. have created EDIT-101, a genome-modifying therapy that uses an AAV5 vector to deliver CEP290 gRNA and *Staphylococcus aureus* Cas9 to restore vision in LCA10. This strategy successfully edited the CEP290 gene in humanized CEP290 mice (Maeder et al., 2019), offering a path forward for the development of LCA10 gene therapies.

3.8.4 Muscular Dystrophy

Duchenne muscular dystrophy (DMD) is an X-linked recessive neuromuscular illness distinguished by major muscle atrophy, cardiomyopathy, and high creatinine kinase levels (Darras et al., 2015). It is among the most prevalent type of childhood muscular dystrophy, affecting around 1 within 3,500–5,000 males (Moat et al., 2013). Typically, symptoms begin in infancy and worsen around teenage years, resulting in reduced mobility and a 20- to 40-year decrease in longevity (Guiraud et al., 2015). Although corticosteroids can help control symptoms, there is still no treatment for DMD. The DMD gene, which is triggered via multiple alterations, encodes the largest known human gene, dystrophin. Located on the X chromosome (Xp21.2-p21.1), dystrophin extends around 2.3 megabases (Hamosh et al., 2005). The DMD gene is susceptible to several sorts of alterations because of its size, including duplications, point mutations, and massive deletions (Bladen et al., 2015). The majority of alterations result in a frameshift, which truncates the dystrophin protein and renders it non-functional, as well as disrupting the open reading frame (ORF). The dystrophin gene has "hotspots," where mutations are more likely to occur. They are composed of exons 45–55, which are associated with the redundant rod domain, and exons 2–10, which are linked to the actin-binding domain (Echigoya et al., 2018). Correcting mutations in these exons is a common focus of efforts for maximizing cross-patient compatibility. Although these therapies are not curative in nature, the current standard of care entails an integrated strategy, including physiotherapy, cardiomyopathy medical care, and corticosteroids, to improve muscular strength and preserve functionality (Echigoya et al., 2018; Birnkrant et al., 2018). The optimal dosing strategy remains uncertain to balance treatment efficacy with side effects, making molecular therapies targeting the root cause of DMD increasingly attractive.

Through localized or systemic delivery of CRISPR-Cas components via AAV, current investigations have used NHEJ to circumvent the premature stop codons in exon 23, resuming dystrophin synthesis in neonatal and adult mice (Long et al., 2016; Tabebordbar et al., 2016). Corresponding to this, systemic AAV administration of CRISPR components produced 3 to 90% dystrophin restoration in skeletal muscle at 8 weeks, with varied effectiveness by muscle type and better muscle histology (Amoasii et al., 2018). CRISPR-Cas-induced NHEJ has also been used to treat DMD in a dog model. In a mouse model, ABE administered locally through intramuscular injection of a trans-splicing AAV has also been utilized to treat DMD (Ryu et al., 2018). These findings highlight the possibility of using modification of genes to treat DMD in individuals. Myasthenia, amyotrophy, and hypotonia are the hallmarks of congenital muscular dystrophy type 1A (MDC1A), a disorder of the neuromuscular system that usually first appears at birth or after the birth of a child. Mutations causing lack of functionality in laminin subunit alpha 2 (LAMA2), the gene encoding laminin-α2, give rise to MDC1A. In a study conducted by Ronald and his co-workers, in order to upregulate laminin subunit alpha 1 (LAMA1), which codes for laminin-α1, a protein that is structurally identical to laminin-α2, they employed CRISPR technology. In an MDC1A animal model, this upregulation reduces muscle atrophy and paralysis, providing a unique, mutation-independent method of treating the disease (Kemaladewi et al., 2019). A number of approaches are either licensed or being researched for the treatment of DMD, such as gene therapies, stop codon read-through medications, and antisense oligonucleotides (ASOs) (Duan et al., 2021). While some therapeutic results have been observed, the ephemeral nature of these techniques restricts their use in clinical settings. The versatility and accuracy of CRISPR-Cas9 make it a promising intervention with long-term potential advantages.

3.8.5 Hereditary Hearing Loss

Hereditary hearing loss is a genetic condition that can be inherited and may present at birth (congenital) or develop later in life (late-onset). It arises from mutations in specific genes crucial for the auditory system's development and function. These genes may affect various auditory system components, ear structure, or cochlear hair cell function. Hereditary hearing loss accounts for approximately 50–60% of childhood deafness cases, representing a significant public health issue (Olusanya and Newton, 2007). It is one of the most frequent congenital disorders, with an incidence of 1 in 500 to 1 in 1,000 infants (Hoffman and Kaplan, 2002). Around 80% of hereditary hearing loss cases are autosomal recessive; meaning, many affected individuals have parents who are carriers but do not have hearing loss themselves (Bayazit and Yılmaz, 2006). Traditional treatments primarily address symptoms rather than the underlying genetic causes. A frequent remedy for sensorineural hearing loss is the use of hearing aids, which enhance hearing. Depending on the user's preferences and level of hearing loss, these devices are available in a variety of versions, including behind-the-ear (BTE), in-the-ear (ITE), in-the-canal (ITC), and completely in-the-canal (CIC) (Ekberg et al., 2014). Although hearing aids benefit many, they do not cure the underlying cause of hearing loss. Cochlear implants may be used for those with significant or severe sensorineural hearing loss, directly stimulating the auditory nerve and resulting in substantial hearing improvement. Bone-anchored hearing aids (BAHA) use bone conduction rather than air conduction to transfer sound for people with conductive hearing loss or single-sided deafness. Despite these traditional approaches significantly improving quality of life, they have limitations and do not address hereditary hearing loss's genetic basis. Through direct targeting and correction of genetic defects, the groundbreaking gene-editing technique CRISPR-Cas9 exhibits significant promise for the treatment of hereditary hearing loss. Guide RNA is used by CRISPR-Cas9 to precisely identify defective genes, and the Cas9 enzyme makes targeted DNA cuts to enable cell repair of the abnormality (Charlier et al., 2021). In preclinical studies, this approach has successfully repaired genes linked to hereditary hearing loss, such as myosin VIIA (MYO7A) and transmembrane channel like 1 (TMC1), demonstrating potential for restoring hearing function (Farooq et al., 2020). Research continues to improve delivery methods to the inner ear, minimize off-target effects, and address ethical concerns related to gene editing (Nourbakhsh et al., 2021). Although still in its early stages, CRISPR-Cas9 holds promise for curing genetic hearing loss beyond the limitations of current symptom-focused therapies (Taiber and Avraham, 2019).

3.8.6 Dystrophic Epidermolysis Bullosa

A serious hereditary disorder called dystrophic epidermolysis bullosa (DEB) is characterized by exceedingly delicate skin that tears and boils quickly at the slightest friction (Hou et al., 2023). Type VII collagen is encoded by the collagen type VII alpha 1 chain (COL7A1) gene, and mutations in this gene cause the condition. Without functional type VII collagen, skin layers fail to adhere properly, leading to severe blistering, persistent sores, and an increased risk of skin cancer. DEB can vary in severity from mild to life-threatening, significantly impacting quality of life. There is no known cure for DEB's genetic cause, so traditional treatments focus mainly on managing symptoms and preventing complications (Georgiadis, 2016). Symptomatic management includes meticulous wound care to avoid further skin damage, using non-adhesive dressings like hydrocolloid or silicone-coated mesh to minimize trauma during changes. Pain management often requires both topical and systemic analgesics. In severe cases, surgical interventions may be necessary to release contractures and scarring, though these come with risks, such as infection and poor wound healing. Earlier gene therapy attempts involved inserting functional genes into patient cells using viral vectors to introduce healthy COL7A1 gene copies (Niti et al., 2023). However, challenges such as insertional mutagenesis and immune responses diminished these strategies' effectiveness. Despite these efforts, traditional care remains largely palliative, focusing on improving life quality rather than curing the condition. CRISPR-Cas9 technology represents a significant advancement in DEB treatment, offering a potential cure by directly targeting and correcting the genetic abnormalities causing the disease. Researchers can now precisely edit the COL7A1 gene within living cells using CRISPR-Cas9, correcting mutations and restoring functional type VII collagen production (Hainzl et al., 2017). The functioning of type VII collagen has also been restored by using gRNA and Cas9 enzyme (Webber et al., 2016). This technique has been applied both *in vivo* and *ex vivo*, with ex vivo methods involving editing patient-derived cells outside the body before reintroducing them (Bhattacharjee et al., 2022). This targeted therapy shows promise in preclinical research and early clinical trials. *In vivo* techniques, where CRISPR is directly administered to affected tissues, are still experimental but could lead to less-invasive treatments (Niemiec and Howard, 2020). Despite its transformative potential, CRISPR use in DEB treatment faces challenges, including delivery efficiency, off-target effects, and long-term stability of modified cells. Ethical considerations, especially regarding germline editing, must also be addressed. Nevertheless, CRISPR-Cas9 represents a remarkable advancement, shifting DEB treatment from symptom management to the potential for genetic corrections and long-term cures.

3.8.7 Primary Hyperoxaluria Type 1 (PH1)

The metabolic disease known as primary hyperoxaluria type 1 (PH1) is triggered by mutations in the AGXT gene, which codes for the hepatic enzyme alanine-glyoxylate aminotransferase (AGXT). This enzyme is essential for minimizing the accumulation of oxalate by changing glyoxylate into glycine. In PH1, the liver's capacity to metabolize glyoxylate is compromised by an AGXT deficit, leading to an overabundance of oxalate formation. These excesses form calcium oxalate crystals, leading to nephrolithiasis, nephrocalcinosis, and if untreated, systemic oxalosis. This progression ultimately causes severe renal failure and multi-organ dysfunction (Cochat, 1999). Traditional treatment strategies for PH1 focus on managing symptoms and slowing disease progression rather than addressing the underlying genetic defect. These treatments include increasing fluid intake, using potassium or sodium citrate to alkalinize urine, reducing dietary oxalate, and supplementing with pyridoxine for mutations responsive to it. Citrate is also used to prevent the formation of calcium oxalate stones. For advanced cases, intensive dialysis and combined liver-kidney transplantation are employed to manage both oxalate overproduction and renal failure and to restore normal AGXT function (Cochat and Rumsby, 2013).

Recent developments in gene-editing technologies, such as CRISPR-Cas9 and CRISPR-Cpf1, offer promising new treatments for PH1. CRISPR-Cas9 aims to correct AGXT gene mutations by delivering the editing components directly to liver cells, thereby restoring AGT activity and reducing oxalate production. CRISPR-Cpf1 facilitates multiplex gene editing by targeting both the Hao1 and Ldha genes, resulting in decreased oxalate levels and improved kidney health in preclinical models (Zheng et al., 2023). One study demonstrated that CRISPR-Cas9-mediated substrate reduction therapy (SRT) targeting glycolate oxidase effectively prevented oxalate overproduction and kidney damage in mice without causing toxicity (Zabaleta et al., 2018). Another study showed that liver-specific CRISPR-Cas9 inhibition of lactate dehydrogenase (LDH) significantly reduced LDH levels, urine oxalate, and kidney damage in PH1 and primary hyperoxaluria type 3 (PH3) mice following a single AAV8 vector dose (Martinez-Turrillas et al., 2022). These innovative approaches, though still experimental, offer promising alternatives for addressing the genetic basis of PH1 and other metabolic disorders associated with toxic metabolites.

3.9 CONCLUSION

A revolutionary development in genetic editing, the CRISPR-Cas9 system provides a variety of techniques for accurate gene alteration. Different delivery methods, like protein delivery, plasmid-mediated distribution, and direct insertion of Cas9 and sgRNA, each have their own benefits and drawbacks. Applications of CRISPR-Cas9 in various genetic diseases demonstrate the technology's promise to treat genetic ailments. The technology has drawbacks despite its potential, such as problems with delivery efficiency, off-target impacts, and moral dilemmas. Upgrading the therapeutic potential of CRISPR-Cas9 and resolving these obstacles require ongoing study and development. CRISPR-Cas9 has the potential to transform genetic medicine as technology develops, providing fresh hope for efficient therapies and cures for hereditary illnesses that were thought to be incurable.

REFERENCES

Amoasii, Leonela, John C. W. Hildyard, Hui Li, Efrain Sanchez-Ortiz, Alex Mireault, Daniel Caballero, Rachel Harron et al. "Gene editing restores dystrophin expression in a canine model of Duchenne muscular dystrophy." *Science* 362, no. 6410 (2018): 86–91.

Anderson, Matthew P., Richard J. Gregory, Simon Thompson, David W. Souza, Sucharita Paul, Richard C. Mulligan, Alan E. Smith, and Michael J. Welsh. "Demonstration that CFTR is a chloride channel by alteration of its anion selectivity." *Science* 253, no. 5016 (1991): 202–205.

Anzalone, Andrew V., Peyton B. Randolph, Jessie R. Davis, Alexander A. Sousa, Luke W. Koblan, Jonathan M. Levy, Peter J. Chen et al. "Search-and-replace genome editing without double-strand breaks or donor DNA." *Nature* 576, no. 7785 (2019): 149–157.

Bangham, Jenny. "Making the 'genetic counsellor' in the UK, 1980–1995." *Medical Humanities* 49, no. 2 (2023): 248–259.

Barrangou, Rodolphe, Christophe Fremaux, Hélène Deveau, Melissa Richards, Patrick Boyaval, Sylvain Moineau, Dennis A. Romero, and Philippe Horvath. "CRISPR provides acquired resistance against viruses in prokaryotes." *Science* 315, no. 5819 (2007): 1709–1712.

Bayazit, Yildirim A., and Metin Yılmaz. "An overview of hereditary hearing loss." *ORL* 68, no. 2 (2006): 57–63.

Baylis F., McLeod M. "First-in-human Phase 1 CRISPR gene editing cancer trials: Are we ready?" *Current Gene Therapy* 17, no. 4 (2017): 309–319. https://doi.org/10.2174/1566523217666171121165935. PMID: 29173170; PMCID: PMC5769084.

Bhattacharjee, Gargi, Nisarg Gohil, Khushal Khambhati, Indra Mani, Rupesh Maurya, Janardhan Keshav Karapurkar, Jigresh Gohil et al. "Current approaches in CRISPR-Cas9 mediated gene editing for biomedical and therapeutic applications." *Journal of Controlled Release* 343 (2022): 703–723.

Biagioni, Alessio, Anna Laurenzana, Francesca Margheri, Anastasia Chillà, Gabriella Fibbi, and Mario Del Rosso. "Delivery systems of CRISPR/Cas9-based cancer gene therapy." *Journal of Biological Engineering* 12 (2018): 1–9.

Birnkrant, David J., Katharine Bushby, Carla M. Bann, Susan D. Apkon, Angela Blackwell, David Brumbaugh, Laura E. Case et al. "Diagnosis and management of Duchenne muscular dystrophy, part 1: Diagnosis, and neuromuscular, rehabilitation, endocrine, and gastrointestinal and nutritional management." *The Lancet Neurology* 17, no. 3 (2018): 251–267.

Bladen, Catherine L., David Salgado, Soledad Monges, Maria E. Foncuberta, Kyriaki Kekou, Konstantina Kosma, Hugh Dawkins et al. "The TREAT-NMD DMD Global Database: Analysis of more than 7,000 Duchenne muscular dystrophy mutations." *Human Mutation* 36, no. 4 (2015): 395–402.

Bortesi, Luisa, and Rainer Fischer. "The CRISPR/Cas9 system for plant genome editing and beyond." *Biotechnology Advances* 33, no. 1 (2015): 41–52.

Carroll, Dana. "Genome engineering with zinc-finger nucleases." *Genetics* 188, no. 4 (2011): 773–782.

Chadwick, Alexandra C., Niklaus H. Evitt, Wenjian Lv, and Kiran Musunuru. "Reduced blood lipid levels with in vivo CRISPR-Cas9 base editing of ANGPTL3." *Circulation* 137, no. 9 (2018): 975–977.

Chadwick, Alexandra C., Xiao Wang, and Kiran Musunuru. "In vivo base editing of PCSK9 (proprotein convertase subtilisin/kexin type 9) as a therapeutic alternative to genome editing." *Arteriosclerosis, Thrombosis, and Vascular Biology* 37, no. 9 (2017): 1741–1747.

Chakraborty, Chiranjib, Seong L. Teoh, and Srijit Das. "The smart programmable CRISPR technology: A next generation genome editing tool for investigators." *Current Drug Targets* 18, no. 14 (2017): 1653–1663.

Charlier, Jeremy, Robert Nadon, and Vladimir Makarenkov. "Accurate deep learning off-target prediction with novel sgRNA-DNA sequence encoding in CRISPR-Cas9 gene editing." *Bioinformatics* 37, no. 16 (2021): 2299–2307.

Cheng, Seng H., Richard J. Gregory, John Marshall, Sucharita Paul, David W. Souza, Gary A. White, Catherine R. O'Riordan, and Alan E. Smith. "Defective intracellular transport and processing of CFTR is the molecular basis of most cystic fibrosis." *Cell* 63, no. 4 (1990): 827–834.

Cho, Galaxy Y., Yazeed Abdulla, Jesse D. Sengillo, Sally Justus, Kellie A. Schaefer, Alexander G. Bassuk, Stephen H. Tsang, and Vinit B. Mahajan. "CRISPR-mediated ophthalmic genome surgery." *Current Ophthalmology Reports* 5 (2017): 199–206.

Cochat, Pierre, and Gill Rumsby. "Primary hyperoxaluria." *New England Journal of Medicine* 369, no. 7 (2013): 649–658.

Cochat, Pierre. "Primary hyperoxaluria type 1." *Kidney International* 55, no. 6 (1999): 2533–2547.

Cohen, Jonathan, Alexander Pertsemlidis, Ingrid K. Kotowski, Randall Graham, Christine Kim Garcia, and Helen H. Hobbs. "Low LDL cholesterol in individuals of African descent resulting from frequent nonsense mutations in PCSK9." *Nature Genetics* 37, no. 2 (2005): 161–165.

Cong, Le, F. Ann Ran, David Cox, Shuailiang Lin, Robert Barretto, Naomi Habib, Patrick D. Hsu et al. "Multiplex genome engineering using CRISPR/Cas systems." *Science* 339, no. 6121 (2013): 819–823.

Coppieters, Frauke, Ingele Casteels, Françoise Meire, Sarah De Jaegere, Sally Hooghe, Nicole Van Regemorter, Hilde Van Esch et al. "Genetic screening of LCA in Belgium: Predominance of CEP290 and identification of potential modifier alleles in AHI1 of CEP290-related phenotypes." *Human Mutation* 31, no. 10 (2010): E1709-E1766.

Darras, Basil T., Caroline C. Menache-Starobinski, Veronica Hinton, and Louis M. Kunkel. "Dystrophinopathies." *Neuromuscular Disorders of Infancy, Childhood, and Adolescence* (2015): 551–592.

Den Hollander, Anneke I., Ronald Roepman, Robert K. Koenekoop, and Frans P. M. Cremers. "Leber congenital amaurosis: Genes, proteins and disease mechanisms." *Progress in Retinal and Eye Research* 27, no. 4 (2008): 391–419.

Ding, Qiurong, Alanna Strong, Kevin M. Patel, Sze-Ling Ng, Bridget S. Gosis, Stephanie N. Regan, Chad A. Cowan, Daniel J. Rader, and Kiran Musunuru. "Permanent alteration of PCSK9 with in vivo CRISPR-Cas9 genome editing." *Circulation Research* 115, no. 5 (2014): 488–492.

Doudna, Jennifer A., and Emmanuelle Charpentier. "The new frontier of genome engineering with CRISPR-Cas9." *Science* 346, no. 6213 (2014): 1258096.

Duan, Dongsheng, Nathalie Goemans, Shin'ichi Takeda, Eugenio Mercuri, and Annemieke Aartsma-Rus. "Duchenne muscular dystrophy." *Nature Reviews Disease Primers* 7, no. 1 (2021): 13.

Echigoya, Yusuke, Kenji Rowel Q. Lim, Akinori Nakamura, and Toshifumi Yokota. "Multiple exon skipping in the Duchenne muscular dystrophy hot spots: Prospects and challenges." *Journal of Personalized Medicine* 8, no. 4 (2018): 41.

Ekberg, Katie, Caitlin Grenness, and Louise Hickson. "Addressing patients' psychosocial concerns regarding hearing aids within audiology appointments for older adults." *American Journal of Audiology* 23, no. 3 (2014): 337–350.

Fan, Zhiqiang, Iuri Viotti Perisse, Calvin U. Cotton, Misha Regouski, Qinggang Meng, Chaim Domb, Arnaud J. Van Wettere et al. "A sheep model of cystic fibrosis generated by CRISPR/Cas9 disruption of the CFTR gene." *JCI Insight* 3, no. 19 (2018).

Farooq, Rimsha, Khadim Hussain, Muhammad Tariq, Ali Farooq, and Muhammad Mustafa. "CRISPR/Cas9: Targeted genome editing for the treatment of hereditary hearing loss." *Journal of Applied Genetics* 61 (2020): 51–65.

Firth, Amy L., Tushar Menon, Gregory S. Parker, Susan J. Qualls, Benjamin M. Lewis, Eugene Ke, Carl T. Dargitz et al. "Functional gene correction for cystic fibrosis in lung epithelial cells generated from patient iPSCs." *Cell Reports* 12, no. 9 (2015): 1385–1390.

Friedland, Ari E., Yonatan B. Tzur, Kevin M. Esvelt, Monica P. Colaiácovo, George M. Church, and John A. Calarco. "Heritable genome editing in C. elegans via a CRISPR-Cas9 system." *Nature Methods* 10, no. 8 (2013): 741–743.

Fuentes, Christina Marie, and David Vernon Schaffer. "Adeno-associated virus-mediated delivery of CRISPR-Cas9 for genome editing in the central nervous system." *Current Opinion in Biomedical Engineering* 7 (2018): 33–41.

Gaj, Thomas, Charles A. Gersbach, and Carlos F. Barbas. "ZFN, TALEN, and CRISPR/Cas-based methods for genome engineering." *Trends in Biotechnology* 31, no. 7 (2013): 397–405.

Georgiadis, Christos. "Development of Gene Therapy for Recessive Dystrophic Epidermolysis Bullosa." PhD diss., UCL (University College London), 2016.

Gidding, Samuel S., and Norrina B. Allen. "Cholesterol and atherosclerotic cardiovascular disease: A lifelong problem." *Journal of the American Heart Association* 8, no. 11 (2019): e012924.

Gratz, Scott J., C. Dustin Rubinstein, Melissa M. Harrison, Jill Wildonger, and Kate M. O'Connor-Giles. "CRISPR-Cas9 genome editing in Drosophila." *Current Protocols in Molecular Biology* 111, no. 1 (2015): 31–2.

Guiraud, Simon, Annemieke Aartsma-Rus, Natassia M. Vieira, Kay E. Davies, Gert-Jan B. van Ommen, and Louis M. Kunkel. "The pathogenesis and therapy of muscular dystrophies." *Annual Review of Genomics and Human Genetics* 16, no. 1 (2015): 281–308.

Gupta, Darshana, Oindrila Bhattacharjee, Drishti Mandal, Madhab Kumar Sen, Dhritiman Dey, Adhiraj Dasgupta, Tawsif Ahmed Kazi et al. "CRISPR-Cas9 system: A new-fangled dawn in gene editing." *Life Sciences* 232 (2019): 116636.

Hainzl, Stefan, Patricia Peking, Thomas Kocher, Eva M. Murauer, Fernando Larcher, Marcela Del Rio, Blanca Duarte et al. "COL7A1 editing via CRISPR/Cas9 in recessive dystrophic epidermolysis bullosa." *Molecular Therapy* 25, no. 11 (2017): 2573–2584.

Hamosh, Ada, Alan F. Scott, Joanna S. Amberger, Carol A. Bocchini, and Victor A. McKusick. "Online Mendelian Inheritance in Man (OMIM), a knowledgebase of human genes and genetic disorders." *Nucleic Acids Research* 33, no. suppl_1 (2005): D514–D517.

Hoffman, Julien I. E., and Samuel Kaplan. "The incidence of congenital heart disease." *Journal of the American College of Cardiology* 39, no. 12 (2002): 1890–1900.

Hou, Ping-Chen, Nathalie Del Agua, Su M. Lwin, Chao-Kai Hsu, and John A. McGrath. "Innovations in the treatment of dystrophic epidermolysis bullosa (DEB): Current landscape and prospects." *Therapeutics and Clinical Risk Management* (2023): 455–473.

Hsu, Patrick D., Eric S. Lander, and Feng Zhang. "Development and applications of CRISPR-Cas9 for genome engineering." *Cell* 157, no. 6 (2014): 1262–1278.

Inouye, Michael, Samuli Ripatti, Johannes Kettunen, Leo-Pekka Lyytikäinen, Niku Oksala, Pirkka-Pekka Laurila, Antti J. Kangas et al. "Novel loci for metabolic networks and multi-tissue expression studies reveal genes for atherosclerosis." *PLoS Genetics* 8, no. 8 (2012): e1002907.

Ishino, Yoshizumi, Hideo Shinagawa, Kozo Makino, Mitsuko Amemura, and Atsuo Nakata. "Nucleotide sequence of the iap gene, responsible for alkaline phosphatase isozyme conversion in Escherichia coli, and identification of the gene product." *Journal of Bacteriology* 169, no. 12 (1987): 5429–5433.

Jansen, Ruud, Jan D. A. van Embden, Wim Gaastra, and Leo M. Schouls. "Identification of genes that are associated with DNA repeats in prokaryotes." *Molecular Microbiology* 43, no. 6 (2002): 1565–1575.

Jason, S. L., and Kosuke Yusa. "Genome-wide CRISPR-Cas9 screening in mammalian cells." *Methods* 164 (2019): 29–35.

Jørgensen, Anders Berg, Ruth Frikke-Schmidt, Børge G. Nordestgaard, and Anne Tybjærg-Hansen. "Loss-of-function mutations in APOC3 and risk of ischemic vascular disease." *New England Journal of Medicine* 371, no. 1 (2014): 32–41.

Kemaladewi, Dwi U., Prabhpreet S. Bassi, Steven Erwood, Dhekra Al-Basha, Kinga I. Gawlik, Kyle Lindsay, Elzbieta Hyatt et al. "A mutation-independent approach for muscular dystrophy via upregulation of a modifier gene." *Nature* 572, no. 7767 (2019): 125–130.

Koenekoop, Robert K. "An overview of Leber congenital amaurosis: A model to understand human retinal development." *Survey of Ophthalmology* 49, no. 4 (2004): 379–398.

Komor, Alexis C., Yongjoo B. Kim, Michael S. Packer, John A. Zuris, and David R. Liu. "Programmable editing of a target base in genomic DNA without double-stranded DNA cleavage." *Nature* 533, no. 7603 (2016): 420–424.

Kumar, Pankaj, Jolly Radhakrishnan, Maksud A. Chowdhary, and Philip F. Giampietro. "Prevalence and patterns of presentation of genetic disorders in a pediatric emergency department." In *Mayo Clinic Proceedings*, vol. 76, no. 8, pp. 777–783. Elsevier, 2001.

Kumaran, Neruban, Anthony T. Moore, Richard G. Weleber, and Michel Michaelides. "Leber congenital amaurosis/early-onset severe retinal dystrophy: Clinical features, molecular genetics and therapeutic interventions." *British Journal of Ophthalmology* 101, no. 9 (2017): 1147–1154.

Landrum, Melissa J., Jennifer M. Lee, Mark Benson, Garth Brown, Chen Chao, Shanmuga Chitipiralla, Baoshan Gu et al. "ClinVar: Public archive of interpretations of clinically relevant variants." *Nucleic Acids Research* 44, no. D1 (2016): D862–D868.

Larsson, Nils-Göran. "Somatic mitochondrial DNA mutations in mammalian aging." *Annual Review of Biochemistry* 79, no. 1 (2010): 683–706.

Lee, E.-Chiang, Urvi Desai, Gennady Gololobov, Seokjoo Hong, Xiao Feng, Xuan-Chuan Yu, Jason Gay et al. "Identification of a new functional domain in angiopoietin-like 3 (ANGPTL3) and angiopoietin-like 4 (ANGPTL4) involved in binding and inhibition of lipoprotein lipase (LPL)." *Journal of Biological Chemistry* 284, no. 20 (2009): 13735–13745.

Li, Huanzheng, Jesse Slone, Lin Fei, and Taosheng Huang. "Mitochondrial DNA variants and common diseases: A mathematical model for the diversity of age-related mtDNA mutations." *Cells* 8, no. 6 (2019): 608.

Li, Mingkun, Roland Schröder, Shengyu Ni, Burkhard Madea, and Mark Stoneking. "Extensive tissue-related and allele-related mtDNA heteroplasmy suggests positive selection for somatic mutations." *Proceedings of the National Academy of Sciences* 112, no. 8 (2015): 2491–2496.

Liu, Jiaqi, Yangzhong Zhou, Xiaolong Qi, Jia Chen, Weisheng Chen, Guixing Qiu, Zhihong Wu, and Nan Wu. "CRISPR/Cas9 in zebrafish: An efficient combination for human genetic diseases modeling." *Human Genetics* 136, no. 1 (2017a): 1–12.

Liu, Chang, Li Zhang, Hao Liu, and Kun Cheng. "Delivery strategies of the CRISPR-Cas9 gene-editing system for therapeutic applications." *Journal of Controlled Release* 266 (2017b): 17–26.

Long, Chengzu, Leonela Amoasii, Alex A. Mireault, John R. McAnally, Hui Li, Efrain Sanchez-Ortiz, Samadrita Bhattacharyya, John M. Shelton, Rhonda Bassel-Duby, and Eric N. Olson. "Postnatal genome editing partially restores dystrophin expression in a mouse model of muscular dystrophy." *Science* 351, no. 6271 (2016): 400–403.

Luo, Ji. "CRISPR/Cas9: From genome engineering to cancer drug discovery." *Trends in Cancer* 2, no. 6 (2016): 313–324.

Maeder, Morgan L., Michael Stefanidakis, Christopher J. Wilson, Reshica Baral, Luis Alberto Barrera, George S. Bounoutas, David Bumcrot et al. "Development of a gene-editing approach to restore vision loss in Leber congenital amaurosis type 10." *Nature Medicine* 25, no. 2 (2019): 229–233.

Mali, Prashant, Luhan Yang, Kevin M. Esvelt, John Aach, Marc Guell, James E. DiCarlo, Julie E. Norville, and George M. Church. "RNA-guided human genome engineering via Cas9." *Science* 339, no. 6121 (2013): 823–826.

Mao, Zefeng, Ruipeng Chen, Xiaojuan Wang, Zixuan Zhou, Yuan Peng, Shuang Li, Dianpeng Han et al. "CRISPR/Cas12a-based technology: A powerful tool for biosensing in food safety." *Trends in Food Science & Technology* 122 (2022): 211–222.

Martinez-Turrillas, Rebeca, Angel Martin-Mallo, Saray Rodriguez-Diaz, Natalia Zapata-Linares, Paula Rodriguez-Marquez, Patxi San Martin-Uriz, Amaia Vilas-Zornoza et al. "In vivo CRISPR-Cas9 inhibition of hepatic LDH as treatment of primary hyperoxaluria." *Molecular Therapy Methods & Clinical Development* 25 (2022): 137–146.

Mashiba, Shinichi, Youichiro Wada, Motohiro Takeya, Akira Sugiyama, Takao Hamakubo, Akio Nakamura, Noriko Noguchi et al. "In vivo complex formation of oxidized α1-antitrypsin and LDL." *Arteriosclerosis, Thrombosis, and Vascular Biology* 21, no. 11 (2001): 1801–1808.

McInnes, Roderick R., and Huntington F. Willard. *Thompson & Thompson Genetics in Medicine*. Philadelphia: WB Saunders Company, 1991.

Moat, Stuart J., Donald M. Bradley, Rachel Salmon, Angus Clarke, and Louise Hartley. "Newborn bloodspot screening for Duchenne muscular dystrophy: 21 years experience in Wales (UK)." *European Journal of Human Genetics* 21, no. 10 (2013): 1049–1053.

Mojica, Francisco J. M., César Díez-Villaseñor, Jesús García-Martínez, and Elena Soria. "Intervening sequences of regularly spaced prokaryotic repeats derive from foreign genetic elements." *Journal of Molecular Evolution* 60 (2005): 174–182.

Musunuru, Kiran, James P. Pirruccello, Ron Do, Gina M. Peloso, Candace Guiducci, Carrie Sougnez, Kiran V. Garimella et al. "Exome sequencing, ANGPTL3 mutations, and familial combined hypolipidemia." *New England Journal of Medicine* 363, no. 23 (2010): 2220–2227.

Nair, Jayaprakash K., Jennifer L. S. Willoughby, Amy Chan, Klaus Charisse, Md Rowshon Alam, Qianfan Wang, Menno Hoekstra et al. "Multivalent N-acetylgalactosamine-conjugated siRNA localizes in hepatocytes and elicits robust RNAi-mediated gene silencing." *Journal of the American Chemical Society* 136, no. 49 (2014): 16958–16961.

Nasrallah, Ali, Eric Sulpice, Farah Kobaisi, Xavier Gidrol, and Walid Rachidi. "CRISPR-Cas9 technology for the creation of biological avatars capable of modeling and treating pathologies: From discovery to the latest improvements." *Cells* 11, no. 22 (2022): 3615.

Newby, Gregory A., Jonathan S. Yen, Kaitly J. Woodard, Thiyagaraj Mayuranathan, Cicera R. Lazzarotto, Yichao Li, Heather Sheppard-Tillman et al. "Base editing of haematopoietic stem cells rescues sickle cell disease in mice." *Nature* 595, no. 7866 (2021): 295–302.

Niemiec, Emilia, and Heidi Carmen Howard. "Ethical issues related to research on genome editing in human embryos." *Computational and Structural Biotechnology Journal* 18 (2020): 887–896.

Niti, Argyrw, Georgios Koliakos, and Anna Michopoulou. "Stem cell therapies for epidermolysis bullosa treatment." *Bioengineering* 10, no. 4 (2023): 422.

Niu, Yuyu, Bin Shen, Yiqiang Cui, Yongchang Chen, Jianying Wang, Lei Wang, Yu Kang et al. "Generation of gene-modified cynomolgus monkey via Cas9/RNA-mediated gene targeting in one-cell embryos." *Cell* 156, no. 4 (2014): 836–843.

Nourbakhsh, Aida, Brett M. Colbert, Eric Nisenbaum, Aziz El-Amraoui, Derek M. Dykxhoorn, Karl Russell Koehler, Zheng-yi Chen, and Xue Z. Liu. "Stem cells and gene therapy in progressive hearing loss: The state of the art." *Journal of the Association for Research in Otolaryngology* 22 (2021): 95–105.

Nussbaum, Robert L., Roderick R. McInnes, and Huntington F. Willard. *Thompson & Thompson Genetics in Medicine E-Book: Thompson & Thompson Genetics in Medicine E-Book*. Elsevier Health Sciences, 2015.

Olusanya, Bolajoko O., and Valerie E. Newton. "Global burden of childhood hearing impairment and disease control priorities for developing countries." *The Lancet* 369, no. 9569 (2007): 1314–1317.

Park, So Hyun, and Gang Bao. "CRISPR/Cas9 gene editing for curing sickle cell disease." *Transfusion and Apheresis Science* 60, no. 1 (2021): 103060.

Paul, Bijoya, and Guillermo Montoya. "CRISPR-Cas12a: Functional overview and applications." *Biomedical Journal* 43, no. 1 (2020): 8–17.

Peng, Rongxue, Guigao Lin, and Jinming Li. "Potential pitfalls of CRISPR/Cas9-mediated genome editing." *The FEBS Journal* 283, no. 7 (2016): 1218–1231.

Perrault, Isabelle, Jean-Michel Rozet, Sylvie Gerber, Imad Ghazi, Corinne Leowski, Dominique Ducroq, Eric Souied, Jean-Louis Dufier, Arnold Munnich, and Josseline Kaplan. "Leber congenital amaurosis." *Molecular Genetics and Metabolism* 68, no. 2 (1999): 200–208.

Rees, Holly A., and David R. Liu. "Base editing: Precision chemistry on the genome and transcriptome of living cells." *Nature Reviews Genetics* 19, no. 12 (2018): 770–788.

Retterer, Kyle, Jane Juusola, Megan T. Cho, Patrik Vitazka, Francisca Millan, Federica Gibellini, Annette Vertino-Bell et al. "Clinical application of whole-exome sequencing across clinical indications." *Genetics in Medicine* 18, no. 7 (2016): 696–704.

Rodríguez, Amaia, Victoria Catalan, Javier Gomez-Ambrosi, Socorro García-Navarro, Fernando Rotellar, Víctor Valentí, Camilo Silva et al. "Insulin-and leptin-mediated control of aquaglyceroporins in human adipocytes and hepatocytes is mediated via the PI3K/Akt/mTOR signaling cascade." *The Journal of Clinical Endocrinology & Metabolism* 96, no. 4 (2011): E586–E597.

Rodríguez-Rodríguez, Diana Raquel, Ramiro Ramirez-Solis, Mario Alberto Garza-Elizondo, Maria de Lourdes Garza-Rodriguez, and Hugo Alberto Barrera-Saldana. "Genome editing: A perspective on the application of CRISPR/Cas9 to study human diseases." *International Journal of Molecular Medicine* 43, no. 4 (2019): 1559–1574.

Rodwell, C., and S. Ayme, eds. *Report on the State of the Art of Rare Disease Activities in Europe Part II: Key Developments in the Field of Rare Diseases in Europe in 2013* (2014), 1–90.

Ruan, Guo-Xiang, Elizabeth Barry, Dan Yu, Michael Lukason, Seng H. Cheng, and Abraham Scaria. "CRISPR/Cas9-mediated genome editing as a therapeutic approach for Leber congenital amaurosis 10." *Molecular Therapy* 25, no. 2 (2017): 331–341.

Ruan, Jinxue, Hiroyuki Hirai, Dongshan Yang, Linyuan Ma, Xia Hou, Hong Jiang, Hongguang Wei et al. "Efficient gene editing at major CFTR mutation loci." *Molecular Therapy-Nucleic Acids* 16 (2019): 73–81.

Russell, Stephen, Jean Bennett, Jennifer A. Wellman, Daniel C. Chung, Zi-Fan Yu, Amy Tillman, Janet Wittes et al. "Efficacy and safety of voretigene neparvovec (AAV2-hRPE65v2) in patients with RPE65-mediated inherited retinal dystrophy: A randomised, controlled, open-label, phase 3 trial." *The Lancet* 390, no. 10097 (2017): 849–860.

Ryan, Owen W., Snigdha Poddar, and Jamie H. D. Cate. "CRISPR–Cas9 genome engineering in Saccharomyces cerevisiae cells." *Cold Spring Harbor Protocols* 2016, no. 6 (2016): pdb-prot086827.

Ryu, Seuk-Min, Taeyoung Koo, Kyoungmi Kim, Kayeong Lim, Gayoung Baek, Sang-Tae Kim, Heon Seok Kim et al. "Adenine base editing in mouse embryos and an adult mouse model of Duchenne muscular dystrophy." *Nature Biotechnology* 36, no. 6 (2018): 536–539.

Schaefer, Andrew M., Robert W. Taylor, Douglass M. Turnbull, and Patrick F. Chinnery. "The epidemiology of mitochondrial disorders—past, present and future." *Biochimica et Biophysica Acta (BBA)-Bioenergetics* 1659, no. 2–3 (2004): 115–120.

Schwank, Gerald, Bon-Kyoung Koo, Valentina Sasselli, Johanna F. Dekkers, Inha Heo, Turan Demircan, Nobuo Sasaki et al. "Functional repair of CFTR by CRISPR/Cas9 in intestinal stem cell organoids of cystic fibrosis patients." *Cell Stem Cell* 13, no. 6 (2013): 653–658.

Shao, Yanjiao, Yuting Guan, Liren Wang, Zhongwei Qiu, Meizhen Liu, Yuting Chen, Lijuan Wu et al. "CRISPR/Cas-mediated genome editing in the rat via direct injection of one-cell embryos." *Nature Protocols* 9, no. 10 (2014): 2493–2512.

Shen, Bin, Jun Zhang, Hongya Wu, Jianying Wang, Ke Ma, Zheng Li, Xueguang Zhang, Pumin Zhang, and Xingxu Huang. "Generation of gene-modified mice via Cas9/RNA-mediated gene targeting." *Cell Research* 23, no. 5 (2013): 720–723.

Speicher, Michael R., Arno G. Motulsky, and Stylianos E. Antonarakis, eds. *Vogel and Motulsky's Human Genetics*. Berlin, Heidelberg: Springer Berlin Heidelberg, 2010.

Stella, Stefano, Pablo Alcón, and Guillermo Montoya. "Structure of the Cpf1 endonuclease R-loop complex after target DNA cleavage." *Nature* 546, no. 7659 (2017): 559–563.

Sweeney, Meredith O., Terri L. McGee, Eliot L. Berson, and Thaddeus P. Dryja. "Low prevalence of lecithin retinol acyltransferase mutations in patients with Leber congenital amaurosis and autosomal recessive retinitis pigmentosa." *Molecular Vision* 13 (2007): 588–593.

Tabebordbar, Mohammadsharif, Kexian Zhu, Jason K. W. Cheng, Wei Leong Chew, Jeffrey J. Widrick, Winston X. Yan, Claire Maesner et al. "In vivo gene editing in dystrophic mouse muscle and muscle stem cells." *Science* 351, no. 6271 (2016): 407–411.

Taiber, Shahar, and Karen B. Avraham. "Genetic therapies for hearing loss: Accomplishments and remaining challenges." *Neuroscience Letters* 713 (2019): 134527.

Varela, Malena Daich, Thales Antonio Cabral de Guimaraes, Michalis Georgiou, and Michel Michaelides. "Leber congenital amaurosis/early-onset severe retinal dystrophy: Current management and clinical trials." *British Journal of Ophthalmology* 106, no. 4 (2022): 445–451.

Vermersch, Eva, Charlène Jouve, and Jean-Sébastien Hulot. "CRISPR/Cas9 gene-editing strategies in cardiovascular cells." *Cardiovascular Research* 116, no. 5 (2020): 894–907.

Wang, Hui, Xia Wang, Xuan Zou, Shan Xu, Hui Li, Zachry Tore Soens, Keqing Wang et al. "Comprehensive molecular diagnosis of a large Chinese Leber congenital amaurosis cohort." *Investigative Ophthalmology & Visual Science* 56, no. 6 (2015): 3642–3655.

Ward-Caviness, Cavin K., Lucas M. Neas, Colette Blach, Carol S. Haynes, Karen LaRocque-Abramson, Elizabeth Grass, Z. Elaine Dowdy et al. "A genome-wide trans-ethnic interaction study links the PIGR-FCAMR locus to coronary atherosclerosis via interactions between genetic variants and residential exposure to traffic." *PLoS One* 12, no. 3 (2017): e0173880.

Webber, Beau R., Mark J. Osborn, Amber N. McElroy, Kirk Twaroski, Cara-lin Lonetree, Anthony P. DeFeo, Lily Xia et al. "CRISPR/Cas9-based genetic correction for recessive dystrophic epidermolysis bullosa." *NPJ Regenerative Medicine* 1, no. 1 (2016): 1–11.

Weleber, Richard G., Mark E. Pennesi, David J. Wilson, Shalesh Kaushal, Laura R. Erker, Lauren Jensen, Maureen T. McBride et al. "Results at 2 years after gene therapy for RPE65-deficient Leber congenital amaurosis and severe early-childhood–onset retinal dystrophy." *Ophthalmology* 123, no. 7 (2016): 1606–1620.

Wong, Alan S. L., Gigi C. G. Choi, Cheryl H. Cui, Gabriela Pregernig, Pamela Milani, Miriam Adam, Samuel D. Perli et al. "Multiplexed barcoded CRISPR-Cas9 screening enabled by CombiGEM." *Proceedings of the National Academy of Sciences* 113, no. 9 (2016): 2544–2549.

Wyman, Claire, and Roland Kanaar. "DNA double-strand break repair: All's well that ends well." *Annual Review of Genetics* 40, no. 1 (2006): 363–383.

Xu, Yu-Xin, Valeska Redon, Haojie Yu, William Querbes, James Pirruccello, Abigail Liebow, Amy Deik et al. "Role of angiopoietin-like 3 (ANGPTL3) in regulating plasma level of low-density lipoprotein cholesterol." *Atherosclerosis* 268 (2018): 196–206.

Yin, Shuming, Lie Ma, Tingting Shao, Mei Zhang, Yuting Guan, Liren Wang, Yaqiang Hu et al. "Enhanced genome editing to ameliorate a genetic metabolic liver disease through co-delivery of adeno-associated virus receptor." *Science China Life Sciences* (2020): 1–13.

Zabaleta, Nerea, Miren Barberia, Cristina Martin-Higueras, Natalia Zapata-Linares, Isabel Betancor, Saray Rodriguez, Rebeca Martinez-Turrillas et al. "CRISPR/Cas9-mediated glycolate oxidase disruption is an efficacious and safe treatment for primary hyperoxaluria type I." *Nature Communications* 9, no. 1 (2018): 5454.

Zaidi, Arslan A., Peter R. Wilton, Marcia Shu-Wei Su, Ian M. Paul, Barbara Arbeithuber, Kate Anthony, Anton Nekrutenko, Rasmus Nielsen, and Kateryna D. Makova. "Bottleneck and selection in the germline and maternal age influence transmission of mitochondrial DNA in human pedigrees." *Proceedings of the National Academy of Sciences* 116, no. 50 (2019): 25172–25178.

Zhan, Tianzuo, Niklas Rindtorff, Johannes Betge, Matthias P. Ebert, and Michael Boutros. "CRISPR/Cas9 for cancer research and therapy." In *Seminars in Cancer Biology*, vol. 55, pp. 106–119. Academic Press, 2019.

Zhang, Dandan, Zhenxiang Li, Bingyu Yan, and Jian-Feng Li. "A novel RNA-guided RNA-targeting CRISPR tool." *Science China. Life Sciences* 59, no. 8 (2016): 854.

Zheng, Rui, De-Xin Zhang, Yan-Jiao Shao, Xiao-Liang Fang, Lei Yang, Ya-Nan Huo, Da-Li Li, and Hong-Quan Geng. "Multiplex gene editing reduces oxalate production in primary hyperoxaluria type 1." *Zoological Research* 44, no. 6 (2023): 993.

Zhong, Hua, Yiyun Chen, Yumei Li, Rui Chen, and Graeme Mardon. "CRISPR-engineered mosaicism rapidly reveals that loss of Kcnj13 function in mice mimics human disease phenotypes." *Scientific Reports* 5, no. 1 (2015): 8366.

Zhu, Shiyou, Yuexin Zhou, and Wensheng Wei. "Genome-wide CRISPR/Cas9 screening for high-throughput functional genomics in human cells." *Innate Antiviral Immunity: Methods and Protocols* (2017): 175–181.

4 CRISPR-Cas Systems in Agriculture

Farkhanda Jabeen, Noreen Aasim, and Farheen Shafique

4.1 INTRODUCTION

The rapid development in gene-editing technologies has set off a new stage for personalized farming, with clustered regularly interspaced short palindromic repeats (CRISPR) and CRISPR-associated protein 9 (Cas9) systems leading this transition. CRISPR when combined with Cas9 has become one of the most effective and adaptable instruments for precise genome editing. CRISPR-Cas9, first described as an adaptive immune mechanism in bacteria, has been converted into a genome editing tool that allows DNA alterations with high precision (Satheesh et al., 2019). This technology works by recruiting the Cas9 enzyme to a specific DNA sequence using an RNA molecule which acts as the guide, and then it aids in cutting the DNA so another DNA-consisting genetic material can either be inserted or deleted (Zhu et al., 2020). CRISPR-Cas9 is providing unique openings to improve a variety of crop traits, such as higher yields, disease resistance, and tolerance to the environment in the agricultural sector. So, while with traditional plant breeding methods it often requires years of cross-breeding and selection by humans for plants to get even marginally better at resisting pests or growing in dry conditions, (or both), CRISPR-Cas9 can do the same thing, but faster, better, and with much higher accuracy. The technology has succeeded to modify staple food crops like rice, maize, and wheat, paving the way to solving global issues of food security/food safety–related challenges, including climate change and deficiencies in nutritional attributes. Take rice, for example; the recently reported strain of rice with enhanced grain yield and resistance to bacterial blight (a devastating global disease) has been developed using CRISPR-Cas9 technology (Satheesh et al., 2019; Li et al., 2021).

However, despite the significant potential for agriculture, the use of CRISPR-Cas systems raises critical ethical, ecological, and regulatory concerns. Yet this requires careful evaluation of the possibility of off-target effects on other genes or the emergence of new variations (Zhu et al., 2020). Further, the ethical considerations associated with developing CRISPR-Cas9-based genetically engineered food organisms should be subject of a continued debate and appropriate societal deliberation. These benefits and concerns are carefully weighed by this research—the trade-off between the ability to exploit this technology so we can continue to advance agriculturally, alongside our conserved biodiversity and good health, is a key topic of ongoing research and policy discussions. This chapter explores the mechanisms, uses, and potential impact on agriculture occurring with the CRISPR-Cas9 system.

DOI: 10.1201/9781003537151-4

4.2 CRISPR-CAS9 GENE-EDITING SYSTEM IN PLANTS

The CRISPR technology is an emerging technique and has been used for plant and animal genome editing. The CRISPR-Cas9 system was originally found in bacteria, as their natural defense mechanism against viruses. It is used to cleave targeted double-stranded breaks (DSBs) in DNA by the creation of single guide RNAs (sgRNAs), designed to recognize specific site in the viral DNA. Now, this mechanism is modified for use in animals and plants, where the system is used to make DSBs in the targeted DNA. Following that, it allows either non-homologous end joining (NHEJ) or homology-directed repair (HDR) to fix the induced break when needed (Zetsche et al., 2015b; Satheesh et al., 2019).

The CRISPR-Cas9 system involves a Cas9 protein, which is directed to the DNA by an sgRNA. After the Cas9–sgRNA complex is bound to the DNA, the Cas9 protein will cleave the DNA at the targeted site. Secondly, as indicated, the telomere will be delineated by a precise cleavage, which will be sealed by cellular repair mechanisms, which can give rise to mutational lesions, depending on whether the repair is error-prone (NHEJ) or correct (HDR) (Satheesh et al., 2019).

The development of genetic diversity in agricultural crops is crucial for promoting sustainable farming practices by enhancing crop resilience, yield, and nutritional value (Podevin et al., 2013). Traditional approaches, such as ethyl methanesulfonate (EMS) mutagenesis and γ-radiation, depend on random mutagenesis, which can lead to unpredictable genomic alterations. In contrast, genome editing technologies have transformed this area by providing a more accurate method for introducing genetic variation at specific genomic sites. Sequence-specific nucleases (SSNs) are among the most commonly utilized tools in genome editing, designed to create DNA DSBs at targeted locations within the genome (Sun et al., 2016). These breaks trigger the plant's repair processes, which can be utilized to introduce specific mutations or integrate new genetic material. The incorporation of transfer DNAs (T-DNA) in plants via Agrobacterium-mediated DNA transfer marks a significant advancement in genome editing techniques (Hsu et al., 2014). These T-DNAs are preferentially integrated at sites of induced DNA DSBs, representing a notable improvement over previous methodology (Hsu et al., 2014). In earlier approaches, T-DNA that contained homologous regions to the target integration site resulted in low-frequency insertions (Sternberg et al., 2014). This development has facilitated the more effective integration of genetic material at precise genomic locations (Sternberg et al., 2014). Nevertheless, achieving HDR-mediated gene insertion continues to pose challenges in plant genome editing. To date, the majority of genome editing initiatives in plants have relied on the NHEJ pathway for gene knockout, often resulting in frameshift mutations or substantial deletions, as NHEJ is more prevalent than HDR (Feng et al., 2014; Zhou et al., 2014). Successful instances of gene insertion via HDR in plants remain relatively uncommon (Feng et al., 2014; Zhou et al., 2014).

Unlike conventional mutagenesis, which often yields a combination of advantageous and detrimental mutations, SSNs facilitate targeted modifications, thereby reducing unintended genetic alterations. The emergence of SSNs, such as CRISPR-Cas9, transcription activator-like effector nucleases (TALENs), and zinc finger nucleases (ZFNs), enables precise gene editing, accelerating crop enhancement and addressing some of the challenges posed by earlier methods (Sun et al., 2016; Li et al., 2016). With these advanced technologies, researchers can directly alter genes linked to critical traits, like disease resistance, drought tolerance, and nutrient content, thereby fostering more sustainable and efficient agricultural systems. As these technologies continue to advance, they present a viable solution to the increasing demands of global food security while minimizing ecological impact. DSBs caused by SSNs are repaired through the cell's intrinsic DNA repair pathways, primarily NHEJ and HDR. NHEJ is the predominant repair mechanism, where DNA ligase IV plays a crucial role in directly reconnecting the separated DNA ends. However, if the ends have been altered by nucleases or if there are missing bases, the repair may not be flawless, resulting in small insertions or deletions (indels) at the site of the break. This can lead to frameshift mutations or gene knockouts, which are often utilized to disrupt specific genes for functional analysis. Conversely, HDR provides a more accurate repair method, utilizing a template to direct the repair process, thereby ensuring precision. Typically, during cell division, the sister chromatid acts as the template for HDR, facilitating error-free repair. In genome editing contexts, researchers can introduce custom DNA templates aimed at recoding specific loci or inserting new genetic elements flanked by homologous regions. This capability to deliver precise genetic information through HDR significantly enhances the potential for targeted gene modification, the addition of advantageous traits, or the introduction of new functional components, thereby expanding the scope of genome editing in plants. By integrating SSNs with HDR, it is possible to achieve highly precise genetic alterations, enabling more efficient engineering of plants with desired characteristics.

Currently, the Cas9 enzyme is the main focus within CRISPR-based genome editing; nonetheless, advancements in CRISPR techniques have introduced proteins like Cas12a (formerly Cpf1) that distinguish themselves from Cas9. Cas12a is not only smaller in size, which simplifies delivery into plant cells, but also exhibits a different cutting mechanism that provides greater flexibility in genome editing (Zetsche et al., 2015a). Cas12a leaves behind sticky ends through cuts that are staggered instead of the blunt-ended, DSBs in DNA made by Cas9. This distinct cleavage pattern is gaining attention because it may facilitate more precise gene insertions through HDR, enhancing the accuracy and efficiency of genetic modifications in plants (Tang et al., 2019). It also reveals that Cas12a has special and extraordinary targeting ability. Together with the RNA guiding strand, Cas12a carries a different protospacer adjacent motif (PAM) sequence from Cas9, which addresses a wider choice of targetable genomic locations. This versatility allows researchers to access previously untargetable regions of the plant genome, providing plant breeders with a broader toolkit for improving crop traits (Liu et al., 2019). Moreover, the smaller size of Cas12a makes it more suitable for use in delivery vectors, like the adeno-associated virus (AAV) and other plant-compatible systems, potentially increasing the efficiency of genome editing in diverse plant species (Bubeck and Grimm, 2024).

Apart from Cas12a, successful inventions of base editing and prime editing utilize the advancement of plant genome editing technology. CRISPR-based base editing is the manipulation of tissues by using modified CRISPR systems to induce point mutations without generating double-strand breaks. For example, cytosine base editors (CBEs) and adenine base editors (ABEs) enable the conversion of C-G to T-A and A-T to G-C base pairs, respectively, allowing for precise changes at the nucleotide level (Gaudelli et al., 2018; Komor et al., 2017). These tools can be valuable in plant breeding because they make it possible to treat single nucleotide polymorphism (SNP) that can have an impact on yields, resistance to diseases, and stress conditions.

Prime editing is another emerging technique that enables the introduction of targeted insertions, deletions, and all 12 possible base-to-base conversions without requiring double-strand breaks or donor DNA templates (Anzalone et al., 2019). Prime editors apply a reverse transcriptase with a new guide RNA to identify the site and the kind of alteration. This precision reduces the risk of unintended mutations and can be employed to introduce or correct mutations that improve plant phenotypes, such as enhancing nutrient content or conferring resistance to pests (Zong et al., 2022). Combining these new, cutting-edge technologies with the regular CRISPR systems can assist plant researchers to create better modifications that can advance crop improvement programs. The lesser chance of off-target effects and being able to introduce specific changes without depending on DSBs can improve the efficacy and safety of plant genome editing (Ren et al., 2018). These advancements point to a better future of plant bioengineering, where experts have several genetic tools at their fingertips for modifying plant genes for food security and sustainable agriculture.

Undertaking the regulatory landscapes that vary among countries poses a crucial difficulty for CRISPR-Cas9 technology. In some regions, genome-edited crops are subject to the same regulations as genetically modified organisms (GMOs), while others distinguish CRISPR-edited crops as non-GMO, provided no foreign DNA is introduced (Zetsche et al., 2017; Satheesh et al., 2019). For instance, in the United States, CRISPR-edited crops are sometimes easier to bring to market since the United States department of agriculture (USDA) does not regulate them, unless they include foreign DNA (Ishii, 2017). However, CRISPR-edited crops face the same strict constraints as GMOs in the European Union (EU), which slows down their commercialization (Puchta, 2017).

The way people view CRISPR-edited plants significantly affects their approval. In places that battled against genetically modified crops, resistance could prevent similar hurdles for CRISPR-edited produce if people are not informed about the key differences between the technologies. Transparent communication about the safety, benefits, and risks of CRISPR technology is necessary to build public trust and encourage the widespread adoption of gene-edited crops (Jaganathan et al., 2018). Efforts are being made by scientists and policymakers to promote science-based regulations that can help fast-track the commercialization of CRISPR-edited crops while addressing public concerns about food safety and environmental impacts (Puchta, 2017).

4.3 APPLICATIONS IN PLANT GENOME EDITING

Recently, genome editing technologies, including CRISPR-Cas9, have gained recognition as a potential tool for improving desired characteristics in agricultural plants (Ercolano and Wang, 2023; Jaganathan et al., 2018). Various agronomic traits, including nutritional content, stress tolerance, and production, may be enhanced by the targeted modification of specific genes (Khatodia et al., 2016). As an example, research has shown that CRISPR-Cas9 may improve wheat grain quality, improve soybean oil content, and make rice more drought-tolerant (Rai et al., 2023; Peng et al., 2020).

4.3.1 Genome Editing for Sustainable Agricultural Improvement

Genome editing technologies, especially CRISPR-Cas9, are being explored as a potential approach to improve desirable characteristics in agricultural plants (Ercolano and Wang, 2023; Jaganathan et al., 2018). The developments in crop growth facilitated by genome editing technologies have the potential to promote stronger and more resilient agricultural systems, therefore contributing in addressing global concerns, such as food shortages and environmental degradation. Researchers can enhance agricultural productivity and minimize dependence on harmful pesticides by developing crop varieties with better characteristics that are more adaptable to adverse environmental conditions, which promotes environmentally sustainable agriculture (Roberts and Mattoo, 2018).

CRISPR-Cas9 has been used in rice cultivation to enhance resistance to bacterial blight and alter plant development responses to environmental stress (Zetsche et al., 2015b). Researchers have focused on genes such as OsSWEET13, which is associated with resistance to bacterial blight, therefore lowering the disease's effect on the production of rice (Ancydiana et al., 2022). In maize, genes linked to drought tolerance and nitrogen-use efficiency have been edited to improve the crop's performance in nutrient-poor soils and under water-limiting conditions (Shi et al., 2017).

CRISPR-Cas9 is advantageous because it allows for the precise and targeted editing of specific genes without introducing foreign DNA, distinguishing it from traditional transgenic methods (Zetsche et al., 2015b). The reduced risk of unintended off-target effects enhances safety and acceptance for this method as a way to get better agricultural products (Satheesh et al., 2019; Zetsche et al., 2015b). Recent developments have supported multiplex gene editing, permitting simultaneous gene modification of various features, such as productivity, stress tolerance, and adjustment to disease resistance (Xie et al., 2015), which has clearly upgraded many crop types.

The innovations occurring in crop enhancement derived from genome editing technologies form opportunities to construct sustainable and resilient agricultural systems to tackle global issues, such as food security and environmental problems. Researchers aim to improve agricultural productivity, reduce dependence on toxic pesticides, and enhance sustainable farming practices that withstand adverse environmental situations through the cultivation of varied crop strains with modified traits (Roberts and Mattoo, 2018). With the ability to exactingly edit genes in plants utilizing CRISPR-Cas9, scientists can newly explore approaches to enhance crop improvement by customizing important trait aspects essential for increasing agricultural sustainability. Improving yield along with stress resilience is important, as well as modifying metabolic pathways to result in crops with elevated nutritional profiles, like heightened vitamin or mineral levels (Husaini, 2022).

The use of genome editing techniques in plants can enhance crop resistance to diseases and pests, thus decreasing chemical pesticide demand while encouraging eco-friendly agricultural methods. Researchers can improve crop varieties resistant to viral, bacterial, and fungal infections, as well as insect invasions, by modifying genes involved in plant–pathogen interactions. These changes generate higher productivity, reduce crop losses, and reduce the use of hazardous substances that are beneficial to the long-term sustainability and adaptability of agricultural systems (Ur Rahim et al., 2021). Today, CRISPR-Cas9 technology enables the precise modification of plant genomes, leading to an emergence of a specialized domain within plant development. Such developments make researchers come up with new solutions to existing problems in today's agriculture, which include food production, climatic change, and environmental degradation (Khan et al., 2017; Zhang et al., 2017).

4.3.2 Pathogen and Pest Resistance

The development of better resistance to pests and diseases is one of the main uses of genome editing in plants. Researchers have produced crop varieties with higher resistance and tolerance to a broad range of pathogens, including viruses, bacteria, and fungus, by carefully identifying and deactivating or changing the genes engaged in plant sensitivity to diseases (Borrelli et al., 2018; Zaidi et al., 2016). Also, genome editing has made it possible to produce agricultural plants with enhanced insect pest resistance, therefore preventing the requirement for chemical pesticides (Dhaliwal and Uchimiya, 1999). By applying these developments in the production of disease-resistant and pest-resistant crops, agricultural systems might become much more sustainable, therefore limiting dependence on harmful chemical pesticides while encouraging more ecologically friendly farming methods. This method is especially helpful in areas where new diseases and pests seriously threaten the harvest, as genome editing helps create crop variants specifically designed to meet these challenges (Roberts et al., 2021).

Furthermore, this technique helps reduce the usage of pesticides, consequently conserving natural ecosystems and protecting biodiversity. Considering this, a major focus of developing sustainable productive agricultural systems is on seeking for plant resistance to pests and diseases via genome editing. The principal emphasis of CRISPR-Cas9 application in plant genome editing is the improvement of agronomic characteristics. One relevant case is that the use of CRISPR-Cas9 has resulted in the development of rice types with improved drought tolerance and wheat varieties with better resistance to fungal pathogens (Ricroch et al., 2017; Eş et al., 2019; Khatodia et al., 2016).

4.3.3 Enhancing Nutritional Content and Food Quality in Crops

CRISPR-Cas9 has been used to optimize yield and improve stress resistance, but the most interesting applications have been directed toward modifying plant metabolic pathways toward crops of improved nutritional value. This comprises biofortification measures intent to enhance the amounts of vital vitamins and minerals in commonly consumed crops, which go a long way in combating malnutrition as well as complications related to nutrient-poor diets across the world. For instance, the utilization of CRISPR-Cas9 has contributed to enhancing the vitamin A content in rice, known as Golden Rice, to address the shortage of vitamin A in regions where rice is the basic staple food (Khatodia et al., 2016). Likewise, this technology enriches various grains, such as wheat and maize, with a higher content of iron and zinc, respectively; these are essential micronutrients that are fundamental for human consumption (Erpen-Dalla Corte et al., 2019). In the same manner, CRISPR-Cas9 has brought about positive developments to food quality since it makes it possible to introduce changes in desirable characteristics of crops, including taste, texture, and shelf life. For instance, researchers have applied this brevity to lower the content of bitter alkaloids in some vegetables, leading to their improved uptake and market value (Eş et al., 2019). It has also been applied in manipulating genes that control ripening, hence slowing fruit and vegetable shelf life, thus reducing post-harvest losses and food wastage (Camerlengo et al., 2022).

These advancements enhance agricultural productiveness, not only in the sense of increasing the crop yield, but also in crafting food that is superior in both quality and nutrition and consumers' tastes. Moreover, the possibility to breed plants with certain desirable characteristics promotes building more sustainable and adaptive agricultural practices. Increased crop resilience from developmental biotechnologies has the capability to counter environmental stress and provide better nutritional returns, solving the major food insecurity and health and sustainable environment issues (Ricroch et al., 2017). As it helps create plants yielding food rich in nutrients and plants that are resistant to climate change, it is a critical technology which defines the future of the world's agriculture (Eş et al., 2019; Erpen-Dalla Corte et al., 2019; Camerlengo et al., 2022).

Further, the system has been used in modifying the metabolic pathways in plants used in getting vitamin-laden crops, besides enhancing the quality, taste, and texture of food crops. Conventional breeding in today's advanced research by using CRISPR-Cas9 technology in crop improvement can create enhanced and sustainable agricultural outputs that meet the food demand in addition to combating the effect of ecological depletion. Through the deletion of some specific genes conclusively in plants, new breeding procedure can be started (Eş et al., 2019; Erpen-Dalla Corte et al., 2019; Camerlengo et al., 2022). These sorts of outcomes can help in increasing crop yield, improving diet availability, and reducing pesticides hazardous to crops, thus increasing sustainability and eco-friendliness in farming (Sharma and Prabhakar, 2014; Zhou et al., 2014).

By using CRISPR-Cas9 technology in wheat, the research community has created resistance to numerous diseases with disabling specific susceptibility genes. The gene knockout of Mildew Locus O (MLO) has been linked to better protection from powdery mildew, a frequent and devastating fungal disease that significantly lowers wheat yields worldwide (Wang et al., 2014). The factors that made the plant susceptible to the pathogen and created wheat strains that provided enhanced natural protection against it have been removed. Wheat genome editing is advancing through the use of CRISPR-Cas9 to boost resistance against Fusarium head blight (FHB), a critical fungal disease influencing wheat grain quality and safety by producing mycotoxins. To improve resistance against FHB in wheat varieties, research studies have targeted and modified defense-related genes. This genetic alteration boosts disease protection and lowers the need for chemical sprays. **Table 4.1** shows a list of CRISPR-Cas9-edited crops.

TABLE 4.1
Examples of CRISPR-Cas9-Edited Crops

Crop	Targeted Trait	Outcome	References
Wheat	Powdery mildew resistance	Knockout of *MLO* gene for disease resistance	Wang et al., 2014
Rice	Drought tolerance	Enhanced tolerance to drought	Xu et al., 2020
Tomato	Fruit shape	Modified fruit morphology	Brooks et al., 2014
Maize	Nutrient content (phytic acid)	Reduced phytic acid levels for better digestibility	Liang et al., 2014
Cassava	Virus resistance	Increased resistance to cassava mosaic virus	Odipio et al., 2017

In rice farming, CRISPR-Cas9 serves as an effective strategy to manage varied environmental risks. One important use is improving defenses against bacterial diseases. Engineered rice varieties have been developed with increased resistance to bacterial blight by changing the genes linked to Xanthomonas resistance 13 (Xa13) susceptibility. CRISPR-Cas9 excels at producing specific and inheritable genetic alterations and is thus a popular method for boosting crop enhancement plans. This led to swift progress in the creation of many crop varieties featuring various positive characteristics. EXCRISPR-Cas9 technology has produced variations in rice that feature improved traits, like plant height and grain quality, while bolstering resistance to pests. Rice plants with enhanced grain size and yield were developed through gene editing of Grain Size 3 (GS3) and Dense and Erect Panicle 1 (DEP1) genes, followed by targeting SWEET14 for bacterial resistance. These innovations show how CRISPR-Cas9 helps modify crops to tackle particular agricultural challenges and proves to be a vital asset for global food stability. CRISPR-Cas9 technology has also been explored to improve nutritional levels in crops. Ongoing studies and uses of CRISPR-Cas9 in wheat and rice promise to resolve critical problems in agriculture, including outbreaks of diseases and environmental challenges. The ability to manipulate plant genomes with precision through CRISPR-Cas9 presents new opportunities for cultivating crops that are durable and fruitful.

4.4 CRISPR-CAS9-MEDIATED MOLECULAR BREEDING ACCELERATING CROP QUALITY

For molecular breeding, CRISPR-Cas9 has certainly proven to be a valuable tool, enabling the rapid improvement of crop quality in aspects like texture, flavor, physical appearance, and nutritional content. With this gene-editing approach, plant breeders can introduce desired features into crops more effectively since it offers the precise and controlled modification of a specific gene (Zetsche et al., 2015b; Satheesh et al., 2019).

4.4.1 Improving Crop Physical Appearance

Marketability is critically dependent on physical appearance, which affects consumer preferences and agricultural profitability. Advances in physical characteristics, such as color, size, and shape, are also being made with CRISPR Cas9. CRISPR-Cas9 has been used in tomato plants to manipulate genome genes influencing fruit morphology, including the formation of a uniform size and increased color (Satheesh et al., 2019). Also, rice plants have been selectively modified to enhance their physical characteristics, such as the number of tillers and the angle of their leaves, which leads to increased crop output and desirable appearance (Zetsche et al., 2017). In the same way, rice plants have been tested with intentional modifications in order to enhance plant architecture. These adjustments include changes in the number of tillers and the angle of the leaves, both of which lead to improved crop productivity and aesthetic appeal (Zetsche et al., 2017). Because of such modifications, the economic value of crops has significantly grown. This is because breeders are now able to customize the physical characteristics of crops to the preferences of customers and to the efficiency of production. For instance, the modification of genes that are involved in the size and shape of tomatoes has resulted in improved uniformity, which has had a direct influence on both the market value and the convenience of processing (Brooks et al., 2014). **Table 4.2** shows the crops that have been modified for their physical appearance by the CRISPR-Cas9 approach.

4.4.2 Improving Crop Texture Quality

The importance of texture within the consumer experience and quality processing has a positive effect on the appeal of fruits and vegetables. The development in choice gene editing, especially through the application of the CRISPR-Cas9 technique, has made it possible to edit the genes that determine the formation of cell walls and, ultimately, the firmness and texture of various crops. For example, studies have successfully applied gene editing to apples and potatoes to reduce the activity of polyphenol oxidase (PPO), a key enzyme that contributes to browning and texture degradation during storage (Divya et al., 2024). Through precise manipulation of the identified PPO genes, engineered varieties that do not only delay browning but also enhance the digestibility of the produce have been developed. This enhancement extends the shelf life of the products and meets the consumer demand for fresh-looking, aesthetically

TABLE 4.2
Examples of CRISPR-Cas9 Modifications for Physical Appearance

Crop	Trait	Gene Targeted	Outcome	References
Tomato	Fruit size and shape	Gene-regulating morphology	Uniform size, improved shape	Brooks et al. (2014)
Rice	Plant architecture	Genes affecting tillering and leaves	Optimized plant structure for higher yield	Zetsche et al. (2017)

pleasing produce, aligning with retailers' preferences for items that maintain their visual appeal (Kumar et al., 2020). In the case of wheat, CRISPR-Cas9 has been employed to modify genes responsible for the gluten structure, resulting in enhanced dough elasticity and improved quality for baking applications (Satheesh et al., 2019). Gluten level, in the food industry, thus calls for precise regulation because the texture determines customer experience of food, especially in the production of specialty bread. A study by Liu et al. (2020) highlights how the manipulation of gluten-related genes can produce wheat varieties that meet specific texture requirements for different types of bread, catering to diverse consumer preferences.

Furthermore, the capability to refine texture-associated genes by CRISPR-Cas9 offers substantial benefits in fulfilling industry standards and customer demands, especially for crops necessitating certain textures for processing and commercial viability (Zetsche et al., 2015b). Modifications to genes associated with pectin production in tomatoes have resulted in alterations in firmness and juiciness, producing types more appropriate for fresh consumption or processing into sauces (Tovar-Perez et al., 2020). It can also be done in other crops, which bring us to the diverse potential uses and advantages of CRISPR technology. In rice, for example, genetic modifications aimed at altering starch granule size and distribution have resulted in improvements in cooking texture and quality, catering to preferences for specific rice varieties like jasmine and basmati (Chen et al., 2021). Such improvements have highlighted the role of CRISPR-Cas9 tools in improving on the organoleptic nature of an array of horticultural crops, thus promoting the welfare of both the consumer and the producer.

4.4.3 Improving Palatability

Taste and *flavor* can be defined as palatability and is a key factor that enhances crop value that directly affects consumers' consumer acceptability. Improvement of this attribute requires the creation of changes in sugars, acid, and volatile compound profiles in various crops. The latest breakthrough in enhancing these traits has been defined by the use of the CRISPR-Cas9 technology. For instance, Zetsche et al. (2015b) demonstrated that CRISPR-Cas9 can effectively modify the biosynthesis pathways of key flavor compounds in crops such as tomatoes and strawberries. Since sweetness, the level of acidity, and the level of flavor volatility are greatly influenced by the genes of crops and products being sold, crisp editing of these specific traits ensures consumers move to the products.

Besides, CRISPR can also be used effectively for the elimination of undesirable characteristics, like the bitterness in cucumbers and eggplants. By knocking out genes responsible for producing bitter compounds, scientists have developed varieties that are more appealing to consumers (Satheesh et al., 2019). These gene edits not only increase the sensory quality of the crops but also increase their salability, which is in accordance with the changes that consumers want to see on the market, in terms of foods that taste better.

Within rice production science, CRISPR-Cas9 has been used to alter firmness and taste of cooked rice by changing the biosynthetic pathways of starch. These precise genetic alterations contribute to better grain quality, satisfying consumer preferences for specific rice varieties, such as aromatic and sticky rice (Zetsche et al., 2015b). Apart from that, Li et al. (2018) investigated the possible application of CRISPR-Cas9 to improve the nutritional composition of rice by augmenting necessary amino acid levels, hence attracting health-conscious customers. It was shown by Ahn et al. (2020) that genes linked with the production of anthocyanin in apples may be effectively modified, resulting in fruit that is more appealing and delicious. Because of the antioxidant properties of anthocyanins, this augmentation not only improves the visible appearance of the fruit, making it more attractive to the eye, but also increases the fruit's nutritional value. Additionally, Wang et al. (2021) highlighted the potential of CRISPR-Cas9 in developing disease-resistant crops, which can improve overall crop health and yield. Better yields are, of course, healthier yields, which means tastier and more attractive produce for consumers. The fact that CRISPR technology is versatile adds to its potentiality of modifying the appearance, taste, and aroma of crops, thus contributing toward higher acceptability and successful presence in the market.

4.4.4 Biofortification of Nutrient Elements

Biofortification involves the process of increasing the nutritional quality of crops, for instance, improving the vitamin or mineral composition by using either genetic techniques or conventional breeding. This has become a useful approach in tackling conventional nutritional problems, such as malnutrition, and the inherent low nutrient density of staple foods prepared by a large populace in the developing world, where their ability to access a pallet of diets is often restrained. Over the last few years, genetic engineering, especially by use of biotechnology methods, such as CRISPR-Cas9, has been developed to enhance the nutritional value of crops to fight malnutrition in the society.

Another popular and precise genetic tool, called CRISPR-Cas9, has been used efficiently to increase the level of micronutrients, like iron, zinc, and vitamin A, in different crops. Probably one of the most famous examples of the use of genetic biofortification is the creation of the Golden Rice, a glutinous rice that has been developed through genetic engineering which contains provitamin A (beta-carotene), that is, vitamin A in the human body, and can aid in curbing vitamin A deficiency, which results in easily preventable blindness and immune deficiencies, most especially among children of the third world countries. The utilization of CRISPR-Cas9 has facilitated the improvement of the biochemical route, regulating provitamin A production in rice, substantially increasing its concentrations (Zetsche et al., 2015b). This advancement shows how genetic engineering might tackle public health issues by improving the nutritional quality of crops that are essential (Tang et al., 2013).

Besides improving positive nutrients, CRISPR-Cas9 has been applied effectively for decreasing negative nutrients, which means antinutritional factors that negatively affect nutrient bioavailability in the human body. For example, in maize, the gene which encodes the production of phytic acid, an anti-nutrient which binds to minerals such as calcium, zinc, and iron, thus hindering their absorption for digestion by humans has been studied. Through modification of this gene, studies have effectively reduced phytic acid concentrations in maize, hence enhancing mineral absorption and improving its nutritional value for consumers (Liang et al., 2014; Shi et al., 2007).

TABLE 4.3
Examples of CRISPR-Cas9-Mediated Biofortification

Crop	Nutrient Enhanced	Gene Targeted	Outcome	References
Rice	Provitamin A	PSY, CRTI	Increased vitamin A content	Paine et al., 2005
Maize	Phytic acid reduction	ZmIPK	Lowered phytic acid, enhanced mineral absorption	Liang et al., 2014
Wheat	Iron and zinc content	NAS, Ferritin	Higher micronutrient levels	Satheesh et al., 2019

CRISPR-Cas9 has been used to improve the nutritional content of certain staple crops. In wheat, gene editing to enhance the levels of micronutrients, including iron and zinc, in the grain has been used. This biofortification attempt is significant in eliminating the iron and zinc deficiencies that plague millions of its users, especially from areas where wheat is a staple food. Deficiency of both iron and zinc leads to anemia, which results from coordination deficit and has its major effect on biochemical and immunological processes among children. By increasing the concentration of these micronutrients in wheat, scientists are helping to alleviate "hidden hunger"—a form of malnutrition where people may have enough calories but lack essential nutrients (Satheesh et al., 2019; Gupta et al., 2020). **Table 4.3** shows the outcome of nutrient-enhanced crops with their respective target genes.

4.5 ADVANTAGES AND DISADVANTAGES OF THE EXTENSIVE APPLICATION OF CRISPR-CAS SYSTEM IN AGRICULTURE

The potential of the CRISPR-Cas9 system is huge for making positive changes in agriculture because it can replace genes as the solution to some of the greatest concerns of crop growers, like higher yields, greater nutritional content, and increased resistance to diseases and unfavorable conditions. This gene-editing technology is particularly valuable in addressing climate change, food security, and the nutritional needs of a growing global population (Zetsche et al., 2015b). On the same note, some of the drawbacks include the fact that there are many complications linked with the use of this technology when it is implemented in almost all fields.

4.5.1 Potential Advantages

The potential advantages of the CRISPR-Cas system in the agriculture sector are mentioned in what follows.

4.5.1.1 More Crop Yield and Durability

CRISPR-Cas9 technology is an innovative tool used to edit the gene responsible for many important characteristics, such as drought tolerance, disease resistance, and nutrient use efficiency, among others. These genetic changes in a big way improve the "drought tolerance" of crops and such other factors, ensuring better production despite adverse conditions. With the increasing world population and climate change enhancement, increasing crop durability and productivity is of paramount importance for food security. More so, CRISPR-Cas9 has been successfully used in some of the world's major crops, like maize, rice, and wheat, to enhance their performance under stress. For example, maize, the third globally consumed cereal, has been genetically altered to make it more drought-resistant. Different studies have identified and edited key genes involved in the plant's stress response pathways, improving its ability to conserve water and survive in arid environments (Liang et al., 2014; Shi et al., 2017). The same tactics have also been applied in rice, where CRISPR technology has enhanced the ability of rice plants to endure both drought and low-quality saline groundwater availability for irrigational purposes. It is especially critical for rice production in Southeast (SE) Asia, where climate change consequences may decrease water supply and soil fertility (Wang et al., 2018).

Besides, the CRISPR-Cas9 has also be used to increase the disease resistance ability of crops. Some diseases are very destructive to the crops and may take a very short time to render farmers and food security prone. Due to research, crops that are resistant to the recurrent diseases are being planted after genes that make one be prone to diseases have been manipulated. For example, CRISPR has been used to knock out susceptibility genes in wheat, making the plants more resistant to fungal diseases like powdery mildew and rust (Wang et al., 2014; Zhang et al., 2017). This is an effective method of combating fungal pathogens since it decreases the amount of chemical fungicides used in agricultural practices which have negative impacts on the environment. Similarly, the CRISPR-Cas9 has been utilized in enhancing nutrient use by the crop plants especially nitrogen use efficiency. Nutrient like nitrogen is very essential for the growth of plants, but the over use of nitrogenous compounds results in pollution load on soil and water bodies. In maize, they have applied CRISPR in improving nitrogen use efficiency through muting genes associated with nitrogen transport and metabolism. This allows crops to maintain high yields with reduced fertilizer inputs, promoting more sustainable agricultural practices (Shi et al., 2017). By reducing the reliance on chemical fertilizers, CRISPR technology helps lower greenhouse gas emissions associated with fertilizer production and application, contributing to climate change mitigation (Xu et al., 2019).

Furthermore, CRISPR has been used for enhancing phosphorus use efficiency in plants more so in crops. Phosphorus is another essential nutrient, but like potassium (K^+), phosphorus (P^{++}) is tightly bound to forms that are unavailable to plant uptake. By editing genes involved in phosphorus metabolism, scientists have enhanced the ability of crops like maize and rice to utilize phosphorus more efficiently, reducing the need for phosphorus fertilizers (Chen et al., 2020). These innovations do not only increase the yields through better crop germination and growth but also assist in preserving such scarce resources like phosphorus.

4.5.1.2 Enhanced Nutritional Content

CRISPR-Cas9 has revolutionized the biofortification of crops by enabling the alteration of the nutritional density of crops to feed the ever-expanding population, particularly in areas with low nutrition quality. This technology enables the manipulation of the genes that are involved in the production and storage of these nutrients, making this a solution to malnutrition. Of the resulting and previously discussed applications, the most well-known example is the production of Golden Rice—a type of rice having enhanced provitamin A (beta-carotene). Night blindness can easily be prevented when enough vitamin A is supplied through diet or supplements, since lack of vitamin A in the diet is a major health concern in developing countries with high mortality rate, especially among the children. Golden Rice, due to the traits inherited from maize and bacteria, is capable of synthesizing efficient amounts of provitamin A occupied in the grains of the rice, which can be further turned into vitamin A in the human body. This innovation has the potential to address widespread vitamin A deficiency, particularly in Asia and Africa, where rice is a dietary staple (Zhu et al., 2020; Tang et al., 2013).

Iron deficiency anemia is another global health issue; it is present in over 1.6 billion people, more in female and children from third world countries. Some of these solutions include iron-enriched wheat, which can be created by utilizing CRISPR-Cas9. By editing genes involved in iron transport and storage, researchers have significantly increased the iron content in wheat grains, providing a more accessible source of this vital nutrient (Chen et al., 2024). This new improved wheat variety eradicates the problem of iron deficiency anemia, which leads to impaired physical and mental development, low productivity, and high mortality among pregnant women. Additionally, CRISPR has been used to enhance the zinc content in staple crops like wheat and rice, targeting another key micronutrient that is essential for immune function and overall growth (Garg et al., 2018).

Apart from boosting the quantity of useful nutrients, the method of applying CRISPR-Cas9 technology has been used to decrease dangerous elements in foods, thus increasing food security while making some foods more available to people who have special diets. An example of this is the extraction of allergens from peanuts, a product which accounts for some of the severest allergies. By targeting and editing the genes responsible for producing allergenic proteins, researchers have developed hypoallergenic peanuts, which could potentially reduce the risk of life-threatening allergic reactions (Liang et al., 2014). This innovation could play a tremendous role in reducing the risks of consuming peanuts to millions of people around the world.

Likewise, CRISPR has been applied to alter wheat in order to decrease or exclude gluten content, for the consumption of which people with celiac disease should not suffer from their disease—*celiac disease* is an autoimmune disorder caused by the reaction of the human (http://aboriginesofaustralia.com/scholar-essay-writing-help/) body to gluten. Scientists have deprived genetically modified wheat of certain genes that code for gluten proteins, thus developing low-gluten or gluten-free wheat. These innovations can improve the quality of life of individuals with celiac disease or gluten sensitivity, offering them more dietary options without compromising their health (Sanchez-Leon et al., 2018).

In addition, CRISPR-Cas9 has been deployed to increase the protein, iron, and vitamin A content of cassava, a staple in the sub-Saharan region. Cassava is naturally low in essential micronutrients, but by using CRISPR to increase its content of beta-carotene and iron, researchers are working to develop varieties that can help alleviate widespread deficiencies in these nutrients among populations that rely heavily on cassava for their daily caloric intake (Adeyemo et al., 2019). These examples show how CRISPR has been applied to enhance the nutritional value of staple crops as well as to make crops nutrient-dense in regions of the world where they are produced.

Finally, it can be noted that the use of CRISPR-Cas9 technology opens up good opportunities in the field of biofortification and can significantly improve the nutritional value of crops, as well as reduce toxic compounds. Focusing on micronutrients such as vitamin A, iron and zinc, allergens, and gluten, CRISPR has opened the door for improved, healthy, and nutritional crops that respond to health problems across the globe, particularly in the third world. Thus, there still remains a vast area of investigation of CRISPR-driven biofortification in enhancing food security and the health of the population, since the concept has the ingredients for the longtime resolution to the problem of malnutrition and foodborne diseases.

4.5.1.3 Reduced Food Waste

CRISPR-Cas9 technology offers promising solutions to one of the most significant global challenges: food waste. Given the ability to edit genes that decide when fruits and vegetables go rotten or ripe, CRISPR helps preserve the freshness of such crops, minimizing spoilage. The current statistics hold that a third of all food produced worldwide is wasted, and food waste is a major component of total food production losses. Most of this waste results from spoilage, especially in the developing world, where most fruits and vegetables destined for export are processed and shipped. Mentioning the fact that sometimes the crops are harvested but they cannot be consumed for months, due to perishability, CRISPR-Cas9 provides a genetic modification that can actually reduce the spoiling rates. This area has witnessed one of the major exemplary applications of CRISPR, where companies have developed techniques in non-browning apples and mushrooms. Scientists have applied CRISPR to edit the PPO gene, which encodes for the polyphenol oxidase enzyme, involved in the browning process in fruits or severed apparatus of vegetables on exposure to air. By knocking out the PPO gene, scientists have been able to prevent browning in apples and mushrooms, extending their shelf life and improving their marketability (Waltz, 2016). These non-browning crops not only appeal to consumers due to their enhanced appearance but also contribute to reducing food waste at both the retail and consumer levels, where significant losses occur due to spoilage and unappealing aesthetics (Kang et al., 2021). Not only apples and mushrooms but also other crops have been genetically modified using CRISPR-Cas9 to reduce ripening and increase shelf life. For instance, tomatoes, which are very sensitive to early maturation and rotting, have been genetically edited to manage the genes which control the hormone ethylene useful in ripening crops. By reducing ethylene production, CRISPR-edited tomatoes can stay fresh for a longer period after harvest, reducing post-harvest losses and allowing for more flexible transportation and storage options (Li et al., 2018). Due to its ability to increase efficiency of product distribution in organizations, thereby minimizing wastage of resources, especially in areas where some species cannot stand the high temperatures during transportation, this technology has the potential to solve the problem of food wastage in supply chains.

In bananas, which are a staple fruit in many tropical countries, where conditions such as high temperature and humidity are conducive to spoilage, CRISPR technology has also been used. This fruit is very sensitive to ripening and, hence, undergoes the process at a very high rate, more especially when in transit. Researchers are targeting genes associated with the ripening process to generate bananas that mature more slowly, so prolonging their shelf life and enhancing their suitability for exports and long-term preservation (KB et al., 2024). This is expected to significantly reduce banana loss in producing areas, while enhancing food security by protecting the crop. In addition to specific crops, CRISPR could reduce food waste, which is critical to global security of food. Post-harvest losses significantly threaten food security in several developing nations, particularly in rural regions, where access to refrigeration is limited. CRISPR may enhance the storage life of crops and minimize degradation, hence minimizing food shortages and optimizing food delivery in areas with inadequate infrastructure (Kim et al., 2020).

4.5.2 Risks and Ethical Concerns

4.5.2.1 Off-Target Effects

Such risk of off-target effects among CRISPR-Cas9 technology is one of the significant concerns. These off-target mutations occur when the Cas9 protein cuts DNA at unintended or undesirable sites within the DNA, including mutations that interfere with crop performance as well as more pathologically, mutations that compromise food safety. CRISPR is very respected for its precision, but potential off-target effects in agriculture are not ideal, as these off-target effects could impair both plant and consumer consequences (Gaudelli et al., 2018; Tsai and Joung, 2016). Off-target mutations in crops could have a detrimental effect on agricultural productivity by somehow causing phenotypes that are detrimental. Unexpected shifts may adversely affect yield, disease resistance, or environmental adaptation, so lowering the planned benefits of the initial gene edits. Concerns exist that off-target mutations may compromise food safety by introducing toxic chemicals, allergic compounds, or other unwanted traits that might endanger consumer health (Hsu et al., 2013). For this reason, any genetic modification of CRISPR-edited crops must be rigorously tested and validated to ensure the crop's safety and stability. To mitigate the risk of off-target effects, high-fidelity Cas9 variants with less off-target activity and greater specificity have been developed by researchers. Among these variants, SpCas9-HF1 has been engineered to minimize off-target cuts by increasing DNA recognition and cleavage precision. Other high-fidelity Cas9 variants, such as SpCas9-HF1, have been shown to reduce unintended genetic changes without losing the efficacy of CRISPR-based gene editing in crops (Kleinstiver et al., 2016; Zetsche et al., 2015b).

To further reduce the possibility of off-target mutations appearing, many techniques have been proposed. Modifications in guide RNA design, for instance, might ensure the gRNA binds correctly to the targeted DNA sequence, decreasing the possibility of off-target binding and enhancing the precision of CRISPR editing (Tycko et al., 2016). Computational techniques that forecast probable off-target locations via sequence similarity have been essential in enhancing CRISPR accuracy, enabling researchers to foresee and alleviate the danger of adverse effects of off-target (Kim et al., 2015). Moreover, paired nickases, using two Cas9 proteins to generate single-stranded DNA breaks rather than double-strand breaks, have been devised as an alternate approach that may reduce the risk of unwanted variations (Ran et al., 2013). An alternative approach has been, and continues to be, the development of next-generation CRISPR systems, including CRISPR-Cpf1 (commonly called CRISPR-Cas12a), that enable enhanced precision of genome editing. CRISPR-Cpf1 differs from Cas9 in both its structure and mechanism of DNA cleavage, and it has been shown to have fewer off-target effects in some studies (Zetsche et al., 2015a).

4.5.2.2 Environmental Effect

Introducing transgenic plants into the natural environment could negatively impact the ecosystem. Sustainable agricultural solutions are possible with CRISPR-Cas9 technology, but the unexpected effects of genetically modified crops in ecosystems must be acknowledged. However, gene flow from CRISPR-edited crops to the wild and other non-target species via cross-pollination is the biggest concern. This might release altered genes in wild populations, affecting ecosystems unintentionally (Puchta, 2017). One important example of a concern of this sort is the possibility of transferring herbicide resistance genes from CRISPR-edited crops to weedy species. If herbicide resistance traits spread to wild plants, they could provide the basis for "superweeds"—weeds resistant to common weed killers—that are more difficult to control. Herbicide-resistant crops generated via classical genetic modification and breed selection have increased agricultural weed control problems due to gene flow to closely related weeds (Nandula et al., 2019). Superweeds reduce agricultural yields and need stronger, more hazardous pesticides, causing additional damage to the ecosystem (Kuiken, 2017).

There are also worries about CRISPR-edited crops causing harm to biodiversity—not just through gene flow. Genetic modifications aimed to boost certain traits, together with pest resistance, may adversely affect non-target organisms—including helpful insects, soil fungi, or other wildlife. For example, transgenic crops which produce a broader range of plant-based pesticides may have an impact on populations of insects, which may subsequently destabilize food chains and lead to a decline in the biodiversity throughout ecosystems (Hokanson et al., 2019). One of the potential consequences of this is that insect populations may be affected. On the other hand, these unanticipated results highlight the need for performing extensive ecological assessments before extensive use of CRISPR-edited crops.

Furthermore, the use of CRISPR-modified crops could impact agricultural practices and adversely affect ecosystems. Monoculture farming—cultivating large areas of land with a single crop—may be effective for pest- and disease-resistant varieties. Monocultures enhance efficiency but limit genetic diversity, resulting in ecosystems more prone to pests, diseases, and climate change (Altieri and Nicholls, 2018). However, the depletion of agricultural variety may affect ecological balance and resilience. Rigorous supervision and surveillance are essential in preventing these hazards. CRISPR-edited crops must undergo thorough risk assessments before obtaining authorization for commercial use, as mandated by regulatory frameworks that require such evaluations. This involves assessing gene transfer to wild relatives, superweeds, and non-target species (Jaganathan et al., 2018). After CRISPR-modified crops are integrated into agriculture, they must be monitored for their effects on the environment.

Several methods have been developed to decrease gene flow as well as potential hazards of CRISPR-edited crops. First, apply genetic containment technology to stop modified gene transmission into the wild. Researchers have used gene-driven suppression mechanisms to prevent wild populations from transferring modified genes (Esvelt et al., 2014). Sterile CRISPR-edited crops might reduce gene flow by preventing genetic exchange between cultivated and indigenous species (DiCarlo et al., 2015). Along with other sustainable agricultural approaches, CRISPR-Cas9 might be used to minimize impact on the environment. A possible scenario are integrated pest management (IPM) techniques, which are designed to minimize environmental impacts using genetic modifications combined with ecological measures, including rotated crops, biological pest control, and less chemical inputs (Lefebvre et al., 2020). Adopting such extensive CRISPR technology could increase agricultural output and minimize the negative ecological impacts.

4.5.2.3 Socioeconomics and Equity

Emerging technologies like CRISPR-Cas9 are revolutionary for agriculture but also seem to offer a disproportionate benefit to large-scale, resource-rich farmers and agribusinesses at the expense of smallholder farmers and developing nations. Because of their advantage in terms of access to financial and technological resources, large agricultural corporations are better equipped to adopt, develop, and commercialize CRISPR technology. This could widen the gap between these major players and smaller, resource-poor farmers, who may lack the means to integrate this cutting-edge technology into their agricultural practices (Ishii, 2017). If such events are not prevented, the inequality could increase significantly, because smallholder farmers in developing countries where a large proportion of smallholders are already struggling because they lack access to agricultural inputs and advanced technologies, will suffer the most.

In order to avoid exacerbating socioeconomic disparities, it is essential to ensure that CRISPR technology is available to farmers in low-income regions. Unless proper interventions, such as public funding or support programs, are instituted, agricultural productivity in developing economies may stagnate as resource-rich farmers move through CRISPR-driven innovations. By facilitating access to CRISPR for smallholder farmers, these innovations could help address critical agricultural challenges, such as pest resistance and climate resilience, rather than reinforcing existing inequalities (Satheesh et al., 2019).

Intellectual property rights are also another very important equity issue. Currently, CRISPR biotechnology patents are available to large agribusinesses, who then gain control over the use of these technologies. Such a concentration of ownership of patents on CRISPR could thwart public sector and smallholder farmer access to CRISPR technology to the extent that they do not have the ability to reap benefits from the advancement. Patent monopolization by large corporations may also drive up the costs of these technologies, making them inaccessible to those who need them the most (Jasanoff, 2018). To achieve this level of CRISPR benefits sharing across differing sectors of society, the need to address these intellectual property challenges is paramount.

4.5.2.4 Public and Regulatory Support

There are still considerable regulatory challenges surrounding the use of CRISPR-edited plants and crops, and these are largely due to differences in standard-setting across the world, and this affects the pace and manner in which CRISPR-edited crops can be produced and marketed. If a country, for example, the EU, already labels crops that have been edited using CRISPR as GMOs, these crops can still be subjected to safety analyses and risk assessment. Such regulations entail procedural environmental and health clearance procedures that may take considerable time before a CRISPR crop proceeds to the market, which is a major setback for developers both financially and in terms of time. The European Court of Justice ruled in 2018 that gene-edited crops should fall under the same strict regulations as GMOs, a decision that has been met with both support and criticism from scientists and policymakers (Puchta, 2017; Callaway, 2018). The main problem motivating these regulations are the potential side consequences that CRISPR editing may have on the environment and organisms, therefore the human body.

On the other hand, countries such as the United States have relatively liberal regulatory environment. Based on USDA rules, products of CRISPR editing are not GMO if no extra foreign gene is inserted in the process of editing. This classification greatly lessens the amount of red tape that CRISPR crops have to navigate around, freeing them from the year-long approval process that traditional GMO crops must go through. This approach is based on the premise that crops edited with CRISPR, when used for small, targeted genetic changes, pose no greater risks than conventional breeding methods (Zetsche et al., 2017; Waltz, 2016). Other countries, including Canada, Australia, and Brazil, have adopted similar measures, though there are slight difference in the criteria used in distinguishing between gene-edited crops and traditional GMOs. Consumer acceptance of crops containing CRISPR is just as important in the success of the technology as well. Manufactured product consumer resistance to GMO crops has been evidenced especially in the regions of Europe, where anti-GMO campaigns politically dominate. Consequently, the application of CRISPR technology has the same challenges because issues like food safety, environmental effects, and ethics have not been withdrawn from discourses. However, unlike GMOs, CRISPR does not always involve the introduction of foreign genes, which could make it more palatable to consumers if these differences are effectively communicated (Jaganathan et al., 2018). **Table 4.4** shows the comparative potential benefits and risks of CRISPR-Cas 9.

In the case of the release of CRISPR-edited crops, such an enhanced transparency is necessary because the public must be trusting the government and commercial actors who will oversee the agricultural production using such plants. Engaging the public through education, media, and stakeholder involvement can help address concerns about the potential risks and highlight the benefits, such as improved crop resilience and reduced reliance on pesticides (Rao et al., 2019). Studies have shown that when people are provided with clear and scientifically accurate information about gene-editing technologies, their acceptance tends to increase (Rose, 2018). Moreover,

TABLE 4.4
Comparison of Potential Benefits and Risks of CRISPR-Cas9 Applications

Aspect	Potential Benefits	Potential Risks	References
Crop resilience	Enhanced drought and disease resistance	Risk of unintended off-target effects	Menz et al., 2020
Nutritional quality	Biofortification of crops	Potential long-term ecological impacts	Zhu et al., 2020
Food security	Increased crop yield and efficiency	Economic barriers for smallholder farmers	Jaganathan et al., 2018
Environmental sustainability	Reduced need for chemical inputs	Gene flow to wild relatives	Puchta, 2017

bureaucratic organizations and the scientific committees should work together in order to bring the necessary control to the application of CRISPR technique, and its consequences, including the question of influence on the further development of biodiversity, as well as food sovereignty.

4.6 CHALLENGES AND FUTURE PERSPECTIVES

Unresolved issues must be addressed prior to the complete utilization of CRISPR-Cas9 in agricultural transformation. The obstacles may be classified into three categories: technological and biological challenges related to the system, legal difficulties aimed at addressing the issues, and sociological challenges affecting the community in which the issues arise. The future use of CRISPR-Cas9 in agriculture will have to grow from the added development, acceptance by many people, and the enactment of enabling laws.

4.6.1 Technical Challenges

4.6.1.1 Off-Target Mutations

One of the biggest issues scientists find is that CRISPR-Cas9 may cause off-target effects and unintended modifications of genetically mobile areas, which in turn may have an effect on the independent growth and advancement of the plant. In order to overcome this problem, Cas9-enhanced variants, like SpCas9-HF1 and eSpCas9, have been generated, which mimic Cas9 protein with approachable off-target impacts (Kim et al., 2020). Additionally, computational tools and machine learning models are designed to predict off-target sites with greater accuracy, further minimizing risks (Zhu et al., 2020; Kim et al., 2024).

4.6.1.2 Gene Delivery Systems

Introducing CRISPR-Cas9 components into plant cells is still a problem, especially for plants such as wheat and corn, because these plants are hard to transform after the introduction of the compound (Zhang et al., 2017). Traditional methods, such as

Agrobacterium-mediated transformation and biolistic particle delivery, have limitations in their effectiveness across various plant species (Gao et al., 2020). Some of the emerging approaches have been identified as nanoparticles, electroporation, and viral vectors targeting the improvement of loading and incorporation of CRISPR in plant genome systems (Kis et al., 2019).

4.6.1.3 Understanding Complex Traits

The application of CRISPR-Cas9 has been greatly successful at editing single genes of interest, and since many agri-economically important traits, including yield, drought, and disease resistance, can be polygenes. Altering these intricate features necessitates concurrent adjustments to multiple systems of genes (Li, 2020). Multiplex gene-editing methodologies, enabling simultaneous modification of several genes, are being advanced to address this difficulty (Brooks et al., 2014). This method is anticipated to facilitate the effective alteration of polygenic characteristics, enhancing overall crop performance.

4.6.2 Future Perspectives

4.6.2.1 Advancements in Precision Editing and New CRISPR Variants

There is hope for better utilization of today's CRISPR-Cas9 for agriculture, and it is dependent on the further enhancements of more sophisticated genome editing tools. Other variants include Cas12a (Cpf1) and Cas13, which have slightly different target specificities, thus have relatively few collateral effects on the genome (Zetsche et al., 2015b). Also, base editing methods which have recently opened the possibility of direct nucleotide conversion without causing DSBs are also new prospects in plant genome editing (Gaudelli et al., 2018). These developments will only enrich CRISPR technology and enhance the utility of these applications.

4.6.2.2 Multiplexing for Complex Trait Improvement

Multiplex gene editing, the prospect of which arose in the present work, is undoubtedly beneficial for enhancing intricate characteristics, like drought tolerance, disease resistance, and yield (Svitashev et al., 2015). This enhancement of polygenic traits is possible because during one editing session, researchers can modify several the key genes. Recent developments in the application of bioinformatics and computational biology enable researchers to discover important gene regulatory networks that can accelerate breeding improvement through gene-editing techniques (Jiang et al., 2013).

4.6.2.3 Implementing CRISPR for Small-Scale Farmers

Making CRISPR technology available to smallholder farmers, especially in the developing world, is a steadily emerging concern. To democratize the CRISPR system, research institutions and allied groups in the public sector, as well as international organizations, target crops, including cassava, maize, and millet (Mao et al., 2019). The ability to create affordable and easily accessible CRISPR systems makes these aspects of improvement relevant to farmers with limited resources in all parts of the world.

4.6.2.4 Combination with Other Technologies

it can be expected that future use of the CRISPR tool for agricultural purposes will be developed with the combination of precision agriculture, artificial intelligence, machine learning, or other exponential technologies. By anticipating off-target effects, evaluating intricate features, and offering insights regarding crop growth, AI-driven systems can maximize CRISPR studies (Yang and Chen, 2023). Furthermore, when applied together with digital farming technologies, which include drones, smart sensors, and real-time data analysis, CRISPR will improve crop performance as well as responsiveness to environmental conditions (Tang et al., 2019).

4.6.2.5 Public Engagement and Education

Again, if the public is to embrace the technology, people must be taken through an education-and-engagement process on issues pertaining to food security, environmental effects, and ethics behind this novel technology (Kumar et al., 2019). Scientists, policymakers, and civil society organizations must work together to present accurate, evidence-based information concerning the use of CRISPR's advantages and drawbacks. In this way, all the stakeholders can guarantee public acceptance of the technology and the responsible further evolution of CRISPR (Khatodia et al., 2016).

4.7 CONCLUSION

The CRISPR-Cas9 system is now widely recognized as a revolution in plant genome editing; its precision means that desirable crop characteristics can now be targeted to address global concerns in agriculture. The uses it has been found to have in the areas of increasing crop production; protecting crops against pests, diseases, and environmental factors; and increasing crop nutritional value are all hoped to be a game-changer when it comes to food production and food security. Making precise changes to genes, CRISPR-Cas9 has helped create plants that could survive drought and pests and have greater nutrient density, solutions that will be vital with climate change and food scarcity. Risk issues of unintended exposure, environmental degradation, and injustice and inequalities require essentials social control measures and conflict resolution. The likelihood of off-target effects and the propagation of the altered genes to neighboring wild-type species make the need for constant control and evaluation of possible risks inherent. Also, these CRISPR technologies are not equally available with smallholder farmers in the developing world still grappling with adaption.

REFERENCES

Adeyemo, Temitayo, Paul Amaza, Victor Okoruwa, Vincent Akinyosoye, Kabir Salman, and Adebayo Abass. 2019. "Determinants of Intensity of Biomass Utilization: Evidence from Cassava Smallholders in Nigeria." *Sustainability (Switzerland)* 11 (9).

Ahn, Chang Ho, Mummadireddy Ramya, Hye Ryun An, Pil Man Park, Yae Jin Kim, Su Young Lee, and Seonghoe Jang. 2020. "Progress and Challenges in the Improvement of Ornamental Plants by Genome Editing." *Plants* 9 (6): 1–13.

Altieri, Miguel, and Clara Nicholls. 2018. *Biodiversity and Pest Management in Agroecosystems*. CRC Press.

Ancydiana, P. G., A. Shanthinie, T. Arulganesh, Yaiphabi Kumam, K. K. Kumar, L. Arul, E. Kokiladevi, S. Varanavasiappan, S. Manonmani, and D. Sudhakar. 2022. "Targeted Editing of OsSWEET13, a Bacterial Leaf Blight Susceptible Gene in Rice Using CRISPR Tool." *Electronic Journal of Plant Breeding* 13 (3): 772–779.

Anzalone, Andrew V., Peyton B. Randolph, Jessie R. Davis, Alexander A. Sousa, Luke W. Koblan, Jonathan M. Levy, Peter J. Chen et al. 2019. "Search-and-Replace Genome Editing without Double-Strand Breaks or Donor DNA." *Nature* 576.

Borrelli, Virginia M. G., Vittoria Brambilla, Peter Rogowsky, Adriano Marocco, and Alessandra Lanubile. 2018. "The Enhancement of Plant Disease Resistance Using CRISPR/Cas9 Technology." *Frontiers in Plant Science* 9 (August).

Brooks, Christopher, Vladimir Nekrasov, Zachary B. Lipppman, and Joyce Van Eck. 2014. "Efficient Gene Editing in Tomato in the First Generation Using the Clustered Regularly Interspaced Short Palindromic Repeats/CRISPR-Associated9 System." *Plant Physiology* 166 (3): 1292–1297.

Bubeck, Felix, and Dirk Grimm. 2024. "When Size Matters: A Novel Compact Cas12a Variant for in vivo Genome Editing." PLoS Biology 22 (7): e3002637. doi: 10.1371journal.pbio.300263

Callaway, Ewen. 2018. "CRISPR Plants Now Subject to Tough GM Laws in European Union." *Nature* 560 (7716): 16–16. https://doi.org/10.1038/d41586-018-05814-6.

Camerlengo, Francesco, Arianna Frittelli, and Riccardo Pagliarello. 2022. "CRISPR towards a Sustainable Agriculture." *Encyclopedia* 2 (1): 538–558.

Chen, Feng, Lu Chen, Zhao Yan, Jingyuan Xu, Luoluo Feng, Na He, Mingli Guo et al. 2024. "Recent Advances of CRISPR-Based Genome Editing for Enhancing Staple Crops." *Frontiers in Plant Science* 15 (September).

Chen, Jiawen, Erica Hawkins, and David Seung. 2021. "Towards Targeted Starch Modification in Plants." *Current Opinion in Plant Biology* 60: 102013. Elsevier Ltd.

Chen, Li, Qian Hao Zhu, and Kerstin Kaufmann. 2020. "Long Non-Coding RNAs in Plants: Emerging Modulators of Gene Activity in Development and Stress Responses." *Planta* 252 (5): 1–14. Springer Berlin Heidelberg.

Corte, Lígia Erpen Dalla, Lamiaa M. Mahmoud, Tatiana S. Moraes, Zhonglin Mou, Jude W. Grosser, and Manjul Dutt. 2019. "Development of Improved Fruit, Vegetable, and Ornamental Crops Using the CRISPR/Cas9 Genome Editing Technique." *Plants* 8 (12).

Dhaliwal, Harcharan S., and Hirofumi Uchimiya. 1999. "Genetic Engineering for Disease and Pest Resistance in Plants." *Plant Biotechnology* 16 (4): 255–261.

DiCarlo, James E., Alejandro Chavez, Sven L. Dietz, Kevin M. Esvelt, and George M. Church. 2015. "Safeguarding CRISPR-Cas9 Gene Drives in Yeast." *Nature Biotechnology* 33 (12): 1250–1255.

Divya, K., Makeshkumar Thangaraj, and N. Krishna Radhika. 2024. "CRISPR/Cas9: An Advanced Platform for Root and Tuber Crops Improvement." *Frontiers in Genome Editing* 5: 1242510. doi: 10.3389/fgeed.2023

Ercolano, Maria Raffaella, and Kejian Wang. 2023. "Editorial: Targeted Genome Editing for Crop Improvement." *Frontiers in Plant Science* 14 (February): 2022–2024.

Eş, Ismail, Mohsen Gavahian, Francisco J. Marti-Quijal, Jose M. Lorenzo, Amin Mousavi Khaneghah, Christos Tsatsanis, Sotirios C. Kampranis, and Francisco J. Barba. 2019. "The Application of the CRISPR-Cas9 Genome Editing Machinery in Food and Agricultural Science: Current Status, Future Perspectives, and Associated Challenges." *Biotechnology Advances* 37 (3): 410–421. Elsevier Inc.

Esvelt, Kevin M., Andrea L. Smidler, Flaminia Catteruccia, and George M. Church. 2014. "Concerning RNA-Guided Gene Drives for the Alteration of Wild Populations." *ELife* 3 (July): 1–21.

Feng, Zhengyan, Yanfei Mao, Nanfei Xu, Botao Zhang, Pengliang Wei, Dong-Lei Yang, Zhen Wang et al. 2014. "Multigeneration Analysis Reveals the Inheritance, Specificity, and Patterns of CRISPR/Cas-Induced Gene Modifications in Arabidopsis." *Proceedings of the National Academy of Sciences* 111 (12): 4632–4637.

Gao, Huirong, Mark J. Gadlage, H. Renee Lafitte, Brian Lenderts, Meizhu Yang, Megan Schroder, Jeffry Farrell et al. 2020. "Superior Field Performance of Waxy Corn Engineered Using CRISPR–Cas9." *Nature Biotechnology* 38 (5): 579–581.

Garg, Monika, Natasha Sharma, Saloni Sharma, Payal Kapoor, Aman Kumar, Venkatesh Chunduri, and Priya Arora. 2018. "Biofortified Crops Generated by Breeding, Agronomy, and Transgenic Approaches Are Improving Lives of Millions of People around the World." *Frontiers in Nutrition* 5 (February).

Gaudelli, Nicole M., Alexis C. Komor, Holly A. Rees, Michael S. Packer, Ahmed H. Badran, David I. Bryson, and David R. Liu. 2018. "Publisher Correction: Programmable Base Editing of A•T to G•C in Genomic DNA without DNA Cleavage." *Nature* 559 (7714): E8.

Gupta, Om Prakash, Vanita Pandey, Ritu Saini, Sneh Narwal, Vipin Kumar Malik, Tushar Khandale, Sewa Ram, and Gyanendra Pratap Singh. 2020. "Identifying Transcripts Associated with Efficient Transport and Accumulation of Fe and Zn in Hexaploid Wheat (T. Aestivum L.)." *Journal of Biotechnology* 316: 46–55. Elsevier B.V.

Hokanson, K. J., C. A. Mendoza, and K. J. Devito. 2019. "Interactions Between Regional Climate, Surficial Geology, and Topography: Characterizing Shallow Groundwater Systems in Subhumid, Low-Relief Landscapes." *Water Resources Research* 55 (1): 284–297.

Hsu, Patrick D., Eric S. Lander, and Feng Zhang. 2014. "Development and Applications of CRISPR-Cas9 for Genome Engineering." *Cell* 157 (6): 1262–1278.

Hsu, Patrick D., David A. Scott, Joshua A. Weinstein, F. Ann Ran, Silvana Konermann, Vineeta Agarwala, Yinqing Li et al. 2013. "DNA Targeting Specificity of RNA-Guided Cas9 Nucleases." *Nature Biotechnology* 31 (9): 827–832. Nature Publishing Group.

Husaini, Amjad M. 2022. "High-Value Pleiotropic Genes for Developing Multiple Stress-Tolerant Biofortified Crops for 21st-Century Challenges." *Heredity* 128 (6): 460–472. Springer US.

Ishii, Tetsuya, and Motoko Araki. 2017. "A Future Scenario of the Global Regulatory Landscape Regarding Genome-Edited Crops." *GM Crops and Food* 8 (1): 44–56.

Jaganathan, Deepa, Karthikeyan Ramasamy, Gothandapani Sellamuthu, Shilpha Jayabalan, and Gayatri Venkataraman. 2018. "CRISPR for Crop Improvement: An Update Review." *Frontiers in Plant Science* 9 (July): 1–17.

Jasanoff, Sheila. 2018. "Just Transitions: A Humble Approach to Global Energy Futures." *Energy Research & Social Science* 35: 11–14.

Jiang, Wenzhi, Huanbin Zhou, Honghao Bi, Michael Fromm, Bing Yang, and Donald P. Weeks. 2013. "Demonstration of CRISPR/Cas9/SgRNA-Mediated Targeted Gene Modification in Arabidopsis, Tobacco, Sorghum and Rice." *Nucleic Acids Research* 41 (20): 1–12.

KB, Soni, Pritam Ramesh Jadhav, and Swapna Alex. 2024. "Genetic Improvement of Banana." In *Genetic Engineering of Crop Plants for Food and Health Security: Volume 1,* 305–329. Springer Nature Singapore.

Kang, Hui, Ting Ting Zhang, Lu Lu Fu, Yu Xin Yao, Chun Xiang You, Xiao Fei Wang, and Yu Jin Hao. 2021. "The Apple MdCOP1-Interacting Protein 1 Negatively Regulates Hypocotyl Elongation and Anthocyanin Biosynthesis." *BMC Plant Biology* 21 (1): 1–11.

Khan, Aftab A., Ashraf El-Sayed, Asma Akbar, Arianna Mangravita-Novo, Shaheen Bibi, Zunaira Afzal, David J. Norman, and Gul Shad Ali. 2017. "A Highly Efficient Ligation-Independent Cloning System for CRISPR/Cas9 Based Genome Editing in Plants." *Plant Methods* 13 (1): 1–13. BioMed Central.

Khatodia, Surender, Kirti Bhatotia, Nishat Passricha, S. M. P. Khurana, and Narendra Tuteja. 2016. "The CRISPR/Cas Genome-Editing Tool: Application in Improvement of Crops." *Frontiers in Plant Science* 7 (April): 1–13.

Kim, Daesik, Sangsu Bae, Jeongbin Park, Eunji Kim, Seokjoong Kim, Hye Ryeong Yu, Jinha Hwang, Jong Il Kim, and Jin Soo Kim. 2015. "Digenome-Seq: Genome-Wide Profiling of CRISPR-Cas9 Off-Target Effects in Human Cells." *Nature Methods* 12 (3): 237–243. Nature Publishing Group.

Kim, Nahye, Hui Kwon Kim, Sungtae Lee, Jung Hwa Seo, Jae Woo Choi, Jinman Park, Seonwoo Min, Sungroh Yoon, Sung-Rae Cho, and Hyongbum Henry Kim. 2020. "Prediction of the Sequence-Specific Cleavage Activity of Cas9 Variants." *Nature Biotechnology* 38 (11): 1328–1336. doi: 10.1038/s41587-020-0537-9

Kim, Nahye, Sungchul Choi, Sungjae Kim, Myungjae Song, Jung Hwa Seo, Seonwoo Min, Jinman Park, Sung-Rae Cho, and Hyongbum Henry Kim. 2024. "Deep Learning Models to Predict the Editing Efficiencies and Outcomes of Diverse Base Editors." *Nature Biotechnology* 42 (3): 484–497.

Kis, András, Éva Hamar, Gergely Tholt, Rita Bán, and Zoltán Havelda. 2019. "Creating Highly Efficient Resistance against Wheat Dwarf Virus in Barley by Employing CRISPR/Cas9 System." *Plant Biotechnology Journal* 17 (6): 1004–1006.

Kleinstiver, Benjamin P., Vikram Pattanayak, Michelle S. Prew, Shengdar Q. Tsai, Nhu T. Nguyen, Zongli Zheng, and J. Keith Joung. 2016. "High-Fidelity CRISPR-Cas9 Nucleases with No Detectable Genome-Wide off-Target Effects." *Nature* 529 (7587): 490–495. Nature Publishing Group.

Komor, Alexis C., Kevin T. Zhao, Michael S. Packer, Nicole M. Gaudelli, Amanda L. Waterbury, Luke W. Koblan, Y. Bill Kim, Ahmed H. Badran, and David R. Liu. 2017. "Improved Base Excision Repair Inhibition and Bacteriophage Mu Gam Protein Yields C:G-to-T:A Base Editors with Higher Efficiency and Product Purity." *Science Advances* 3 (8).

Kuiken, Todd. 2017. "Vigilante Environmentalism: Are Gene Drives Changing How We Value and Govern Ecosystems?." In *Gene Editing, Law, and the Environment*, edited by Irus Braverman, 1st ed., 94–111. Taylor & Francis Group.

Kumar, Krishan, Geetika Gambhir, Abhishek Dass, Amit Kumar Tripathi, Alla Singh, Abhishek Kumar Jha, Pranjal Yadava, Mukesh Choudhary, and Sujay Rakshit. 2020. "Genetically Modified Crops: Current Status and Future Prospects." *Planta* 251 (4): 1–27. Springer Berlin Heidelberg.

Kumar, Rakesh, Amandeep Kaur, Ankita Pandey, H. M. Mamrutha, and G. P. Singh. 2019. "CRISPR-Based Genome Editing in Wheat: A Comprehensive Review and Future Prospects." *Molecular Biology Reports* 46 (3): 3557–3569.

Lefebvre, Véronique, Nathalie Boissot, and Jean-Luc Gallois. 2020. "Host Plant Resistance to Pests and Pathogens, the Genetic Leverage in Integrated Pest and Disease Management." In *Integrated Pest and Disease Management in Greenhouse Crops*, 259–283. Cham: Springer International Publishing.

Li, Bin, Xi Du, Yunyan Fei, Fangquan Wang, Yang Xu, Xia Li, Wenqi Li et al. 2021. "Efficient Breeding of Early-Maturing Rice Cultivar by Editing PHYC via CRISPR/Cas9." *Rice* 14 (1): 4–7.

Li, Chengdao. 2020. "Breeding Crops by Design for Future Agriculture." *Journal of Zhejiang University-Science B* 21 (6): 423–425.

Li, Ting, Bo Liu, Chih Ying Chen, and Bing Yang. 2016. "TALEN-Mediated Homologous Recombination Produces Site-Directed DNA Base Change and Herbicide-Resistant Rice." *Journal of Genetics and Genomics* 43 (5): 297–305.

Li, Tingdong, Xinping Yang, Yuan Yu, Xiaomin Si, Xiawan Zhai, Huawei Zhang, Wenxia Dong, Caixia Gao, and Cao Xu. 2018. "Domestication of Wild Tomato is Accelerated by Genome Editing." *Nature Biotechnology* 36 (12): 1160–1163.

Liang, Zhen, Kang Zhang, Kunling Chen, and Caixia Gao. 2014. "Targeted Mutagenesis in Zea Mays Using TALENs and the CRISPR/Cas System." *Journal of Genetics and Genomics* 41 (2): 63–68.

Liu, Jian, Merrin L. Macrae, Jane A. Elliott, Helen M. Baulch, Henry F. Wilson, and Peter J. A. Kleinman. 2019. "Impacts of Cover Crops and Crop Residues on Phosphorus Losses in Cold Climates: A Review." *Journal of Environmental Quality* 48 (4): 850–868.

Liu, Juan, Qiao Li, Huisheng Zhai, Yuhong Zhang, Xingquan Zeng, Yawei Tang, Nyima Tashi, and Zhifen Pan. 2020. "Effects of the Addition of Waxy and Normal Hull-Less Barley Flours on the Farinograph and Pasting Properties of Composite Flours and on the Nutritional Value, Textural Qualities, and in Vitro Digestibility of Resultant Breads." *Journal of Food Science* 85 (10): 3141–3149.

Mao, Yanfei, Jose Ramon Botella, Yaoguang Liu, and Jian Kang Zhu. 2019. "Gene Editing in Plants: Progress and Challenges." *National Science Review* 6 (3): 421–437.

Menz, Jochen, Dominik Modrzejewski, Frank Hartung, Ralf Wilhelm, and Thorben Sprink. 2020. "Genome Edited Crops Touch the Market: A View on the Global Development and Regulatory Environment." *Frontiers in Plant Science* 11 (October): 1–17.

Nandula, Vijay K. 2019. "Herbicide Resistance Traits in Maize and Soybean: Current Status and Future Outlook." *Plants* 8 (9).

Odipio, John, Titus Alicai, Ivan Ingelbrecht, Dmitri A. Nusinow, Rebecca Bart, and Nigel J. Taylor. 2017. "Efficient CRISPR/Cas9 Genome Editing of Phytoene Desaturase in Cassava." *Frontiers in Plant Science* 8 (October): 1–11.

Paine, Jacqueline A., Catherine A. Shipton, Sunandha Chaggar, Rhian M. Howells, Mike J. Kennedy, Gareth Vernon, Susan Y. Wright et al. 2005. "Improving the Nutritional Value of Golden Rice Through Increased Pro-Vitamin A Content." *Nature Biotechnology* 23 (4): 482–487. doi: 10.1038/nbt1082

Peng, Cheng, Ming Zheng, Lin Ding, Xiaoyun Chen, Xiaofu Wang, Xuping Feng, Junmin Wang, and Junfeng Xu. 2020. "Accurate Detection and Evaluation of the Gene-Editing Frequency in Plants Using Droplet Digital PCR." *Frontiers in Plant Science* 11 (December): 1–8.

Podevin, Nancy, Howard V. Davies, Frank Hartung, Fabien Nogué, and Josep M. Casacuberta. 2013. "Site-Directed Nucleases: A Paradigm Shift in Predictable, Knowledge-Based Plant Breeding." *Trends in Biotechnology* 31 (6): 375–383.

Puchta, Holger. 2017. "Applying CRISPR/Cas for Genome Engineering in Plants: The Best Is yet to Come." *Current Opinion in Plant Biology* 36: 1–8.

Rai, Gyanendra Kumar, Danish Mushtaq Khanday, Pradeep Kumar, Isha Magotra, Sadiya M. Choudhary, Rafia Kosser, Raviraj Kalunke et al. 2023. "Enhancing Crop Resilience to Drought Stress through CRISPR-Cas9 Genome Editing." *Plants* 12 (12): 1–14.

Ran, F. Ann, Patrick D. Hsu, Jason Wright, Vineeta Agarwala, David A. Scott, and Feng Zhang. 2013. "Genome Engineering Using the CRISPR-Cas9 System." *Nature Protocols* 8 (11): 2281–2308.

Rao, A. V. R. Kesava, Suhas P. Wani, and K. Srinivas. 2019. "Climate Change Adaptation and Mitigation Strategies for Sustainable Crop Production." In *Agricultural Extension and Sustainable Development Goals*, edited by S. Venku Reddy and M. Suryamani, 239. APH Publishing Corporation.

Ren, Bin, Fang Yan, Yongjie Kuang, Na Li, Dawei Zhang, Xueping Zhou, Honghui Lin, and Huanbin Zhou. 2018. "Improved Base Editor for Efficiently Inducing Genetic Variations in Rice with CRISPR/Cas9-Guided Hyperactive HAID Mutant." *Molecular Plant* 11 (4): 623–626.

Ricroch, Agnès, Pauline Clairand, and Wendy Harwood. 2017. "Use of CRISPR Systems in Plant Genome Editing: Toward New Opportunities in Agriculture." *Emerging Topics in Life Sciences* 1 (2): 169–182.

Roberts, Daniel P., and Autar K. Mattoo. 2018. "Sustainable Agriculture—Enhancing Environmental Benefits, Food Nutritional Quality and Building Crop Resilience to Abiotic and Biotic Stresses." *Agriculture (Switzerland)* 8 (1).

Roberts, Daniel P., Nicholas M. Short, James Sill, Dilip K. Lakshman, Xiaojia Hu, and Michael Buser. 2021. "Precision Agriculture and Geospatial Techniques for Sustainable Disease Control." *Indian Phytopathology* 74 (2): 287–305. Springer India.

Rose, Elaine. 2018. "Identifying Mechanisms of Resistance to Oncolytic Virotherapy in Acute Leukemia through a Genome-Wide CRISPR Screen." University of Ottawa.

Sanchez-León, Susana, Javier Gil-Humanes, Carmen V. Ozuna, María J. Giménez, Carolina Sousa, Daniel F. Voytas, and Francisco Barro. 2018. "Low-Gluten, Nontransgenic Wheat Engineered with CRISPR/Cas9." *Plant Biotechnology Journal* 16 (4): 902–910.

Satheesh, Somisetty V., Marit Rasmussen, and Jason Matthews. 2019. "Applications of CRISPR/Cas9 in Agriculture, Nutrition, Health and Disease." *Norsk Tidsskrift for Ernæring* 17 (2): 30–35.

Sharma, Hari C., and Chandra S. Prabhakar. 2014. *Impact of Climate Change on Pest Management and Food Security. Integrated Pest Management: Current Concepts and Ecological Perspective*. Elsevier Inc.

Shi, Jinrui, Huirong Gao, Hongyu Wang, H. Renee Lafitte, Rayeann L. Archibald, Meizhu Yang, Salim M. Hakimi, Hua Mo, and Jeffrey E. Habben. 2017. "ARGOS8 Variants Generated by CRISPR-Cas9 Improve Maize Grain Yield under Field Drought Stress Conditions." *Plant Biotechnology Journal* 15 (2): 207–216.

Shi, Jinrui, Hongyu Wang, Kathleen Schellin, Bailin Li, Marianna Faller, Johan M. Stoop, Robert B. Meeley, David S. Ertl, Jerry P. Ranch, and Kimberly Glassman. 2007. "Embryo-Specific Silencing of a Transporter Reduces Phytic Acid Content of Maize and Soybean Seeds." *Nature Biotechnology* 25 (8): 930–937.

Sternberg, Samuel H., Sy Redding, Martin Jinek, Eric C. Greene, and Jennifer A. Doudna. 2014. "DNA Interrogation by the CRISPR RNA-Guided Endonuclease Cas9." *Biophysical Journal* 106 (2): 695a.

Sun, Yongwei, Jingying Li, and Lanqin Xia. 2016. "Precise Genome Modification via Sequence-Specific Nucleases-Mediated Gene Targeting for Crop Improvement." *Frontiers in Plant Science* 7: 1928.

Svitashev, Sergei, Joshua K. Young, Christine Schwartz, Huirong Gao, S. Carl Falco, and A. Mark Cigan. 2015. "Targeted Mutagenesis, Precise Gene Editing, and Site-Specific Gene Insertion in Maize Using Cas9 and Guide RNA." *Plant Physiology* 169 (2): 931–945.

Tang, Maozhi, Xiaoyun He, Yunbo Luo, Liyan Ma, Xiaoge Tang, and Kunlun Huang. 2013. "Nutritional Assessment of Transgenic Lysine-Rich Maize Compared with Conventional Quality Protein Maize." *Journal of the Science of Food and Agriculture* 93 (5): 1049–1054.

Tang, Xu, Qiurong Ren, Lijia Yang, Yu Bao, Zhaohui Zhong, Yao He, Shishi Liu et al. 2019. "Single Transcript Unit CRISPR 2.0 Systems for Robust Cas9 and Cas12a Mediated Plant Genome Editing." *Plant Biotechnology Journal* 17 (7): 1431–1445.

Tovar-Pérez, Erik Gustavo, Selene Aguilera-Aguirre, Ulises López-García, Maribel Valdez-Morales, Alma Karina Ibarra-Zurita, Rosa Isela Ortiz-Basurto, and Alejandra Chacón-López. 2020. "Effect of Ultrasound Treatment on the Quality and Contents of Polyphenols, Lycopene and Rutin in Tomato Fruits." *Czech Journal of Food Sciences* 38 (1): 20–27.

Tsai, Shengdar Q., and J. Keith Joung. 2016. "Defining and Improving the Genome-Wide Specificities of CRISPR-Cas9 Nucleases." *Nature Reviews Genetics* 17 (5): 300–312. Nature Publishing Group.

Tycko, Josh, Vic E. Myer, and Patrick D. Hsu. 2016. "Methods for Optimizing CRISPR-Cas9 Genome Editing Specificity." *Molecular Cell* 63 (3): 355–370. Elsevier Inc.

Ur Rahim, Hafeez, Muhammad Qaswar, Misbah Uddin, Cinzia Giannini, Maria Lidia Herrera, and Giuseppina Rea. 2021. "Nano-Enable Materials Promoting Sustainability and Resilience in Modern Agriculture." *Nanomaterials* 11 (8).

Waltz, Emily. 2016. "CRISPR-Edited Crops Free to Enter Market, Skip Regulation." *Nature Biotechnology* 34 (6): 582–582. Nature Publishing Group.

Wang, Jin, Sai Kranthi Vanga, Rachit Saxena, Valérie Orsat, and Vijaya Raghavan. 2018. "Effect of Climate Change on the Yield of Cereal Crops: A Review." *Climate* 6 (2).

Wang, Xiaoyi, Renbo Yu, and Jingyang Li. 2021. "Using Genetic Engineering Techniques to Develop Banana Cultivars with Fusarium Wilt Resistance and Ideal Plant Architecture." *Frontiers in Plant Science* 11 (January).

Wang, Y. B., P. T. Wu, B. A. Engel, and S. K. Sun. 2014. "Application of Water Footprint Combined with a Unified Virtual Crop Pattern to Evaluate Crop Water Productivity in Grain Production in China." *Science of the Total Environment* 497–498: 1–9.

Xie, Kabin, Bastian Minkenberg, and Yinong Yang. 2015. "Boosting CRISPR/Cas9 Multiplex Editing Capability with the Endogenous TRNA-Processing System." *Proceedings of the National Academy of Sciences of the United States of America* 112 (11): 3570–3575.

Xu, Weiya, Wensi Tang, Chunxiao Wang, Linhao Ge, Jianchang Sun, Xin Qi, Zhang He et al. 2020. "SiMYB56 Confers Drought Stress Tolerance in Transgenic Rice by Regulating Lignin Biosynthesis and ABA Signaling Pathway." *Frontiers in Plant Science* 11 (June): 1–14.

Xu, Xiangrui, Kun Cheng, Hua Wu, Jianfei Sun, Qian Yue, and Genxing Pan. 2019. "Greenhouse Gas Mitigation Potential in Crop Production with Biochar Soil Amendment—A Carbon Footprint Assessment for Cross-Site Field Experiments from China." *GCB Bioenergy* 11 (4): 592–605.

Yang, Li, and Jia Chen. 2023. "Expanding Genome Editing Scopes with Artificial Intelligence." *Science Bulletin* 68 (23): 2881–2883.

Zaidi, Syed Shan-e-Ali, Manal Tashkandi, Shahid Mansoor, and Magdy M. Mahfouz. 2016. "Engineering Plant Immunity: Using CRISPR/Cas9 to Generate Virus Resistance." *Frontiers in Plant Science* 7 (November): 1–10.

Zetsche, Bernd, Jonathan S. Gootenberg, Omar O. Abudayyeh, Ian M. Slaymaker, Kira S. Makarova, Patrick Essletzbichler, Sara E. Volz et al. 2015a. "Cpf1 Is a Single RNA-Guided Endonuclease of a Class 2 CRISPR-Cas System." *Cell* 163 (3): 759–771. Elsevier Inc.

Zetsche, Bernd, Sara E. Volz, and Feng Zhang. 2015b. "A Split-Cas9 Architecture for Inducible Genome Editing and Transcription Modulation." *Nature Biotechnology* 33 (2): 139–142.

Zetsche, Bernd, Matthias Heidenreich, Prarthana Mohanraju, Iana Fedorova, Jeroen Kneppers, Ellen M. DeGennaro, Nerges Winblad et al. 2017. Mohanraju et al. Multiplex gene editing by CRISPR–Cpf1 using a single crRNA array. *Nature Biotechnology* 35: 31–34.

Zhang, Kai, Nadia Raboanatahiry, Bin Zhu, and Maoteng Li. 2017. "Progress in Genome Editing Technology and Its Application in Plants." *Frontiers in Plant Science* 8 (February).

Zhou, Huanbin, Bo Liu, Donald P. Weeks, Martin H. Spalding, and Bing Yang. 2014. "Large Chromosomal Deletions and Heritable Small Genetic Changes Induced by CRISPR/Cas9 in Rice." *Nucleic Acids Research* 42 (17): 10903–10914.

Zhu, Kaijie, Xiongjie Zheng, Junli Ye, Qihang Jiang, Hongyan Chen, Xuehan Mei, Eleanore T. Wurtzel, and Xiuxin Deng. 2020. "Building the Synthetic Biology Toolbox with Enzyme Variants to Expand Opportunities for Biofortification of Provitamin a and Other Health-Promoting Carotenoids." *Journal of Agricultural and Food Chemistry* 68 (43): 12048–12057.

Zong, Yuan, Yijing Liu, Chenxiao Xue, Boshu Li, Xiangyang Li, Yanpeng Wang, Ji Li et al. 2022. "An Engineered Prime Editor with Enhanced Editing Efficiency in Plants." *Nature Biotechnology* 40 (9): 1394–1402.

5 CRISPR-Cas Systems in Insects and Other Arthropods

Arjumand John, Amaninder Kaur Riat, Kaisar Ahmad Bhat, and Aarif Ali

5.1 INTRODUCTION

The clustered regularly interspaced short palindromic repeats (CRISPR) and CRISPR-associated protein 9 (Cas9) system has revolutionized genome editing by allowing precise, targeted mutations in Deoxyribonucleic acid (DNA) sequences across a wide range of organisms, from plants to animals (Rao et al., 2022). Originating as a bacterial immune defense against viruses and plasmids, CRISPR-Cas9 relies on a highly specific recognition mechanism that has been harnessed and modified for genome editing (Loureiro and da Silva, 2019). This adaptive immunity system has been engineered to create the "CRISPR-Cas9" system, a tool that enables efficient and highly specific targeted mutagenesis. The core of the CRISPR-Cas9 system involves a single guide RNA (sgRNA) and the Cas9 nuclease (Le Trionnaire et al., 2019). The sgRNA is typically composed of a 20-nucleotide sequence complementary to the target genomic region, paired with a 120-nucleotide universal region that folds to form the sgRNA–Cas9 complex (Jiang and Doudna, 2017; Naeem et al., 2020). Once introduced into a cell, the Cas9–sgRNA complex scans the genome, binding to the target site through RNA-DNA base pairing. Cas9's nuclease activity then generates a double-strand break (DSB) in the DNA, which is recognized and repaired by the cell's natural DNA repair mechanisms: non-homologous end joining (NHEJ) or homology-directed repair (HDR) (Munawar and Ahmad, 2021). NHEJ is often imprecise, resulting in insertions or deletions (InDels) that can induce frameshifts, altering protein function and enabling gene disruption. CRISPR-Cas9 has become widely used in various organisms, including plant and animal models (Ray, 2022). In insects, it has shown notable success in species such as *Drosophila melanogaster (D. melanogaster)*, *Aedes aegypti*, *Bombyx mori*, and *Tribolium castaneum (T. castaneum)*. More recently, it has been extended to non-model insects across diverse orders, like Diptera, Lepidoptera, Orthoptera, Coleoptera, and Hymenoptera (Sun et al., 2017). Insect genome editing often relies on microinjection techniques to deliver the Cas9–sgRNA ribonucleoprotein (RNP) complex into single cells or early embryos (Kouranova et al., 2016). Despite its successes, CRISPR-Cas9 application in certain non-model insects presents challenges. For instance, aphids, a major agricultural pest, have a unique reproductive cycle that complicates gene editing. Aphids

DOI: 10.1201/9781003537151-5

reproduce through viviparous parthenogenesis during warmer seasons and switch to sexual reproduction with the arrival of shorter photoperiods (Moon et al., 2022). This life cycle, along with RNA interference (RNAi) inefficiencies, has hindered functional gene studies in aphids. Consequently, optimizing CRISPR-Cas9 for aphids could provide significant insights into their gene functions and pest management (Matsuda, 2020; Mulot et al., 2016). In another study, CRISPR-Cas9 was successfully applied to the two-spotted spider mite, *Tetranychus urticae* (*T. urticae*), a widespread pest that feeds on a wide range of plants. For the first time in Chelicerata, the CRISPR-Cas9 system was shown to induce targeted gene knockouts in *T. urticae* by disrupting the phytoene desaturase gene, essential for pigmentation. Researchers observed a clear loss of red pigmentation in the eyes and legs, a visible marker of successful gene editing (De Rouck et al., 2024). To improve CRISPR-Cas9 delivery to spider mite eggs, researchers explored various transfection agents, including lipofectamine, branched amphiphilic peptide capsules (BAPC), and endosomal escape agents, such as chloroquine and saponins. This work has paved the way for CRISPR-Cas9's use in other hard-to-transform arthropods, including *Frankliniella occidentalis*. Beyond pest control, the potential of CRISPR-Cas9 in non-model insects offers exciting avenues for agricultural and ecological applications. It holds promise for studying traits linked to pollination efficiency and environmental resilience, which could benefit ecosystem stability and productivity under climate change pressures (Singh et al., 2022). Recent advancements in CRISPR technology, such as prime editing and base editing, have enhanced its precision and minimized off-target effects, further expanding its applicability to organisms with complex genomes or life cycles. These advancements position CRISPR-Cas9 not only as a fundamental tool for basic research but also as a transformative approach for addressing challenges in pest control, conservation, and pollinator health in agricultural ecosystems.

5.2 CRISPR-CAS SYSTEMS IN INSECTS

CRISPR-Cas systems in insects have revolutionized genetic research and practical applications in entomology. Insects, the most diverse group of animals on earth, are vital to various ecological functions, including pollination, decomposition, and serving as a food source for other organisms (Banerjee et al., 2023). Understanding their biology is crucial for effective CRISPR-Cas application, encompassing aspects like reproductive strategies, genomic architecture, developmental stages, and ecological roles (Sun et al., 2017). CRISPR-Cas9 has been extensively utilized in various insects for gene function exploration, pest control, and disease resistance. The flexibility and efficiency of CRISPR-Cas9 have made it the tool of choice for genome editing in insects and other arthropods. Researchers are also exploring newer systems, like CRISPR from Prevotella and Francisella 1 (CRISPR-Cpf1) (Cas12a) and CRISPR-C2c2 (Cas13a), which offer different targeting mechanisms and could further enhance editing precision and efficiency (Rato et al., 2021). Model organisms such as *D. melanogaster* have been instrumental in advancing CRISPR research due to their well-characterized genomes, short life cycles, and ease of breeding. In *Drosophila*, CRISPR-Cas9 has enabled precise gene editing, from creating knockouts to study gene function to inserting specific sequences for visualizing protein

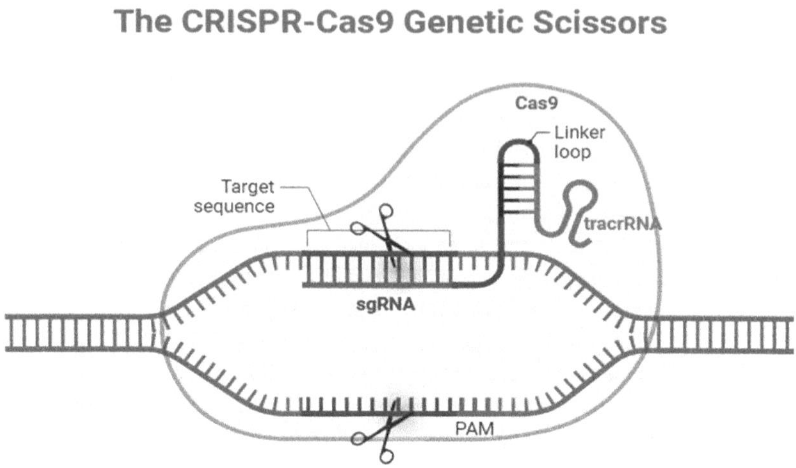

FIGURE 5.1 Components of CRISPR-Cas9. This figure illustrates the CRISPR-Cas9 genome editing system, highlighting the two main components: the Cas9 nuclease enzyme and the single guide RNA (sgRNA). Together, they form a ribonucleoprotein complex that homes in on specific DNA sequences within a genome. Guided by the sgRNA, which provides the target sequence, the Cas9 protein precisely locates and cleaves both strands of DNA at the desired site, creating a double-strand break (DSB). This break triggers cellular repair processes, primarily non-homologous end joining (NHEJ) or homology-directed repair (HDR), which either rejoin or recombine the DNA. NHEJ is an error-prone process that often introduces small insertions or deletions (InDels), potentially disrupting the target gene. HDR, on the other hand, can use a homologous template for accurate repair, enabling precise gene edits. This CRISPR-Cas9 mechanism is widely utilized in research to study gene function, develop disease models, and advance potential therapies.

interactions (Lepesant, 2015). Functional genomics studies using CRISPR interference (CRISPRi) and CRISPR activation (CRISPRa) have allowed to modulate gene expression and investigate gene regulatory networks (Akinci et al., 2021). Recently, a novel CRISPR-like system in eukaryotes, known as the Fanzor system, which uses ωRNAs to target DNA has been discovered. This system, found in various organisms, including algae and clams, could provide new tools for genome editing, with potentially higher precision and less off-target activity compared to current CRISPR systems (Altae-Tran et al., 2023). Additionally, CRISPR-based genetic screens have systematically revealed genes essential for various biological processes. In agricultural pests, CRISPR-Cas systems offer targeted gene modifications to control populations and combat resistance (Rato et al., 2021). Gene drives have been developed to rapidly spread specific genetic traits through pest populations, such as reducing fertility in mosquitoes to combat malaria. Strategies like the sterile insect technique (SIT) leverage CRISPR to produce sterile males that can reduce pest populations when released into the wild. Furthermore, modifying genes related to pest behavior, such as feeding or mating, can mitigate their impact on crops. Beneficial insects, such as bees and butterflies, are essential for pollination and ecosystem health (Macias et al., 2017). CRISPR technology can enhance disease resistance in these insects by modifying immune-related genes and protecting them from pathogens and environmental stresses. Altering the gut microbiome through CRISPR can also improve their health and resilience (Moon et al., 2022). Enhancing pollination efficiency through genetic modifications that affect flight and foraging behavior can lead to more effective pollinators, while nutritional adaptations can improve their health and longevity (Moon et al., 2022). Despite the transformative potential of CRISPR-Cas systems in insects, it is crucial to consider ethical, ecological, and regulatory implications to ensure responsible use. This technology holds promise for advancing our understanding of gene function, developing innovative pest control strategies, and enhancing beneficial insect traits, ultimately benefiting both scientific research and agriculture (Komal et al., 2023).

5.3 CRISPR-CAS SYSTEMS IN OTHER ARTHROPODS

CRISPR-Cas systems in other arthropods, including spiders, mites, and crustaceans, have significantly advanced genetic research and opened new avenues for practical applications. Arthropods, a diverse group of invertebrates, play crucial roles in ecosystems as predators, prey, decomposers, and even as vectors of diseases (Sun et al., 2017). Understanding their biology is essential for the effective use of CRISPR technology, considering factors such as their reproductive strategies, developmental stages, and ecological roles. Researchers have identified a new RNA-guided, DNA-cutting enzyme called Fanzor in eukaryotes, including arthropods. This system, derived from transposable elements, shares similarities with CRISPR-Cas12 but exhibits more extensive interactions with RNA, suggesting potential for precise genome editing (Xu and Li, 2020). The discovery of Fanzor proteins, which

TABLE 5.1
CRISPR-Cas Systems Found in Insects

Insect Species	CRISPR-Cas System Type	Description	Reference(s)
Drosophila melanogaster	Type II (Cas9)	Used extensively in genetic studies	Gratz et al., 2015
Bombyx mori	Type III	Identified with potential for RNA targeting	Wang et al., 2014
Apis mellifera	Type I (Cascade)	Involved in the immune response against viruses	Evans et al., 2006
Tribolium castaneum	Type I (Cascade)	Plays a role in antiviral defense	Kingsolver et al., 2013
Locusta migratoria	Type I (Cascade)	Identified with potential for antiviral defense	Zhang et al., 2015
Anopheles gambiae	Type II (Cas9)	Explored for potential gene drive applications	D'Amato et al., 2024

can be reprogrammed to target specific genome sites, opens new possibilities for efficient genome editing in arthropods and other eukaryotes. In model arthropods, like the two-spotted spider mite (*T. urticae*) and the red flour beetle (*T. castaneum*), CRISPR-Cas systems have enabled precise gene editing to study gene function, investigate developmental processes, and understand evolutionary biology (De Rouck et al., 2024). These organisms have relatively well-characterized genomes, making them suitable for CRISPR applications. For example, gene knockout studies in *T. urticae* have revealed insights into acaricide resistance mechanisms, while in *T. castaneum*, gene editing has been used to study segmentation and development (Sun et al., 2017). In non-model arthropods, CRISPR-Cas systems have also been employed, albeit with greater challenges due to less well-annotated genomes and more complex life cycles. Nevertheless, CRISPR has facilitated functional studies in these species, providing valuable evolutionary insights and enabling the exploration of genetic diversity. Beyond basic research, CRISPR technology in arthropods offers practical benefits, such as controlling pest species and enhancing beneficial traits (Koonin and Makarova, 2019). For instance, in agricultural pests like mites and ticks, CRISPR can target genes responsible for pesticide resistance or feeding behavior, reducing their impact on crops and livestock. In beneficial arthropods like silk-producing spiders, gene editing can enhance silk properties for industrial applications (Singh et al., 2022). Various search algorithms have been developed that could uncover nearly 200 new kinds of CRISPR systems in bacterial genomes. These discoveries include systems that could potentially be adapted for genome editing in arthropods (Moon et al., 2022). The identification of these diverse CRISPR systems suggests a vast, untapped potential for new genome editing tools. The University of Illinois has developed a new tool called "CRISPR-Copies" that accelerates and optimizes genome editing (Ishino et al., 2018). This tool enhances the efficiency of CRISPR-Cas systems, making it more applicable for complex genome editing tasks

TABLE 5.2
CRISPR-Cas Systems Found in Other Arthropods

Arthropod Species	CRISPR-Cas System Type	Description	References
Tetranychus urticae	Type II (Cas9)	Used in studying pest resistance mechanisms	Moon et al., 2022
Limulus polyphemus	Type I (Cascade)	Identified with potential for antiviral defense	Vallespi et al., 2000
Ixodes scapularis	Type II (Cas9)	Investigated for potential applications in tick control	Sharma et al., 2022
Aedes aegypti	Type II (Cas9)	Explored genetic modifications to prevent disease transmission	Kistler et al., 2015
Acyrthosiphon pisum	Type I (Cascade)	Research focusing on understanding aphid–plant interactions	Sanchez-Arcos et al., 2016

in arthropods. The improvements in this tool could significantly impact research and applications in arthropod genomics (De Rouck et al., 2024). Furthermore, CRISPR-Cas systems can be used to study and mitigate the impacts of arthropod-borne diseases by targeting genes in vectors that influence pathogen transmission. While the potential of CRISPR in arthropods is vast, ethical, ecological, and regulatory considerations are paramount to ensure its responsible use (Liu et al., 2021). The technology promises to revolutionize our understanding of arthropod biology, provide innovative solutions for pest management, and enhance beneficial arthropod traits, ultimately contributing to ecological and agricultural sustainability (Bernaola and Holt, 2021).

5.4 TECHNIQUES AND METHODOLOGIES

The techniques and methodologies for employing CRISPR-Cas systems in insects and arthropods involve several crucial steps, starting with the delivery of CRISPR components into target cells (Baci et al., 2022). Effective delivery methods are essential for successful gene editing and can vary widely depending on the species and developmental stage of the organism. In insects, microinjection is a common method, where CRISPR-Cas components (Cas protein and guide RNA) are directly injected into embryos at early developmental stages. This method has been successfully used in model insects, like the *D. melanogaster*, allowing precise control over gene-editing events (Shirai et al., 2022). Electroporation, another technique, involves applying an electric field to create temporary pores in cell membranes through which CRISPR components can enter. This method is particularly useful for larger numbers of cells or for species where microinjection is challenging (Xu et al., 2024). In some arthropods, such as ticks and mites, nanoparticle-mediated delivery is gaining traction, where CRISPR components are encapsulated in nanoparticles that facilitate cellular uptake. Additionally, viral vectors, such as baculoviruses, have been adapted for delivering CRISPR components into certain arthropods, exploiting the natural infection mechanisms of these viruses to achieve high-efficiency gene editing. Optimizing CRISPR-Cas components for specific species is another critical aspect of the methodology. Different species may require different Cas proteins or guide RNA designs to achieve efficient and specific gene editing (Duan et al., 2021). For instance, Cas9 from *Streptococcus pyogenes* is widely used, but other variants, like Cas12a (Cpf1) or Cas13, may be more effective in certain species due to differences in target DNA recognition and cutting mechanisms. The selection of appropriate promoters to drive the expression of CRISPR components is also crucial, as these must be compatible with the host's cellular machinery (Pickar-Oliver and Gersbach, 2019). In insects and arthropods, species-specific promoters can enhance the efficiency and precision of gene editing. Furthermore, guide RNA design is tailored to the target gene and species, ensuring high specificity and minimal off-target effects (Sun et al., 2017). Computational tools and databases have been developed to aid in the design of guide RNAs, taking into account factors like Guanine-Cytosine (GC) content, secondary structure, and potential off-target sites. This optimization process often involves iterative testing and refinement to achieve the desired gene-editing outcomes. Minimizing off-target effects is a significant concern in the application of CRISPR-Cas systems, as unintended edits can lead to undesired phenotypes and complicate the interpretation of results (Tyagi et al., 2020). Several strategies have been developed to reduce off-target effects and improve the specificity of CRISPR-Cas systems. One approach is to use high-fidelity variants of Cas proteins, such as SpCas9-HF1 or eSpCas9, which have been engineered to reduce off-target cleavage while maintaining on-target activity (Guo et al., 2023). These variants achieve greater specificity by reducing the binding affinity for mismatched target sites. Another strategy involves using truncated guide RNAs (tru-gRNAs) with shorter lengths, which can enhance specificity by reducing the likelihood of binding to off-target sequences. Additionally, pairing CRISPR with other technologies, such as base editors or prime editors, can provide more precise editing without introducing DSBs, thereby reducing the potential for off-target effects (Fu et al., 2014). Bioinformatic tools and deep sequencing techniques are employed to predict and identify off-target sites, allowing researchers to design guide RNAs with minimal off-target potential and to validate the specificity of CRISPR-induced edits. By combining these strategies, researchers can enhance the accuracy and reliability of CRISPR-Cas systems in insects and arthropods, paving the way for more precise genetic modifications and a deeper understanding of gene function in these diverse organisms (Tycko et al., 2016).

5.5 APPLICATION OF CRISPR-CAS9 IN INSECTS AND OTHER ARTHROPODS

The application of CRISPR-Cas9 in insects and other arthropods has been systematically reviewed by year and species to illustrate its adoption and evolution in genetic research. Initially, CRISPR-Cas9 was introduced in the model insect *D. melanogaster* and later extended to *Bombyx mori* and several non-model insects, such as flies, mosquitoes, moths, butterflies, and other arthropods (Komal et al., 2023). These studies are organized to facilitate comparison (Table 5.4). As a genetic model, *D. melanogaster* has paved the way for genome editing in other insect species, marking a significant step in insect genetic studies (Singh et al., 2022). In 2013, Gratz and colleagues pioneered the use of CRISPR-Cas9 in *Drosophila*, creating a 4.6 kb deletion in the yellow locus using dual sgRNAs and a single-stranded oligonucleotide donor (ssODNs) template. This study highlighted CRISPR-Cas9's potential for generating "designer flies" tailored to research needs (Gratz et al., 2015). Further developments in 2013 focused on increasing homologous recombination (HR) rates through the reintegration vector approach (Baena-Lopez et al., 2013). Yu et al. explored various HDR strategies using CRISPR-Cas9 and Transcription Activator-Like Effector Nucleases (TALENs), optimizing these techniques for more precise mutagenesis (Yu et al., 2014). The establishment of transgene vectors expressing both Cas9 and sgRNA allowed for targeted gene knockout or knock-in by injecting sgRNA plasmids into transgenic-Cas9 flies (Sun et al., 2017). Studies compared three approaches for HDR: direct injection of donor template and sgRNA into transgenic Cas9 embryos (Gratz et al., 2015), injection of sgRNA into Cas9-expressing transgenic embryos, and injection of all components into non-transgenic individuals (Ge et al., 2016; Gratz et al., 2014). The transgenic

TABLE 5.3
Techniques and Methodologies for Employing CRISPR-Cas Systems in Insects and Arthropods

Technique/Methodology	Description	References
Microinjection	**Insects:** Direct injection of CRISPR-Cas components (Cas protein and guide RNA) into embryos at early developmental stages. **Arthropods:** A similar approach is used where feasible, though often challenging in smaller or more delicate species.	Moon et al., 2022
Electroporation	Application of an electric field to create temporary pores in cell membranes, allowing CRISPR-Cas components to enter cells. Particularly useful for species where microinjection is impractical or inefficient.	Sioson et al., 2021
Nanoparticle-mediated delivery	Utilization of nanoparticles to encapsulate CRISPR-Cas components, facilitating cellular uptake. Effective in certain arthropods, like ticks and mites, where other methods may not be viable.	Duan et al., 2021
Viral vectors	Adaptation of viral vectors (e.g., baculoviruses) for delivering CRISPR-Cas components into arthropods, leveraging natural infection mechanisms for efficient gene editing.	Asmamaw Mengstie 2022
Optimization of CRISPR components	Selection of appropriate Cas proteins (e.g., Cas9, Cas12a, Cas13) and guide RNA designs tailored to the species and target gene.	Bharathkumar et al., 2022
Promoter selection	Choosing species-specific promoters to drive the expression of CRISPR components, ensuring compatibility with host cellular machinery for efficient gene editing.	Kuzmin and Tomilin, 2022
Guide RNA design	Customizing guide RNA sequences for high specificity and minimal off-target effects, using computational tools to optimize design parameters (e.g., GC content, secondary structure).	Konstantakos et al., 2022
Minimizing off-target effects	Strategies include using high-fidelity Cas variants (e.g., SpCas9-HF1, eSpCas9), truncated guide RNAs (tru-gRNAs), and combining CRISPR with base editors or prime editors to reduce unintended edits. Bioinformatic tools and deep sequencing are employed for off-target prediction and validation.	Naeem et al., 2020

approaches demonstrated higher knock-in efficiency (Sun et al., 2017), accelerating *Drosophila* research and broadening CRISPR-Cas9 applications across *Drosophila* species. In addition to whole-organism editing, CRISPR-Cas9 has been applied in *Drosophila* cell lines (Jacinto et al., 2020). For instance, a green fluorescent protein (GFP)-expressing cell line was created to aid functional gene analysis, simplifying CRISPR system optimization for cellular studies (Ceasar et al., 2016). Another notable advancement, the mutagenic chain reaction (MCR) technique enabled autocatalytic mutation to convert heterozygous mutations into homozygous ones (Gantz and Bier, 2015). Similarly, Zimmer et al. used CRISPR-Cas9 to edit the nicotinic acetylcholine receptor (nAChR) gene, resulting in a mutation (G275E) associated with spinosad insecticide resistance, though the resistance level was lower than that of Drosophila nicotinic acetylcholine receptor alpha 6 (Dα6)-null mutants (Zimmer et al., 2016). This indicated the potential of CRISPR to analyze insecticide resistance mechanisms. Innovative methods have also enabled scarless allele replacements, allowing single or multiple allelic substitutions (Lamb et al., 2016). Targeted gene knockouts revealed gene-specific phenotypes, such as mutations in the yellow gene, which affected adult survival rates, depending on sgRNA concentrations (Sun et al., 2017). Building on *D. melanogaster* research, mutations were introduced in the white (w) and sex-lethal (Sxl) genes in *Drosophila suzukii*, affecting reproductive development and highlighting the differences between DNA plasmid and mRNA injection efficacy (Ren et al., 2014; Zimmer et al., 2016). In another study, the succinyl-CoA synthetase/ligase (SCS) gene, essential for energy metabolism, was mutated, resulting in developmental delays and increased mortality under starvation (Quan et al., 2017). The CRISPR-Cas9 system also facilitated research into sex-specific genes and chromatin modifications, such as CLAMP and MSL complexes crucial for balancing X chromosome expression in males (Moschall et al., 2019; Urban et al., 2017). CRISPR-Cas9 has similarly transformed mosquito research. Key species like *Aedes aegypti*, *Anopheles stephensi*, and *Culex quinquefasciatus* are vectors of diseases like malaria, dengue, and Zika, posing global health risks (Chechenova et al., 2017; Kim and Kim, 2017). Traditional insecticides, while effective, cause ecological harm and lead to insecticide resistance (Nebbak et al., 2022; Senthil-Nathan and Sengottayan, 2020). Mosquito reproduction, a blood-dependent process, makes population control challenging. Genome editing in mosquitoes has leveraged CRISPR-Cas9 and alternative tools like zinc-finger nucleases (ZFNs) and TALENs, underscoring the versatility and potential of genome editing for species-specific genetic interventions (Gulia-Nuss et al., 2012; Nourani et al., 2023).

5.6 APPLICATION IN POLLINATION AND ECOLOGICAL STUDIES

The application of CRISPR technology in pollination and ecological studies holds transformative potential for advancing our understanding of pollinator biology, behavior, and resilience to environmental changes (Rai et al., 2023). Recent studies have shown that CRISPR-Cas systems can be used to explore the genetic

TABLE 5.4
Application of the CRISPR-Cas9 System in Insects and Other Arthropods

Species	Target Gene	Application	Method of Delivery	Outcome/Results	References
Drosophila melanogaster	Yellow, white	Gene knockout	Microinjection	Successful gene disruption, observable phenotypic changes	Souza et al., 2022
Bombyx mori	BmBLOS2, BmBLOS3	Gene function study	Electroporation	Targeted mutagenesis, altered silk production	Baci et al., 2022
Aedes aegypti	Kmo	Disease vector control	Embryo microinjection	Disruption of eye pigment gene, visual markers	Chen et al., 2022
Apis mellifera	Vitellogenin	Reproductive and behavioral study	Microinjection	Gene knockdown, effects on reproduction and behavior	Sheng et al., 2023
Tribolium castaneum	Tc'white	Genetic analysis	Embryo injection	Knockout of pigmentation gene, white-eyed phenotype	Rylee et al., 2022
Tetranychus urticae	Actin, vATPase, GPCRs	Pest control	RNA interference	Reduced fitness, potential for pest management strategies	Bensoussan et al., 2020
Bombyx mori	BmPIG-A	Study of glycosylation	Microinjection	Disruption of glycosylation pathway, phenotypic changes	Zhu et al., 2020
Locusta migratoria	Neurotransmitter genes	Neurological function study	Microinjection	Disruption of neurotransmitter function, behavioral changes	Teleanu et al., 2022
Daphnia magna	Ppib	Environmental and evolutionary study	Microinjection	Knockout of target gene, impact on physiology and adaptation	Bojadzija et al., 2019
Parasteatoda tepidariorum	Sp6–9, Sp1	Developmental biology study	Microinjection	Analysis of segmentation and appendage formation	Janssen and Budd, 2020

foundations of traits critical for pollination efficiencies, such as foraging behavior, floral preference, and navigation abilities, which are essential for the productivity of both wild and cultivated plants (Rasheed et al., 2021). Researchers have identified specific genes that influence olfactory, visual, and cognitive functions in pollinators like honeybees and bumblebees, which can now be targeted using CRISPR to understand how genetic modifications impact these critical behaviors (Değirmenci et al., 2020). Additionally, as climate change increasingly threatens pollinator populations, CRISPR provides a promising tool for investigating genes related to temperature tolerance, desiccation resistance, and immune response. For example, editing genes associated with thermal stress tolerance has been proposed as a means to enhance the survival of pollinators under extreme temperature fluctuations, which are becoming more common globally (Erdoğan et al., 2023). Similarly, genes that play a role in pesticide resistance can be studied and potentially modified to help pollinators withstand chemical exposure, offering a genetic-based strategy to mitigate the impact of agricultural practices on these essential insects (Bass et al., 2024). Moreover, beyond studying and modifying individual traits, CRISPR holds potential applications in bolstering pollinator populations against broad environmental challenges. For instance, researchers are now investigating gene-editing approaches to improve bee resistance to diseases such as deformed wing virus and Nosema infections, both of which contribute significantly to bee colony losses worldwide (Movahedi et al., 2023). Modifying genes associated with immune function could strengthen pollinator resilience to pathogens, potentially enhancing pollinator health and stability in ecosystems. Another potential application of CRISPR is in fostering genetic diversity within pollinator populations, which is crucial for adaptation to evolving environmental pressures (Koller et al., 2023). By enabling precise gene edits, CRISPR may one day facilitate the introduction of beneficial genetic variants into wild populations, enhancing their adaptive capacity without compromising their natural behaviors or ecological roles (Jacinto et al., 2020). However, the use of CRISPR in beneficial insects like pollinators raises important ethical considerations. Gene editing in insects that play critical roles in ecosystems could have unintended consequences, both within modified populations and across broader ecological networks. For example, altering genes related to pollination behaviors could impact plant–pollinator interactions, potentially affecting entire plant communities and the animals that depend on them. There is also concern over gene flow from edited individuals to wild populations, which could spread modified traits with unforeseen ecological impacts (Ayanoğlu et al., 2020). To address these issues, researchers advocate for rigorous containment, comprehensive ecological risk assessments, and careful consideration of alternative approaches. Additionally, the prospect of deploying genetically modified pollinators raises questions about ownership, control, and the rights of Indigenous and farming communities who may be impacted by these interventions (Auer, 2008). Engaging these communities in decision-making processes is critical to ethically advancing CRISPR applications in pollinator conservation. As research progresses, a balanced approach that combines scientific innovation with ethical foresight is necessary to ensure that CRISPR technology contributes positively to pollinator health and ecological resilience in a responsible manner (Liu et al., 2021).

5.7 EXPANDING CRISPR APPLICATIONS TO NON-INSECT ARTHROPODS

Expanding CRISPR applications to non-insect arthropods, particularly in groups like chelicerates, has opened new avenues for research and practical applications, though it poses significant challenges due to the unique biology and reproductive characteristics of these species (Liu et al., 2021). Chelicerates, such as mites and spiders, present opportunities for genetic manipulation that could lead to advances in pest control, agricultural productivity, and ecological balance. For instance, the two-spotted spider mite (*T. urticae*), a highly destructive pest across various crops, was recently used in CRISPR research to disrupt genes associated with pigmentation, providing a clear visual marker of successful gene editing (Grbic et al., 2007). This was a significant breakthrough because, for a long time, gene editing in chelicerates lagged behind that of insects due to the difficulty in delivering CRISPR-Cas9 components into their eggs or embryos and ensuring gene expression (Singh et al., 2022). By optimizing injection techniques and using transfection agents like lipofectamine and chloroquine, researchers have now established protocols that improve CRISPR efficiency in *T. urticae* and other challenging arthropod species, including *Frankliniella occidentalis* (De Rouck et al., 2024). The success of these methods marks an essential step in extending CRISPR technology to non-insect arthropods, as the protocols developed could be adapted for other pest species in agriculture, potentially reducing the reliance on chemical pesticides. Another promising area of CRISPR application is in spiders, where researchers are exploring gene functions related to silk production, venom components, and development (Komal et al., 2023). Spider silk, known for its strength and elasticity, has been of great interest for bioengineering and materials science, and CRISPR now offers a means to explore and possibly enhance the genetic basis of silk synthesis. Likewise, venom components in spiders hold potential for pharmaceuticals, and targeted gene editing could help identify and manipulate specific venom proteins with therapeutic applications. However, gene editing in spiders has faced hurdles similar to those seen in mites, primarily due to difficulties in early-stage embryonic injection and achieving stable gene integration (Brankovic et al., 2024). Breakthroughs in micromanipulation and embryo preservation have recently shown promise, leading to some successful gene knockouts that could pave the way for broader CRISPR applications in these species. The potential agricultural and ecological benefits of expanding CRISPR to non-insect arthropods are vast (Zhu, 2022). In agricultural settings, genetic modifications in pest species such as mites could lead to more effective, environmentally friendly pest control strategies. For example, targeting genes associated with pesticide resistance or reproductive traits could offer sustainable pest management solutions (Legros et al., 2021). In ecological research, CRISPR could be used to investigate the roles of various genes in arthropod behaviors, physiology, and interactions within ecosystems, contributing to biodiversity and conservation efforts. However, the unique biological characteristics of non-insect arthropods necessitate specialized CRISPR approaches and careful ecological assessments to ensure that gene modifications do not unintentionally disrupt ecological balance (Komal et al., 2023). The potential for off-target effects and unintended consequences is a crucial consideration, as is the need for ethical oversight when manipulating species that play vital roles in natural ecosystems. Despite these challenges, the extension of CRISPR technology to non-insect arthropods represents an exciting frontier with profound implications for agriculture, ecology, and beyond (Ayanoğlu et al., 2020).

5.8 FUTURE PROSPECTS

The prospects of CRISPR-Cas technology in insects and arthropods are exceptionally promising, driven by continuous advancements in the precision, efficiency, and versatility of gene-editing tools. Recent developments include the creation of high-fidelity Cas variants and base editors that minimize off-target effects, expanding the range of possible genetic modifications (Tyumentseva et al., 2023). These technological improvements pave the way for innovative applications in pest control, such as more targeted gene drives that could suppress or eradicate disease vectors like mosquitoes with greater ecological safety (Wang et al., 2021). In agricultural contexts, CRISPR can be used to engineer pest-resistant crops indirectly by modifying the genetics of insect pests to reduce their virulence or enhance their susceptibility to biocontrol agents (Moon et al., 2022). Moreover, beneficial insects like pollinators can be genetically enhanced to improve their resilience against diseases and environmental stresses, thereby supporting biodiversity and ecosystem health. Long-term goals and visions for CRISPR in this field include creating self-limiting gene drives that naturally degrade after achieving their purpose, ensuring ecological balance, and developing global regulatory frameworks that facilitate the safe and ethical use of genetic technologies (Harvey-Samuel et al., 2023). Ultimately, the integration of CRISPR-Cas technology into insect and arthropod research holds the potential to revolutionize pest management, enhance agricultural productivity, and protect biodiversity, contributing to a sustainable and resilient future (Sun et al., 2017).

5.9 CONCLUSION

CRISPR-Cas systems have revolutionized genetic research and practical applications in insects and arthropods, offering precise tools for targeted genome editing. From model organisms like *D. melanogaster* to agricultural pests such as the two-spotted spider mite (*T. urticae*), CRISPR technology has enabled researchers to dissect gene functions, unravel resistance mechanisms, and explore novel strategies for pest control. Despite challenges in delivery methods and efficiency, advancements in CRISPR techniques continue to expand possibilities for genetic manipulation across diverse species. Ethical considerations and regulatory frameworks remain crucial to ensure responsible use and mitigate potential ecological impacts, emphasizing the need for ongoing research and international collaboration. Looking forward, CRISPR-Cas systems hold promise not only for advancing scientific knowledge but also for addressing global challenges in agriculture, public health, and environmental sustainability.

REFERENCES

Akinci, Ersin, Marisa C. Hamilton, Benyapa Khowpinitchai, and Richard I. Sherwood. 2021. "Using CRISPR to Understand and Manipulate Gene Regulation." *Development* 148 (9): dev182667. https://doi.org/10.1242/dev.182667.

Altae-Tran, Han, Sergey A. Shmakov, Kira S. Makarova, Yuri I. Wolf, Soumya Kannan, Feng Zhang, and Eugene V. Koonin. 2023. "Diversity, Evolution, and Classification of the RNA-Guided Nucleases TnpB and Cas12." *Proceedings of the National Academy of Sciences* 120 (48): e2308224120. https://doi.org/10.1073/pnas.2308224120.

Asmamaw Mengstie, Misganaw. 2022. "Viral Vectors for the in Vivo Delivery of CRISPR Components: Advances and Challenges." *Frontiers in Bioengineering and Biotechnology* 10 (May). https://doi.org/10.3389/fbioe.2022.895713.

Auer, Carol. 2008. "Ecological Risk Assessment and Regulation for Genetically-Modified Ornamental Plants." *Critical Reviews in Plant Sciences* 27 (August): 255–271. https://doi.org/10.1080/07352680802237162.

Ayanoğlu, Fatma Betül, Ayşe Eser Elçin, and Yaşar Murat Elçin. 2020. "Bioethical Issues in Genome Editing by CRISPR-Cas9 Technology." *Turkish Journal of Biology* 44 (2): 110. https://doi.org/10.3906/biy-1912-52.

Baci, Gabriela-Maria, Alexandra-Antonia Cucu, Alexandru-Ioan Giurgiu, Adriana-Sebastiana Muscă, Lilla Bagameri, Adela Ramona Moise, Otilia Bobiş, Attila Cristian Rațiu, and Daniel Severus Dezmirean. 2022. "Advances in Editing Silkworms (Bombyx Mori) Genome by Using the CRISPR-Cas System." *Insects* 13 (1): 28. https://doi.org/10.3390/insects13010028.

Baena-Lopez, Luis, Cyrille Alexandre, Alice Mitchell, Laurynas Pasakarnis, and Jean-Paul Vincent. 2013. "Accelerated Homologous Recombination and Subsequent Genome Modification in Drosophila." *Development (Cambridge, England)* 140 (October). https://doi.org/10.1242/dev.100933.

Banerjee, Sabyasachi, Shruti Gupta, Ritu Raj, Dixit, Uday Kaushal, Gurpreet Kaur, and Satish Krushna Gharde. 2023. "CRISPR-Cas9 Technology: Mechanism and Its Application in The Field of Entomology." *Journal of Advanced Zoology* (December 1). EBSCOhost. https://doi.org/10.17762/jaz.v44i5.2642.

Bass, Chris, Angela Hayward, Bartlomiej J. Troczka, Julian Haas, and Ralf Nauen. 2024. "The Molecular Determinants of Pesticide Sensitivity in Bee Pollinators." *Science of The Total Environment* 915 (March): 170174. https://doi.org/10.1016/j.scitotenv.2024.170174.

Bensoussan, Nicolas, Sameer Dixit, Midori Tabara, David Letwin, Maja Milojevic, Michele Antonacci, Pengyu Jin et al., 2020. "Environmental RNA Interference in Two-Spotted Spider Mite, Tetranychus Urticae, Reveals dsRNA Processing Requirements for Efficient RNAi Response." *Scientific Reports* 10 (November): 19126. https://doi.org/10.1038/s41598-020-75682-6.

Bernaola, Lina, and Jocelyn R. Holt. 2021. "Incorporating Sustainable and Technological Approaches in Pest Management of Invasive Arthropod Species." *Annals of the Entomological Society of America* 114 (6): 673–685. https://doi.org/10.1093/aesa/saab041.

Bharathkumar, Nagaraj, Abraham Sunil, Prabhakar Meera, Sam Aksah, Muthu Kannan, Konda Mani Saravanan, and Thirunavukarasou Anand. 2022. "CRISPR/Cas-Based Modifications for Therapeutic Applications: A Review." *Molecular Biotechnology* 64 (4): 355–372. https://doi.org/10.1007/s12033-021-00422-8.

Bojadzija Savic, Gorenka, Christine Edwards, Enora Briand, Linda Lawton, Claudia Wiegand, and Myriam Bormans. 2019. "Daphnia Magna Exudates Impact Physiological and Metabolic Changes in Microcystis Aeruginosa." *Toxins* 11 (7): 421. https://doi.org/10.3390/toxins11070421.

Branković, Marija, Fatima Zivic, Nenad Grujovic, Ivan Stojadinovic, Strahinja Milenkovic, and Nikola Kotorcevic. 2024. "Review of Spider Silk Applications in Biomedical and Tissue Engineering." *Biomimetics* 9 (3): 169. https://doi.org/10.3390/biomimetics9030169.

Ceasar, S. Antony, Vinothkumar Rajan, Sergey V. Prykhozhij, Jason N. Berman, and S. Ignacimuthu. 2016. "Insert, Remove or Replace: A Highly Advanced Genome Editing System Using CRISPR/Cas9." *Biochimica et Biophysica Acta (BBA)—Molecular Cell Research* 1863 (9): 2333–2344. https://doi.org/10.1016/j.bbamcr.2016.06.009.

Chechenova, Maria B., Sara Maes, Sandy T. Oas, Cloyce Nelson, Kaveh G. Kiani, Anton L. Bryantsev, and Richard M. Cripps. 2017. "Functional Redundancy and Nonredundancy between Two Troponin C Isoforms in Drosophila Adult Muscles." *Molecular Biology of the Cell* 28 (6): 760–770. https://doi.org/10.1091/mbc.E16-07-0498.

Chen, Chujia, Austin Compton, Katerina Nikolouli, Aihua Wang, Azadeh Aryan, Atashi Sharma, Yumin Qi et al., 2022. "Marker-Assisted Mapping Enables Forward Genetic Analysis in Aedes Aegypti, an Arboviral Vector with Vast Recombination Deserts." *Genetics* 222 (3): iyac140. https://doi.org/10.1093/genetics/iyac140.

D'Amato, Rocco, Chrysanthi Taxiarchi, Marco Galardini, Alessandro Trusso, Roxana L. Minuz, Silvia Grilli, Alastair G. T. Somerville et al. 2024. "Anti-CRISPR Anopheles Mosquitoes Inhibit Gene Drive Spread under Challenging Behavioural Conditions in Large Cages." *Nature Communications* 15 (1): 952. https://doi.org/10.1038/s41467-024-44907-x.

De Rouck, Sander, Antonio Mocchetti, Wannes Dermauw, and Thomas Van Leeuwen. 2024. "SYNCAS: Efficient CRISPR/Cas9 Gene-Editing in Difficult to Transform Arthropods." *Insect Biochemistry and Molecular Biology* 165: 104068.

Değirmenci, Laura, Dietmar Geiger, Fábio Ferreira, Alexander Keller, Beate Krischke, Martin Beye, Ingolf Steffan-Dewenter, and Ricarda Scheiner. 2020. "CRISPR/Cas 9-Mediated Mutations as a New Tool for Studying Taste in Honeybees." *Chemical Senses* 45 (September). https://doi.org/10.1093/chemse/bjaa063.

Duan, Li, Kan Ouyang, Xiao Xu, Limei Xu, Caining Wen, Xiaoying Zhou, Zhuan Qin, Zhiyi Xu, Wei Sun, and Yujie Liang. 2021. "Nanoparticle Delivery of CRISPR/Cas9 for Genome Editing." *Frontiers in Genetics* 12 (May): 673286. https://doi.org/10.3389/fgene.2021.673286.

Erdoğan, İbrahim, Birsen Cevher-Keskin, Özlem Bilir, Yiguo Hong, and Mahmut Tör. 2023. "Recent Developments in CRISPR/Cas9 Genome-Editing Technology Related to Plant Disease Resistance and Abiotic Stress Tolerance." *Biology* 12 (7): 1037. https://doi.org/10.3390/biology12071037.

Evans, J. D., K. Aronstein, Y. P. Chen, C. Hetru, J.-L. Imler, H. Jiang, M. Kanost, G. J. Thompson, Z. Zou, and D. Hultmark. 2006. "Immune Pathways and Defence Mechanisms in Honey Bees Apis Mellifera." *Insect Molecular Biology* 15 (5): 645–656. https://doi.org/10.1111/j.1365–2583.2006.00682.x.

Fu, Yanfang, Jeffry D. Sander, Deepak Reyon, Vincent M. Cascio, and J. Keith Joung. 2014. "Improving CRISPR-Cas Nuclease Specificity Using Truncated Guide RNAs." *Nature Biotechnology* 32 (3): 279–284. https://doi.org/10.1038/nbt.2808.

Gantz, Valentino M., and Ethan Bier. 2015. "The Mutagenic Chain Reaction: A Method for Converting Heterozygous to Homozygous Mutations." *Science (New York, N.Y.)* 348 (6233): 442–444. https://doi.org/10.1126/science.aaa5945.

Ge, Daniel Tianfang, Cindy Tipping, Michael H. Brodsky, and Phillip D. Zamore. 2016. "Rapid Screening for CRISPR-Directed Editing of the Drosophila Genome Using White Coconversion." *G3: Genes|Genomes|Genetics* 6 (10): 3197–3206. https://doi.org/10.1534/g3.116.032557.

Gratz, Scott J., C. Dustin Rubinstein, Melissa M. Harrison, Jill Wildonger, and Kate M. O'Connor-Giles. 2015. "CRISPR-Cas9 Genome Editing in Drosophila." *Current Protocols in Molecular Biology* 111 (July): 31.2.1–31.2.20. https://doi.org/10.1002/0471142727.mb3102s111.

Gratz, Scott J., Fiona P. Ukken, C. Dustin Rubinstein, Gene Thiede, Laura K. Donohue, Alexander M. Cummings, and Kate M. O'Connor-Giles. 2014. "Highly Specific and Efficient CRISPR/Cas9-Catalyzed Homology-Directed Repair in Drosophila." *Genetics* 196 (4): 961–971. https://doi.org/10.1534/genetics.113.160713.

Grbic, Miodrag, Abderrahman Khila, Kwang-Zin Lee, Anica Bjelica, Vojislava Grbic, Jay Whistlecraft, Lou Verdon, Maria Navajas, and Lisa Nagy. 2007. "Mity Model: Tetranychus Urticae, a Candidate for Chelicerate Model Organism." *BioEssays: News and Reviews in Molecular, Cellular and Developmental Biology* 29 (5): 489–496. https://doi.org/10.1002/bies.20564.

Gulia-Nuss, Monika, Jai-Hoon Eum, Michael R. Strand, and Mark R. Brown. 2012. "Ovary Ecdysteroidogenic Hormone Activates Egg Maturation in the Mosquito Georgecraigius Atropalpus after Adult Eclosion or a Blood Meal." *The Journal of Experimental Biology* 215 (21): 3758–3767. https://doi.org/10.1242/jeb.074617.

Guo, Congting, Xiaoteng Ma, Fei Gao, and Yuxuan Guo. 2023. "Off-Target Effects in CRISPR/Cas9 Gene Editing." *Frontiers in Bioengineering and Biotechnology* 11 (March): 1143157. https://doi.org/10.3389/fbioe.2023.1143157.

Harvey-Samuel, Tim, Xuechun Feng, Emily M. Okamoto, Deepak-Kumar Purusothaman, Philip T. Leftwich, Luke Alphey, and Valentino M. Gantz. 2023. "CRISPR-Based Gene Drives Generate Super-Mendelian Inheritance in the Disease Vector Culex Quinquefasciatus." *bioRxiv*, June, 2023.06.12.544656. https://doi.org/10.1101/2023.06.12.544656.

Ishino, Yoshizumi, Mart Krupovic, and Patrick Forterre. 2018. "History of CRISPR-Cas from Encounter with a Mysterious Repeated Sequence to Genome Editing Technology." *Journal of Bacteriology* 200 (7): e00580-17. https://doi.org/10.1128/JB.00580-17.

Jacinto, Filipe V., Wolfgang Link, and Bibiana I. Ferreira. 2020. "CRISPR/Cas9-mediated Genome Editing: From Basic Research to Translational Medicine." *Journal of Cellular and Molecular Medicine* 24 (7): 3766–3778. https://doi.org/10.1111/jcmm.14916.

Janssen, Ralf, and Graham E. Budd. 2020. "Expression of the Zinc Finger Transcription Factor Sp6–9 in the Velvet Worm Euperipatoides Kanangrensis Suggests a Conserved Role in Appendage Development in Panarthropoda." *Development Genes and Evolution* 230 (3): 239–245. https://doi.org/10.1007/s00427-020-00661-w.

Jiang, Fuguo, and Jennifer A. Doudna. 2017. "CRISPR–Cas9 Structures and Mechanisms." *Annual Review of Biophysics* 46 (1): 505–529.

Kim, Yea Woon, and AeRi Kim. 2017. "Deletion of Transcription Factor Binding Motifs Using the CRISPR/spCas9 System in the β-Globin LCR." *Bioscience Reports* 37 (4): BSR20170976. https://doi.org/10.1042/BSR20170976.

Kingsolver, Megan B., Zhijing Huang, and Richard W. Hardy. 2013. "Insect Antiviral Innate Immunity: Pathways, Effectors, and Connections." *Journal of Molecular Biology* 425 (24): 4921–4936. https://doi.org/10.1016/j.jmb.2013.10.006.

Kistler, Kathryn E., Leslie B. Vosshall, and Benjamin J. Matthews. 2015. "Genome-Engineering with CRISPR-Cas9 in the Mosquito Aedes Aegypti." *Cell Reports* 11 (1): 51–60. https://doi.org/10.1016/j.celrep.2015.03.009.

Koller, Franziska, Meike Schulz, Matthias Juhas, Andreas Bauer-Panskus, and Christoph Then. 2023. "The Need for Assessment of Risks Arising from Interactions between NGT Organisms from an EU Perspective." *Environmental Sciences Europe* 35 (1): 27. https://doi.org/10.1186/s12302-023-00734-3.

Komal, J., H. R. Desai, Ipsita Samal, Andrea Mastinu, R. D. Patel, P. V. Dinesh Kumar, Prasanta Kumar Majhi, Deepak Kumar Mahanta, and Tanmaya Kumar Bhoi. 2023. "Unveiling the Genetic Symphony: Harnessing CRISPR-Cas Genome Editing for Effective Insect Pest Management." *Plants* 12 (23): 3961. https://doi.org/10.3390/plants12233961.

Konstantakos, Vasileios, Anastasios Nentidis, Anastasia Krithara, and Georgios Paliouras. 2022. "CRISPR–Cas9 gRNA Efficiency Prediction: An Overview of Predictive Tools and the Role of Deep Learning." *Nucleic Acids Research* 50 (7): 3616–3637. https://doi.org/10.1093/nar/gkac192.

Koonin, Eugene V., and Kira S. Makarova. 2019. "Origins and Evolution of CRISPR-Cas Systems." *Philosophical Transactions of the Royal Society B: Biological Sciences* 374 (1772): 20180087. https://doi.org/10.1098/rstb.2018.0087.

Kouranova, Evguenia, Kevin Forbes, Guojun Zhao, Joe Warren, Angela Bartels, Yumei Wu, and Xiaoxia Cui. 2016. "CRISPRs for Optimal Targeting: Delivery of CRISPR Components as DNA, RNA, and Protein into Cultured Cells and Single-Cell Embryos." *Human Gene Therapy* 27 (6): 464–475.

Kuzmin, Andrey A., and Alexey N. Tomilin. 2022. "Building Blocks of Artificial CRISPR-Based Systems beyond Nucleases." *International Journal of Molecular Sciences* 24 (1): 397. https://doi.org/10.3390/ijms24010397.

Lamb, Abigail M., Elizabeth A. Walker, and Patricia J. Wittkopp. 2016. "Tools and Strategies for Scarless Allele Replacement in Drosophila Using CRISPR/Cas9." *Fly* 11 (1): 53–64. https://doi.org/10.1080/19336934.2016.1220463.

Le Trionnaire, Gaël, Sylvie Tanguy, Sylvie Hudaverdian, Florence Gléonnec, Gautier Richard, Bastien Cayrol, Baptiste Monsion, Elodie Pichon, Maelle Deshoux, and Craig Webster. 2019. "An Integrated Protocol for Targeted Mutagenesis with CRISPR-Cas9 System in the Pea Aphid." *Insect Biochemistry and Molecular Biology* 110: 34–44.

Legros, Mathieu, John M. Marshall, Sarina Macfadyen, Keith R. Hayes, Andy Sheppard, and Luke G. Barrett. 2021. "Gene Drive Strategies of Pest Control in Agricultural Systems: Challenges and Opportunities." *Evolutionary Applications* 14 (9): 2162. https://doi.org/10.1111/eva.13285.

Lepesant, Jean-Antoine. 2015. "The promises of neurodegenerative disease modeling." *Comptes Rendus. Biologies* 338 (8–9): 584–592. https://doi.org/10.1016/j.crvi.2015.06.018.

Liu, Wenyi, Luoxi Li, Jianxin Jiang, Min Wu, and Ping Lin. 2021. "Applications and Challenges of CRISPR-Cas Gene-Editing to Disease Treatment in Clinics." *Precision Clinical Medicine* 4 (3): 179–191. https://doi.org/10.1093/pcmedi/pbab014.

Loureiro, Alexandre, and Gabriela Jorge da Silva. 2019. "CRISPR-Cas: Converting a Bacterial Defence Mechanism into a State-of-the-Art Genetic Manipulation Tool." *Antibiotics* 8 (1): 18.

Macias, Vanessa M., Johanna R. Ohm, and Jason L. Rasgon. 2017. "Gene Drive for Mosquito Control: Where Did It Come from and Where Are We Headed?" *International Journal of Environmental Research and Public Health* 14 (9): 1006. https://doi.org/10.3390/ijerph14091006.

Matsuda, Naoki. 2020. "Physiological Study on the Transgenerational Timing Mechanism in an Aphid." https://repository.kulib.kyotou.ac.jp/dspace/bitstream/2433/253117/5/drigk04595.pdf

Moon, Tasfia Tasnim, Ishrat Jahan Maliha, Abdullah Al Moin Khan, Moutoshi Chakraborty, Md Sharaf Uddin, Md Ruhul Amin, and Tofazzal Islam. 2022. "CRISPR-Cas Genome Editing for Insect Pest Stress Management in Crop Plants." *Stresses* 2 (4): 493–514. https://doi.org/10.3390/stresses2040034.

Moschall, Rebecca, Mathias Rass, Oliver Rossbach, Gerhard Lehmann, Lars Kullmann, Norbert Eichner, Daniela Strauss et al., 2019. "Drosophila Sister-of-Sex-Lethal Reinforces a Male-Specific Gene Expression Pattern by Controlling Sex-Lethal Alternative Splicing." *Nucleic Acids Research* 47 (5): 2276–2288. https://doi.org/10.1093/nar/gky1284.

Movahedi, Ali, Soheila Aghaei-Dargiri, Hongyan Li, Qiang Zhuge, and Weibo Sun. 2023. "CRISPR Variants for Gene Editing in Plants: Biosafety Risks and Future Directions." *International Journal of Molecular Sciences* 24 (22): 16241. https://doi.org/10.3390/ijms242216241.

Mulot, Michaël, Sylvaine Boissinot, Baptiste Monsion, Maryam Rastegar, Gabriel Clavijo, David Halter, Nicole Bochet, Monique Erdinger, and Véronique Brault. 2016. "Comparative Analysis of RNAi-Based Methods to down-Regulate Expression of Two Genes Expressed at Different Levels in Myzus Persicae." *Viruses* 8 (11): 316.

Munawar, Nayla, and Aftab Ahmad. 2021. "CRISPR/Cas System: An Introduction." *CRISPR Crops: The Future of Food Security*, 1–35. Springer.

Naeem, Muhammad, Saman Majeed, Mubasher Zahir Hoque, and Irshad Ahmad. 2020. "Latest Developed Strategies to Minimize the Off-Target Effects in CRISPR-Cas-Mediated Genome Editing." *Cells* 9 (7): 1608.

Nebbak, Amira, Lionel Almeras, Philippe Parola, and Idir Bitam. 2022. "Mosquito Vectors (Diptera: Culicidae) and Mosquito-Borne Diseases in North Africa." *Insects* 13 (10): 962. https://doi.org/10.3390/insects13100962.

Nourani, Leila, Akram Abouie Mehrizi, Sakineh Pirahmadi, Zeinab Pourhashem, Elahe Asadollahi, and Babak Jahangiri. 2023. "CRISPR/Cas Advancements for Genome Editing, Diagnosis, Therapeutics, and Vaccine Development for *Plasmodium* Parasites, and Genetic Engineering of *Anopheles* Mosquito Vector." *Infection, Genetics and Evolution* 109 (April): 105419. https://doi.org/10.1016/j.meegid.2023.105419.

Pickar-Oliver, Adrian, and Charles A. Gersbach. 2019. "The Next Generation of CRISPR–Cas Technologies and Applications." *Nature Reviews. Molecular Cell Biology* 20 (8): 490–507. https://doi.org/10.1038/s41580-019-0131-5.

Quan, Xiuming, Yukiko Sato-Miyata, Manabu Tsuda, Keigo Muramatsu, Tsunaki Asano, Satomi Takeo, and Toshiro Aigaki. 2017. "Deficiency of Succinyl-CoA Synthetase α Subunit Delays Development, Impairs Locomotor Activity and Reduces Survival under Starvation in Drosophila." *Biochemical and Biophysical Research Communications* 483 (1): 566–571. https://doi.org/10.1016/j.bbrc.2016.12.105.

Rai, Gyanendra Kumar, Danish Mushtaq Khanday, Pradeep Kumar, Isha Magotra, Sadiya M. Choudhary, Rafia Kosser, Raviraj Kalunke et al., 2023. "Enhancing Crop Resilience to Drought Stress through CRISPR-Cas9 Genome Editing." *Plants* 12 (12): 2306. https://doi.org/10.3390/plants12122306.

Rao, Yuchun, Xi Yang, Chenyang Pan, Chun Wang, and Kejian Wang. 2022. "Advance of Clustered Regularly Interspaced Short Palindromic Repeats-Cas9 System and Its Application in Crop Improvement." *Frontiers in Plant Science* 13 (May). https://doi.org/10.3389/fpls.2022.839001.

Rasheed, Adnan, Rafaqat Ali Gill, Muhammad Umair Hassan, Athar Mahmood, Sameer Qari, Qamar U. Zaman, Muhammad Ilyas et al., 2021. "A Critical Review: Recent Advancements in the Use of CRISPR/Cas9 Technology to Enhance Crops and Alleviate Global Food Crises." *Current Issues in Molecular Biology* 43 (3): 1950. https://doi.org/10.3390/cimb43030135.

Rato, Cláudia, Miguel F. Carvalho, Cristina Azevedo, and Paula Rodrigues Oblessuc. 2021. "Genome Editing for Resistance against Plant Pests and Pathogens." *Transgenic Research* 30 (4): 427–459. https://doi.org/10.1007/s11248-021-00262-x.

Ray, Atish. 2022. "DNA Mutation, Repair, and Recombination." In *Genetics Fundamentals Notes*, 433–490. Springer.

Ren, Xingjie, Zhihao Yang, Jiang Xu, Jin Sun, Decai Mao, Yanhui Hu, Su-Juan Yang et al., 2014. "Enhanced Specificity and Efficiency of the CRISPR/Cas9 System with Optimized sgRNA Parameters in *Drosophila*." *Cell Reports* 9 (3): 1151–1162. https://doi.org/10.1016/j.celrep.2014.09.044.

Rylee, Johnathan C., Alexandra Nin-Velez, Simpla Mahato, Kennedy J. Helms, Michael J. Wade, Gabriel E. Zentner, and Andrew C. Zelhof. 2022. "Generating and Testing the Efficacy of Transgenic Cas9 in Tribolium Castaneum." *Insect Molecular Biology* 31 (5): 543–550. https://doi.org/10.1111/imb.12778.

Sanchez-Arcos, Carlos, Michael Reichelt, Jonathan Gershenzon, and Grit Kunert. 2016. "Modulation of Legume Defense Signaling Pathways by Native and Non-Native Pea Aphid Clones." *Frontiers in Plant Science* 7 (December). https://doi.org/10.3389/fpls.2016.01872.

Senthil-Nathan, Sengottayan. 2020. "A Review of Resistance Mechanisms of Synthetic Insecticides and Botanicals, Phytochemicals, and Essential Oils as Alternative Larvicidal Agents Against Mosquitoes." *Frontiers in Physiology* 10 (February): 1591. https://doi.org/10.3389/fphys.2019.01591.

Sharma, Arvind, Michael N. Pham, Jeremiah B. Reyes, Randeep Chana, Won C. Yim, Chan C. Heu, Donghun Kim et al., 2022. "Cas9-Mediated Gene Editing in the Black-Legged Tick, *Ixodes Scapularis*, by Embryo Injection and ReMOT Control." *iScience* 25 (3): 103781. https://doi.org/10.1016/j.isci.2022.103781.

Sheng, Yifeng, Jiani Chen, Hanyu Jiang, Yueqi Lu, Zhi Dong, Lan Pang, Junwei Zhang, Ying Wang, Xuexin Chen, and Jianhua Huang. 2023. "The Vitellogenin Receptor Gene Contributes to Mating and Host-Searching Behaviors in Parasitoid Wasps." *iScience* 26 (4): 106298. https://doi.org/10.1016/j.isci.2023.106298.

Shirai, Yu, Maria-Dolors Piulachs, Xavier Belles, and Takaaki Daimon. 2022. "DIPA-CRISPR Is a Simple and Accessible Method for Insect Gene Editing." *Cell Reports Methods* 2 (5): 100215. https://doi.org/10.1016/j.crmeth.2022.100215.

Singh, Sanchita, Somnath Rahangdale, Shivali Pandita, Gauri Saxena, Santosh Kumar Upadhyay, Geetanjali Mishra, and Praveen C. Verma. 2022. "CRISPR/Cas9 for Insect Pests Management: A Comprehensive Review of Advances and Applications." *Agriculture* 12 (11): 1896.

Sioson, Victor Aaron, Minjong Kim, and Jinmyoung Joo. 2021. "Challenges in Delivery Systems for CRISPR-Based Genome Editing and Opportunities of Nanomedicine." *Biomedical Engineering Letters* 11 (3): 217–233. https://doi.org/10.1007/s13534-021-00199-4.

Souza, Dariane, Shawn A. Christensen, Ke Wu, Lyle Buss, Kaylin Kleckner, Constance Darrisaw, Paul D. Shirk, and Blair D. Siegfried. 2022. "RNAi-Induced Knockdown of White Gene in the Southern Green Stink Bug (Nezara Viridula L.)." *Scientific Reports* 12 (June): 10396. https://doi.org/10.1038/s41598-022-14620-0.

Sun, Dan, Zhaojiang Guo, Yong Liu, and Youjun Zhang. 2017. "Progress and Prospects of CRISPR/Cas Systems in Insects and Other Arthropods." *Frontiers in Physiology* 8: 608.

Teleanu, Raluca Ioana, Adelina-Gabriela Niculescu, Eugenia Roza, Oana Vladâcenco, Alexandru Mihai Grumezescu, and Daniel Mihai Teleanu. 2022. "Neurotransmitters—Key Factors in Neurological and Neurodegenerative Disorders of the Central Nervous System." *International Journal of Molecular Sciences* 23 (11): 5954. https://doi.org/10.3390/ijms23115954.

Tyagi, Shaily, Karthik Kesiraju, Manjesh Saakre, Maniraj Rathinam, Venkat Raman, Debasis Pattanayak, and Rohini Sreevathsa. 2020. "Genome Editing for Resistance to Insect Pests: An Emerging Tool for Crop Improvement." *ACS Omega* 5 (33): 20674–20683. https://doi.org/10.1021/acsomega.0c01435.

Tycko, Josh, Vic E. Myer, and Patrick D. Hsu. 2016. "Methods for Optimizing CRISPR-Cas9 Genome Editing Specificity." *Molecular Cell* 63 (3): 355–370. https://doi.org/10.1016/j.molcel.2016.07.004.

Tyumentseva, Marina, Aleksandr Tyumentsev, and Vasiliy Akimkin. 2023. "CRISPR/Cas9 Landscape: Current State and Future Perspectives." *International Journal of Molecular Sciences* 24 (22): 16077. https://doi.org/10.3390/ijms242216077.

Urban, Jennifer A., Caroline A. Doherty, William T. Jordan, Jacob E. Bliss, Jessica Feng, Marcela M. Soruco, Leila E. Rieder, Maria A. Tsiarli, and Erica N. Larschan. 2017. "The Essential Drosophila CLAMP Protein Differentially Regulates Non-Coding roX RNAs in Male and Females." *Chromosome Research: An International Journal on the Molecular, Supramolecular and Evolutionary Aspects of Chromosome Biology* 25 (2): 101–113. https://doi.org/10.1007/s10577-016-9541-9.

Vallespi, Maribel G., Luis A. Glaria, Osvaldo Reyes, Hilda E. Garay, Joel Ferrero, and Manuel J. Araña. 2000. "A Limulus Antilipopolysaccharide Factor-Derived Peptide Exhibits a New Immunological Activity with Potential Applicability in Infectious Diseases." *Clinical and Diagnostic Laboratory Immunology* 7 (4): 669–675. https://www.ncbi.nlm.nih.gov/pmc/articles/PMC95932/.

Wang, Guan-Hong, Stephanie Gamez, Robyn R. Raban, John M. Marshall, Luke Alphey, Ming Li, Jason L. Rasgon, and Omar S. Akbari. 2021. "Combating Mosquito-Borne Diseases Using Genetic Control Technologies." *Nature Communications* 12 (1): 4388. https://doi.org/10.1038/s41467-021-24654-z.

Wang, Xin, Shun-ming Tang, and Xing-jia Shen. 2014. "Overview of Research on Bombyx Mori microRNA." *Journal of Insect Science* 14 (October): 133. https://doi.org/10.1093/jis/14.1.133.

Xu, Sen, Swatantra Neupane, Hongjun Wang, Thinh Phu Pham, Marelize Snyman, Trung V. Huynh, and Li Wang. 2024. "Efficient CRISPR Genome Editing and Integrative Genomic Analyses Reveal the Mosaicism of Cas-Induced Mutations and Pleiotropic Effects of Scarlet Gene in an Emerging Model System." *bioRxiv*, January, 2024.01.29.577787. https://doi.org/10.1101/2024.01.29.577787.

Xu, Yuanyuan, and Zhanjun Li. 2020. "CRISPR-Cas Systems: Overview, Innovations and Applications in Human Disease Research and Gene Therapy." *Computational and Structural Biotechnology Journal* 18 (September): 2401–2415. https://doi.org/10.1016/j.csbj.2020.08.031.

Yu, Zhongsheng, Hanqing Chen, Jiyong Liu, Hongtao Zhang, Yan Yan, Nannan Zhu, Yawen Guo et al., 2014. "Various Applications of TALEN- and CRISPR/Cas9-Mediated Homologous Recombination to Modify the Drosophila Genome." *Biology Open* 3 (4): 271–280. https://doi.org/10.1242/bio.20147682.

Zhang, Wei, Jianhong Chen, Nemat Keyhani, Zhengyi Zhang, Sai Li, and Yuxian Xia. 2015. "Comparative Transcriptomic Analysis of Immune Responses of the Migratory Locust, Locusta Migratoria, to Challenge by the Fungal Insect Pathogen, Metarhizium Acridum." *BMC Genomics* 16 (October). https://doi.org/10.1186/s12864-015-2089-9.

Zhu, Feifei, Dong Li, Dandan Song, Shuhao Huo, Shangshang Ma, Peng Lü, Xiaoyong Liu, Qin Yao, and Keping Chen. 2020. "Glycoproteome in Silkworm Bombyx Mori and Alteration by BmCPV Infection." *Journal of Proteomics* 222 (June): 103802. https://doi.org/10.1016/j.jprot.2020.103802.

Zhu, Youmin. 2022. "Advances in CRISPR/Cas9." *BioMed Research International* 2022 (September): 9978571. https://doi.org/10.1155/2022/9978571.

Zimmer, Christoph T., William T. Garrood, A. Mirel Puinean, Manuela Eckel-Zimmer, Martin S. Williamson, T. G. Emyr Davies, and Chris Bass. 2016. "A CRISPR/Cas9 Mediated Point Mutation in the Alpha 6 Subunit of the Nicotinic Acetylcholine Receptor Confers Resistance to Spinosad in *Drosophila Melanogaster*." *Insect Biochemistry and Molecular Biology* 73 (June): 62–69. https://doi.org/10.1016/j.ibmb.2016.04.007.

6 CRISPR-Cas Systems in Cancer Biology and Therapeutics

Gh Jeelani Mir, Aarif Ali, Nisar ul Ashraf, Javeed Iqbal Ahmad Bhat, Showkat Ahmad Ganie, Sheikh Bilal Ahmad, and Rabia Hamid

6.1 UNDERSTANDING CANCER

Cancer is a complex disease characterized by uncontrolled cellular proliferation, the ability to invade nearby tissues, and the potential to metastasize to distant sites in the body (Hanahan and Weinberg, 2000). Cancer represents a condition marked by changes in signaling mechanisms and metabolic functions, resulting in the unchecked growth and survival of altered cells. Prior to categorizing cancers as classical disease entities, it is essential to recognize that they may be viewed as a phenomenon characterized by sequential alterations at the molecular, cellular, tissue, and organ levels, which collectively impact the overall physiology of the organism. Cancer initiates when a cell escapes the normal regulatory mechanisms governing cell division and begins to pursue its own proliferation agenda. All cells resulting from the division of this initial ancestral cell and its descendants exhibit similar inappropriate proliferation. A *tumor*, characterized by a cluster of abnormal cells, can either stay confined to the tissue where it originated, known as in situ cancer, or begin to spread to nearby tissues, which is referred to as invasive cancer. An invasive tumor is deemed malignant, and cells that break away from a malignant tumor can enter the bloodstream or lymphatic system, potentially leading to the development of new tumors in other parts of the body, a process known as metastasis. Tumors pose a significant threat to an individual's life when their growth interferes with the functioning of essential tissues and organs. A significant number of genetic mutations have been documented that can cause normal human cells to transform, resulting in tumor development and the onset of cancer (Mroz and James, 2017). In oncological research, numerous techniques are used; however, none of these methods have proven adequate in creating successful treatment options thus far (Bishayee and Block, 2015).

The initiation of the disease remains one of the most captivating aspects of oncology, an area in which our understanding is still limited (Visvader, 2011). A vast amount of data has been collected, revealing various factors that contribute to the process of transformation. Regrettably, these findings have only scratched the surface of the issue. Without a comprehensive understanding of the disease's onset, it is

DOI: 10.1201/9781003537151-6

difficult to identify the precise targets necessary for effective treatment. The onset of cancer is fueled by a range of genetic and epigenetic changes that provide cancer cells with unique traits, known as the "hallmarks of cancer." These hallmarks are critical for the initiation, growth, and progression of tumors, defining the fundamental features that drive cancer development (Hanahan, 2022).

6.2 UNDERSTANDING CANCER GENOMICS

Cancer is primarily a genetic disease caused by mutations that lead to uncontrolled cell growth and prolonged cell survival (Hanahan and Weinberg, 2011). Cancer is driven by a variety of genetic mutations and alterations that disrupt normal cellular functions. Modifications in these genes can influence oncogenes and tumor suppressor genes, resulting in unregulated cell proliferation and persistence (Vogelstein and Kinzler, 2004).

6.2.1 Key Genomic Alterations

The main types of mutations involved in cancer development include the following:

6.2.1.1 Driver Mutations

These are critical mutations that confer a growth advantage to cancer cells, driving the initiation and progression of the disease. Driver mutations often occur in two main types of genes: oncogenes, which promote cell proliferation and survival, and tumor suppressor genes, which normally work to restrain cell division and induce apoptosis (Vogelstein and Kinzler, 2004). When these mutations occur, they disrupt the normal cell regulatory processes, such as control of the cell cycle, programmed cell death, ultimately causing the cells to become cancerous (Stratton et al., 2009).

6.2.1.2 Passenger Mutations

These mutations do not contribute to cancer development but occur as a consequence of the genomic instability inherent in cancer cells. They accumulate alongside driver mutations but do not directly influence the cancer phenotype.

6.2.1.3 Point Mutations

These mutations reflect a change in a solitary nucleotide base within the DNA sequence. Such a small alteration can have significant consequences, including the activation of oncogenes such as Kirsten rat sarcoma viral oncogene homolog (KRAS) and B-Raf proto-oncogene, serine/threonine kinase (BRAF) or the inactivation of tumor suppressor genes (like TP53) (Stratton et al., 2009). When an oncogene is activated by a point mutation, it can result in the production of proteins that drive excessive cell growth and division. Conversely, a point mutation that inactivates a tumor suppressor gene can prevent the production of proteins that normally inhibit cell proliferation or promote cell death, allowing cells to multiply unchecked.

6.2.1.4 Insertions and Deletions (Indels)

The occurrence of mutations is linked to the insertion or deletion of small DNA segments, which disrupts the coding sequences or regulatory domains of genes.

Indels can cause a shift in the reading frame of the genetic code (frameshift mutations), which often results in the production of malfunctioning proteins (Hanahan and Weinberg, 2011). This disruption can alter cellular processes, promoting cancer development by either enhancing oncogenic pathways or diminishing the protective effects of tumor suppressor genes (Stratton et al., 2009).

6.2.1.5 Copy Number Variations (CNVs)

These genetic alterations involve changes in the number of copies of a particular gene or genomic region. In cancer cells, CNVs often manifest as amplifications of oncogenes such as Myelocytomatosis oncogene (MYC) or deletions of tumor suppressor genes like Phosphatase and tensin homolog (PTEN) (Vogelstein and Kinzler, 2004). This imbalance leads to an overproduction of proteins that promote cell growth and survival, while reducing the availability of proteins that would typically restrain these processes, thereby facilitating tumor progression.

6.2.1.6 Structural Variations

Large-scale rearrangements of the genome, including translocations, inversions, and gene fusions, can significantly impact cancer development. For example, the translocation between chromosomes 9 and 22 creates the Breakpoint Cluster Region-Abelson (BCR-ABL) fusion gene, which produces a hybrid protein with oncogenic properties that drive chronic myeloid leukemia (Stratton et al., 2009). These structural changes can create new, abnormal proteins or disrupt normal gene function, contributing to uncontrolled cell growth and cancer progression (Hanahan and Weinberg, 2011).

6.2.1.7 Epigenetic Alterations

Epigenetic changes, in contrast to genetic mutations, do not modify the DNA sequence directly; instead, they involve alterations in DNA methylation and histone modifications (Jones and Baylin, 2007). By altering the regulation of genes without changing their underlying sequence, epigenetic modifications can lead to cancer development, thereby making them an important area of study in cancer research.

Together, these genetic and epigenetic alterations create a landscape that favors the uncontrolled proliferation and survival of cancer cells, underscoring the complexity and adaptability of cancer as a disease.

6.2.2 Genomic Heterogeneity in Cancer

Cancer exhibits significant genomic heterogeneity, both between different tumors (intertumor heterogeneity) and within a single tumor (intratumor heterogeneity) (Gerlinger et al., 2012). This diversity is driven by several key factors.

6.2.2.1 Clonal Evolution

Cancer cells undergo clonal selection, where cells harboring advantageous mutations proliferate more rapidly than others (Nowell, 1976). Over time, this results in the emergence of multiple subclones within a tumor, each possessing unique genetic alterations, thereby contributing to intratumor heterogeneity (McGranahan and Swanton, 2017).

6.2.2.2 Tumor Microenvironment

In shaping the diversity of the genome, an important role is played by the relationship between cancer cells and the surrounding microenvironment, which includes immune cells, stromal cells, and extracellular matrix components (Quail and Joyce, 2013). These interactions can promote tumor progression and influence the genetic landscape of cancer cells, further enhancing heterogeneity (Junttila and de Sauvage, 2013).

6.2.2.3 Therapeutic Pressure

Exposure to chemotherapy, radiation, or targeted therapies can create selective pressure that favors the survival of resistant cancer cell clones. As a result, treatment-resistant subpopulations can emerge and proliferate, complicating the treatment landscape and contributing to tumor recurrence and progression (Turner and Reis-Filho, 2006).

6.2.3 Clinical Implications

The comprehension of the cancer genomic landscape has significant implications for clinical practice, influencing various aspects of cancer care.

6.2.3.1 Diagnosis and Prognosis

Genomic profiling helps in the identification of particular mutations and biomarkers linked to different cancer types, facilitating more accurate diagnoses and providing valuable prognostic information about the likely course and outcome of the disease (Ding et al., 2008).

6.2.3.2 Targeted Therapies

By pinpointing driver mutations and their associated signaling pathways, it becomes possible to develop targeted therapies that particularly inhibit the cancer-promoting proteins or pathways involved.

6.2.3.3 Personalized Medicine

Utilizing genomic data allows for the customization of cancer treatments to match the unique genetic makeup of an individual's tumor. This personalized treatment aims to increase treatment effectiveness and reduce side effects by tailoring therapeutic strategies to the specific molecular characteristics of each patient's cancer (Garraway et al., 2013).

6.2.3.4 Resistance Mechanisms

Understanding the genetic underpinnings of resistance to therapies can guide the development of combination treatments or new drugs designed to overcome these resistance mechanisms, ultimately improving patient outcomes (Holohan et al., 2013).

6.3 MECHANISMS BY WHICH MUTATIONS DRIVE CANCER DEVELOPMENT

Genetic mutations are central to the process of cancer development as they disrupt key cellular functions and pathways that maintain normal cell behavior. These mutations can lead to malignancy by altering several fundamental biological processes.

6.3.1 Activation of Oncogenes

Oncogenes are mutated or overexpressed versions of normal genes, known as proto-oncogenes. These genes typically regulate cell growth and division. When mutations occur, they can result in the constant activation or overexpression of these genes, leading to uncontrolled cell proliferation and survival. This relentless cellular growth is a hallmark of cancer. Common examples of oncogenes that become activated through mutations include Rat sarcoma (RAS), which controls cell signaling pathways for growth; BRAF, involved in cell division and differentiation; and epidermal growth factor receptor (EGFR), a receptor that, when activated, triggers pathways promoting cell survival and proliferation (Yarden and Sliwkowski, 2001).

6.3.2 Inactivation of Tumor Suppressor Genes

Tumor suppressor genes are crucial for maintaining normal cellular functions, such as regulating the cell cycle, initiating apoptosis (programmed cell death), and ensuring DNA repair and genomic stability. When mutations inactivate these genes, the cell loses its ability to control growth and division properly, resulting in unregulated proliferation and potential transformation into cancer cells. Notable examples of tumor suppressor genes include TP53, often referred to as the "guardian of the genome" due to its role in DNA repair and apoptosis; retinoblastoma 1 (RB1), which regulates progression of cell cycle; and PTEN, which controls cell growth by inhibiting signaling pathways (Li et al., 1997b).

6.3.3 Disruption of DNA Repair Mechanisms

Genes responsible for repairing DNA damage are vital for maintaining genomic integrity. Mutations in these genes can severely impair the DNA repair process, leading to the accumulation of additional mutations throughout the genome. This accumulation increases genomic instability, creating an environment conducive to cancer development. For example, mutations in Breast cancer gene 1 (BRCA1) and Breast cancer gene 2 (BRCA2) genes, which are integral to the homologous recombination repair pathway, lead to inefficient repair of double-stranded DNA breaks, increasing the risk of developing cancers, such as breast and ovarian cancer (Roy et al., 2012).

6.3.4 Evasion of Apoptosis

Apoptosis is a natural cellular process designed to eliminate damaged, dysfunctional, or potentially dangerous cells. Mutations that impair the apoptotic pathways allow cells to evade programmed cell death, even when they carry significant genetic damage. This evasion contributes to tumor growth and survival. For instance, mutations in the TP53 gene, which is crucial for initiating apoptosis, are generally observed in various cancers allowing cells to survive and proliferate unchecked (Vousden and Lane, 2007).

6.3.5 Promotion of Angiogenesis and Metastasis

Mutations may also improve the capacity of cancer cells to develop new blood vessels (angiogenesis) and disseminate to other regions of the body (metastasis). Angiogenesis is crucial for delivering oxygen and nutrients to the swiftly expanding tumor mass. Mutations in genes such as vascular endothelial growth factor A (VEGFA) can lead to the secretion of proteins that promote blood vessel formation around the tumor, thereby sustaining its growth.

6.4 OVERVIEW OF TRADITIONAL CANCER THERAPIES

The approach to cancer treatment has been characterized by significant complexity. The primary aim of cancer treatment is to eradicate or diminish the cancerous growth and to inhibit its progression to other areas of the body. Long-standing treatment options, such as surgery, chemotherapy, and radiotherapy, have been widely implemented.

6.4.1 Surgical Intervention

Surgical intervention is a cornerstone of cancer treatment that involves the physical excision of the tumor mass. During this procedure, not only is the tumor itself removed, but often a margin of surrounding healthy tissue is also taken out to ensure that any microscopic cancer cells that may have spread are also eliminated. This approach is particularly effective for treating localized solid tumors, where the cancer is confined to a specific area (Fisher, 1977). When a tumor is completely resected, surgery can be potentially curative, providing a high chance of long-term remission or even a cure. The effectiveness of surgery diminishes when dealing with metastatic cancers, where cancer cells have spread to other parts of the body beyond the primary site. In such cases, surgery alone is often insufficient because it cannot address the widespread nature of the disease. Furthermore, surgical procedures come with a range of potential risks and complications, such as infection, bleeding, and unintended damage to surrounding healthy tissues or organs. There is also the possibility of postoperative pain and a lengthy recovery period, which can impact the patient's overall quality of life. Thus, while surgical treatment remains a fundamental option in the management of cancer, particularly for early-stage and localized tumors, it must be carefully considered within the broader context of each patient's overall treatment plan (Kufe et al., 2003).

6.4.2 Radiation Therapy

The process of radiation therapy employs high-energy radiation sources, such as X-rays, gamma rays, or proton beams, to inflict DNA damage in cancer cells, ultimately leading to their destruction. This treatment is highly effective for targeting localized tumors, as it can precisely focus radiation on the tumor site, maximizing damage to cancer cells while minimizing exposure to surrounding tissues (Park et al., 2006). Radiation therapy is often used in combination with other treatment modalities, such as surgery, to remove residual cancer cells, or with chemotherapy, to enhance the overall therapeutic effect (Delaney et al., 2005). However, there are notable limitations to radiation therapy. Despite its precision, nearby healthy tissues can still be affected by the radiation, potentially resulting in a range of side effects, from mild skin irritation to more severe complications, depending on the location and dose of radiation (Joiner and van der KogeJ, 2025). Additionally, radiation therapy is generally less effective against cancers that have spread throughout the body, as it is designed to target specific areas rather than treat systemic disease. As a result, its role is often limited to controlling localized tumors or relieving symptoms in cases of advanced metastatic cancer (Halperin et al., 2018).

6.4.3 Chemotherapy

This method involves the use of powerful drugs designed to target and destroy cells that divide rapidly, making it effective in treating both primary tumors and cancers that have metastasized. This systemic approach allows chemotherapy to reach cancer cells throughout the body. However, because these drugs are not selective solely for cancer cells, they also affect healthy cells that undergo frequent division, such as those in the immune system, digestive tract, and hair follicles. This results in a range of side effects, including immunosuppression, gastrointestinal discomfort, and hair loss. Another significant challenge with chemotherapy is that cancerous cells develop drug resistance, reducing the treatment's effectiveness over time and complicating the ongoing management of the disease (Kanno et al., 2021).

6.5 LIMITATIONS OF TRADITIONAL CANCER TREATMENTS

6.5.1 Toxicity and Side Effects

Traditional cancer therapies like chemotherapy and radiation primarily target cells that are rapidly dividing. While this approach can effectively eliminate cancer cells, it also impacts normal, fast-dividing cells, such as those found in the bone marrow, digestive system, and hair follicles. This non-targeted action leads to numerous side effects.

6.5.1.1 Bone Marrow Suppression

The production of blood cells can be reduced, leading to conditions such as anemia (low red blood cell number), heightened susceptibility to infections (low white blood cell number), and bleeding complications (low platelet number).

6.5.1.2 Gastrointestinal Problems

Damage to the lining of the digestive tract by these treatments can lead to nausea, vomiting, diarrhea, and mucositis (inflammation of the mucous membranes).

6.5.1.3 Hair Loss

The high turnover rate of hair follicle cells makes them vulnerable to chemotherapy, often causing hair loss (alopecia).

6.5.1.4 Fatigue

This common side effect arises due to the body's response to both cancer and its treatment, and it is frequently worsened by anemia and metabolic changes. The presence of these side effects can significantly reduce a patient's overall quality of life, limiting the dosage and duration of treatment that can be safely administered.

6.5.2 Resistance

The mechanisms by which cancer cells become resistant to therapies can be diverse, with some being inherent and others developing over time. Spontaneous mutations can occur in cancer cells, leading to changes that render treatments ineffective. Cancer cells can adapt by activating alternative signaling pathways that bypass the blocked pathway. Certain cancer cells can enhance the expression of efflux pumps, which are proteins that actively expel chemotherapy drugs from within the cells. This mechanism lowers the intracellular concentration of the drugs, thereby diminishing their therapeutic effectiveness (Robey et al., 2008). These resistance mechanisms can emerge during treatment, leading to relapse and progression of the disease. Overcoming resistance often requires combining multiple drugs or developing new therapies that target the resistant pathways (Kanno et al., 2021).

6.5.3 Tumor Heterogeneity

Tumors are inherently heterogeneous, consisting of diverse populations of cells that exhibit distinct genetic and phenotypic traits. This heterogeneity manifests on several levels.

6.5.3.1 Intertumoral Heterogeneity

This refers to the differences that exist between tumors in different patients, or even between a primary tumor and its metastases within the same patient. Such variability can lead to distinct tumor characteristics, varying responses to therapies, and differing outcomes or prognoses (Gerlinger et al., 2012).

6.5.3.2 Intratumoral Heterogeneity

Variability exists within a single tumor, where subpopulations of cells may have different mutations, growth rates, and responses to treatment (Burrell et al., 2013). This diversity means that while some cells in a tumor may be susceptible to a particular treatment, others may be resistant. This can lead to incomplete eradication of the tumor and subsequent regrowth of resistant cell populations. Effective treatment strategies must therefore address this heterogeneity, often requiring combination therapies that target multiple pathways simultaneously (McGranahan and Swanton, 2017).

6.5.4 Limited Efficacy in Metastatic Disease

Metastasis, which involves the spread of cancer cells to distant parts of the body, accounts for most cancer-related fatalities. Conventional treatments frequently face challenges in effectively targeting metastatic cancer due to several factors.

6.5.4.1 Physical Barriers

Metastatic tumors may be located in sites that are difficult to reach with surgery or localized radiation therapy (Fidler, 2003).

6.5.4.2 Systemic Spread

Metastatic cancer cells can disseminate through the bloodstream or lymphatic system, establishing new tumors in multiple locations, making it challenging to target all sites effectively (Massague and Obenauf, 2016).

6.5.4.3 Biological Differences

Metastatic cells often exhibit different characteristics compared to the primary tumor, including increased resistance to apoptosis, enhanced migratory abilities, and different interactions with the surrounding tissue microenvironment (Hanahan and Weinberg, 2011). These factors contribute to the difficulty in achieving long-term control or cure in patients with metastatic cancer using traditional therapies.

6.5.5 Tumor Microenvironment

The tumor microenvironment (TME) is composed of a diverse array of non-cancerous cells, including stromal cells, immune cells, and blood vessels, and elements of the extracellular matrix. These components interact closely with cancer cells, playing a crucial role in promoting tumor growth and survival. The TME can significantly affect the effectiveness of treatments in multiple ways:

6.5.5.1 Protective Niche

Stromal cells and other components of the extracellular matrix can create a physical barrier that protects cancer cells from therapeutic agents, making them less accessible to treatment.

6.5.5.2 Signaling Interactions

Cancer cells can attract and modify immune cells to establish an immunosuppressive environment, allowing tumors to grow unchecked and evade immune detection. For example, tumor-associated macrophages (TAMs) can release factors that encourage tumor progression and inhibit anti-tumor immune responses.

6.5.5.3 Metabolic Support

The TME can provide cancer cells with essential nutrients and oxygen, supporting their survival and proliferation even when they are subjected to treatment-induced stress (Bergers and Benjamin, 2003). A comprehensive understanding of the TME is crucial for enhancing the effectiveness of cancer therapies. Approaches such as inhibiting angiogenesis, modulating the immune response, and disrupting stromal cell support are being investigated to improve the efficacy of conventional treatments and to combat therapeutic resistance (Hanahan and Weinberg, 2011).

6.6 ADVANCED AND INNOVATIVE CANCER THERAPIES

In cancer treatment, a significant obstacle is the problem of drug resistance and the effectiveness of delivery systems, which hinder the ability to cure cancer and alleviate its symptoms. However, there are now numerous approved treatment options and medications available. The effectiveness of traditional cancer therapies is often compromised by tumor pathology and the structural irregularities of blood vessels within the tumor tissue. The following outlines various cutting-edge and groundbreaking cancer therapies, along with their respective advantages and challenges.

6.6.1 Stem Cell Therapy

The practice of stem cell therapy utilizes undifferentiated cells found in the bone marrow, which possess the capability to transform into various cell types within the body. This therapeutic approach is recognized as a viable treatment option for cancer, demonstrating both safety and efficacy. At present, the use of stem cells is largely confined to experimental settings, with numerous clinical trials exploring their capacity to regenerate damaged tissues. Mesenchymal stem cells (MSCs), derived from sources such as bone marrow, adipose tissue, and connective tissues, are being actively employed in these investigations.

6.6.2 Pluripotent Stem Cells

Derived from the inner cell mass of embryos, pluripotent stem cells, especially embryonic stem cells (ESCs), are uniquely capable of developing into nearly all cell types, with the exception of those forming the placenta. The landmark discovery of Yamanaka factors in 2006, which allow somatic cells to be reprogrammed into induced pluripotent stem cells (iPSCs), has marked a significant milestone in the field of cell biology (Takahashi and Yamanaka, 2006). This approach avoids the ethical complications associated with embryo destruction, as iPSCs and ESCs share similar traits. Currently, hematopoietic embryonic stem cells (hESCs) and induced pluripotent stem cells (iPSCs) are employed in the generation of effector T cells and natural killer (NK) cells, in addition to the development of anti-tumor vaccines (Ouyang et al., 2019).

6.6.3 Adult Stem Cells

In cancer treatment, adult stem cells (ASCs) that are often used include hematopoietic stem cells (HSCs), mesenchymal stem cells (MSCs), and neural stem cells (NSCs). HSCs, located in the bone marrow, possess the ability to develop into various mature blood cell types. Currently, the only FDA-approved application involves the infusion of HSCs sourced from cord blood for the treatment of multiple myeloma and leukemia (Trujillo-Alonso et al., 2019). Due to their unique biological properties, MSCs are often used in conjunction with other therapeutic strategies for tumor treatment (Lin et al., 2019). With their capability for self-renewal and the production of new neurons and glial cells, NSCs are being explored for their potential in treating primary and metastatic breast tumors, among other diseases (Chu et al., 2020).

6.6.4 Cancer Stem Cells

Through epigenetic mutations, normal stem cells or precursor/progenitor cells can transform into cancer stem cells (CSCs). Their contribution in the treatment of tumor is significant, as they contribute to cancer growth, metastasis, and recurrence, offering potential in the management of solid tumors. Stem cells employ various mechanisms to combat tumors. One such mechanism is the homing process, where HSCs rapidly migrate to specific stem cell niches in the bone marrow, followed by engraftment and differentiation into particular blood cells. Another mechanism is the tumor-tropic effect, where MSCs migrate toward the TME, attracted by factors such as IL-6, CXCL16, CCL-25, and SDF-1, respectively. This migration leads to the differentiation of MSCs within the tumor, aiding in the development of tumor stroma (Sullivan et al., 2020).

6.6.5 Targeted Drug Therapy

This approach involves drugs/substances that function by disrupting the activity of growth molecules, thus inhibiting cancer growth and metastasis. The TM is pivotal in determining tumor initiation and progression. These TM-forming cells engage in complex signaling interactions with cancer cells, fostering an environment that supports robust cellular proliferation. A tremendous interest in exploring TM conditions to facilitate more effective cancer targeting strategies has been found (Adjei and Blanka, 2015). Targeted therapies stand apart from conventional chemotherapy by specifically aiming at cancer cells, which helps protect healthy cells and leads to improved treatment outcomes and a reduction in side effects. Imatinib, approved by the US Food and Drug Administration in 2001, is the first small tyrosine kinase inhibitor (TKI) molecule and has marked a pivotal moment, leading to a surge in the development of targeted therapies (Savage and Antman, 2002). Over the last two decades, there has been a notable rise in the number of targeted cancer drug treatments approved by FDA.

6.6.6 Gene Therapy

The process of gene therapy entails the incorporation of a normal gene into the genome to target and treat a specific genetic disorder. There is growing interest in gene therapy as a viable and promising research domain, offering an alternative approach to cancer treatment by delivering therapeutic genes into cancer cells or tissues. This method aims to induce cell death or inhibit cancer growth. Various strategies are employed, including the activation of beneficial genes and the silencing of harmful or defective ones, leading to the exploration of a diverse range of genes and both viral and non-viral vectors in ongoing studies. The delivery of the thymidine kinase (TK) gene proves to be an effective method for administering the prodrug ganciclovir, facilitating its activation and promoting targeted cytotoxic effects (Freeman et al., 1993). When used in combination with radiotherapy, Gendicine, a recombinant adenovirus that contains the wild-type p53 gene, has exhibited remarkable efficacy in facilitating complete disease regression in cases of head and neck squamous cell carcinoma (Raty et al., 2008).

6.7 POTENTIAL APPLICATIONS OF CRISPR-CAS9 TECHNOLOGY IN CANCER THERAPY

Cancer development and progression are driven by mutations and the improper expression of a range of genes, such as oncogenes, tumor suppressor genes, chemoresistance-related genes, metabolism-associated genes, and genes linked to cancer stem cells. The CRISPR-Cas9 technology holds significant promise for cancer treatment by facilitating accurate gene editing within cancer cells. This capability enables researchers to target specific oncogenes responsible for tumor growth, improve the effectiveness of immunotherapy through the modification of immune cells, discover new therapeutic targets, and create personalized cancer therapies by rectifying genetic mutations specific to an individual's tumor. The key applications of CRISPR-Cas9 technology in cancer treatment are mentioned next.

6.7.1 Gene Knockout/In Studies

The CRISPR-Cas system has revolutionized the field of functional genomics, enabling precise and targeted modifications of the genome. This tool is crucial for studying gene functions and understanding the molecular underpinnings of cancer. Gene knockout and knock-in experiments play a vital role in exploring how specific genes contribute to cancer development, progression, and resistance to therapies.

6.7.1.1 Gene Knockout Studies

Gene knockout involves completely disabling a gene's function, typically achieved by inducing double-strand breaks (DSBs) in DNA. These breaks are repaired by the error-prone non-homologous end joining (NHEJ) pathway, leading to insertions or deletions (indels) that disrupt the gene's coding sequence, effectively knocking it out (Doudna and Charpentier, 2014). Oncogenes exhibit distinct regulatory mechanisms compared to normal genes and have the potential to induce malignant transformation in normal cells. The CRISPR-Cas9 system serves as an effective tool for deletion, expression modulation, and alteration of oncogenes activity, which can subsequently hinder tumor growth. Knockout studies help determine the significance of particular genes in cancer cell survival and proliferation. For example, knocking out the KRAS gene in cells with a KRAS mutation can reveal its role in driving cancer (Shalem et al., 2014). By knocking out single or multiple genes, researchers can explore whether different genes or pathways can compensate for each other's function (Wang et al., 2015). Disrupting genes involved in key signaling pathways allows scientists to uncover the molecular processes that drive cancer progression.

For instance, knocking out genes in the phosphatidylinositol 3-kinase and protein kinase B (PI3K/AKT) pathway can shed light on their roles in promoting cell growth and survival (Barrangou et al., 2007). Knockout studies can identify potential therapeutic targets by showing that certain genes are critical for cancer cell survival, as knocking out Bromodomain containing 4 (BRD4), a key regulator of transcription, has shown promise in inhibiting the growth of various cancers (Kleinstiver et al., 2016). Moreover, creating knockout models allows to test how cancer cells respond to specific drugs, helping to develop targeted therapies that are more effective and have fewer side effects (Cox et al., 2017). These findings suggest that the CRISPR-Cas9 gene-editing system serves as a powerful method for identifying oncogenes and assessing the therapeutic potential of targeting these genes.

In the context of cancer, the inactivation of tumor suppressor genes is a prominent aspect of both its initiation and progression (Chen et al., 2007). Their inactivation, loss, or alteration can trigger the activation of oncogenes, resulting in the onset and advancement of tumors (Morris and Chan, 2015). Notably, the emergence of CRISPR-Cas9 technology has dramatically changed the landscape of cancer research by allowing for the quick validation of tumor suppressor genes in both laboratory and living organism studies.

6.7.1.2 Gene Knock-In Studies

Gene knock-in techniques involve the precise addition of genetic sequences into specific locations within the genome. This is typically achieved through homology-directed repair (HDR) following CRISPR-induced DSBs. Knock-in strategies allow researchers to introduce specific mutations, reporter genes, or therapeutic genes into the genome, providing valuable insights into gene function and therapeutic potential (Richardson et al., 2016). The gene knock-in approach has been useful to create cancer models, study gene functions and therapeutic development, and explored resistance to drugs.

6.7.2 Oncolytic Viral Therapy

Oncolytic viruses (OVs) are specialized viruses that target and eliminate cancer cells while sparing healthy tissues (Zheng et al., 2019). These viruses replicate within the infected cancer cells, generating new viral offspring, and stimulate immune responses that specifically attack uninfected cancer cells (Russel et al., 2012). Consequently, oncolytic virotherapy offers a multifaceted approach to effectively and selectively eradicate cancer cells. CRISPR-Cas9 technology is also being harnessed to create oncolytic viruses. These viruses are engineered to include CRISPR elements that can target and disrupt oncogenes within cancer cells. The viruses that are most frequently researched for therapeutic purposes include mumps, herpes simplex virus (HSV), retroviruses, adenoviruses, and vaccinia viruses (Chulpanova et al., 2018). Modifications are made to these viruses to improve their effectiveness and ability to generate an immune response. In cancer therapy, the efficacy of OVs has been reported by various clinical and preclinical research studies (Au et al., 2005). The OVs engineered by the CRISPR-Cas9 method have boosted immune responses and also showed enhanced tumor selectivity.

6.7.3 Cancer Immunotherapy

The incorporation of the CRISPR-Cas approach into immunotherapy has transformed the arena, enabling the development of highly targeted cancer treatments. By allowing for precise editing of immune cells, CRISPR-Cas significantly boosts the effectiveness and safety of cancer immunotherapies. Immunotherapy has produced effective outcomes in tumor treatment, making it a promising and innovative therapeutic strategy. The CRISPR-Cas9 has widespread applicability, particularly in the development of chimeric antigen receptor T (CAR-T) cells for gene therapy. CAR-T therapy, which involves the modification of patients' T cells to precisely target cancer cells, has been developed as a potential treatment, especially for blood cancers. Traditionally, isolating the CAR-T cells from the individual's T cells was an expensive and time-consuming process and was unsuitable for patients with low T cell counts. The CRISPR-Cas9 technology overcomes these challenges by allowing the engineering of T cells from donors that can be used across multiple patients. By disrupting genes like the T cell receptor (TCR) and beta-2-microglobulin (B2M), scientists have created CAR-T cells that significantly reduce the risk of graft-versus-host disease (GVHD). These modifications enable universal CAR-T cells to be administered to various patients without the risk of immune rejection, providing a more scalable and accessible treatment option (Dimitri et al., 2022). CRISPR-Cas9 also enhances the anti-tumor potency of CAR-T cells by enabling the deletion of inhibitory receptors that tumors use to suppress immune responses. Additionally, CRISPR-Cas allows the introduction of genetic modifications into the CAR-T cells to enhance the metabolic resilience, permitting them to sustain their anti-tumor activity over extended periods. Looking forward, ongoing research and clinical trials are expected to further refine these genetic modifications and explore new targets to improve CAR-T cell therapy.

6.7.4 Amplifying Immune System Potency against Tumors

CRISPR-Cas9 technology extends its impact beyond CAR-T cells by enhancing the broader immune response against tumors. This is achieved by reprogramming various immune cells, such as natural killer (NK) cells and macrophages, to more effectively target and eliminate cancer cells. In NK cell therapy, CRISPR-Cas9 has been instrumental in improving both the functionality and survivability of NK cells. Additionally, editing the CD38 gene reduces NK cell fratricide, which significantly increases their effectiveness in targeting acute myeloid leukemia (AML) (Elahi et al., 2021). Macrophages, another key player in the immune system, can also be genetically modified using CRISPR-Cas9 to overcome tumor-induced immune suppression. By deleting inhibitory receptors or altering signaling pathways, CRISPR-Cas9-modified macrophages can more effectively phagocytose and destroy tumor cells, thereby amplifying the overall anti-tumor immune response. This strategy holds considerable potential for enhancing macrophage-based therapies, especially in cancers that are adept at evading the immune system. The integration of CRISPR-Cas9 into cancer immunotherapy marks a substantial evolution in cancer treatment strategies. By leveraging precise gene-editing capabilities, CRISPR-Cas9 not only improves the effectiveness and safety of existing therapies but also paves the way for novel treatment approaches. As research advances and clinical trials continue, CRISPR-Cas9 is poised to become an integral part of cancer immunotherapy, offering new hope to patients worldwide.

6.7.5 Editing Cancer Epigenome

The CRISPR-Cas9 epigenome tool for editing can be used to address cancers (Nakamura et al., 2021). This procedure comprises of the combination of Cas9 with a transcription activator to enhance gene expression, or a repressor to diminish it. Within the cancerous cells, genetic mutations in tumor suppressor genes and oncogenes have been reported by studies (Hilton et al., 2015). The CRISPR-Cas9 is adept at creating both gain-of-function and loss-of-function mutations, which assists in uncovering the genes responsible for various cancers. By detecting and targeting these genes, it is possible to exert control over the related cancers (Thakore et al., 2015). The Steroid receptor coactivator-1 (SRC-1) gene, a key regulator in estrogen receptor–positive tumors, significantly influences the development of metastatic disease. SRC-1 collaborates with estrogen receptors to manage a web of genes associated with cancer cell differentiation and proliferation (Sarvari et al., 2022). The use of CRISPR-Cas9 to silence SRC-1 results in the diminished expression of genes involved in these critical processes, thus providing an effective means to control breast cancer progression and tumor metastasis (Pei et al., 2020).

6.7.6 High-Throughput CRISPR-Cas Screening for Identifying Cancer Driver Genes

High-throughput screening (HTS) using CRISPR-Cas technology has revolutionized cancer genomics by enabling comprehensive exploration of the genome to uncover cancer driver genes—those critical genes whose mutations promote tumor development and progression. This approach allows systematical identification of essential genes, enhancing our knowledge of cancer biology and paving way for the development of novel therapeutic targets. High-throughput CRISPR screening involves the simultaneous alteration of thousands of genes within a huge cell population. This process relies on libraries of sgRNAs, each precisely designed to target a specific gene, enabling broad or targeted screening across the genome. The success of a CRISPR screen begins with a meticulously designed sgRNA library. Once designed, sgRNAs are synthesized and cloned into vectors suitable for cellular delivery. Lentiviral vectors are frequently used due to their efficiency in transducing cells and their capability to stably merge into the genome of the host, ensuring prolonged sgRNA expression (Wang et al., 2013; Jinek et al., 2012). The sgRNA library is delivered into the selected cancer cells through viral transduction. Over time, cells containing sgRNAs that target genes essential for survival or proliferation will decrease in number, while those targeting non-essential genes will persist or increase if they provide a growth advantage (Marraffini and Sontheimer, 2008). Advanced bioinformatics tools are employed to analyze the sequencing data, identifying sgRNAs that are significantly depleted or enriched compared to the initial library. Depletion indicates that the targeted gene is crucial for survival or proliferation, while enrichment suggests a role in drug resistance or growth advantage (Kim and Kim, 2014). High-throughput CRISPR screens are instrumental in identifying genes that are needed by cancerous cells for their survival and growth. Targeting unique metabolic pathways in cancer cells can reveal vulnerabilities absent in normal cells, offering therapeutic opportunities (Sander and Joung, 2014). Similarly, synthetic lethality screens identify gene pairs where the simultaneous loss leads to cell death, particularly valuable in targeting cancer-specific mutations. In addition, HTS can reveal genes whose loss confers resistance to chemotherapy or targeted therapies. HTS can also uncover pathways involved in adaptive resistance, where cancer cells evolve under drug pressure. Identifying these pathways helps in designing strategies to prevent or delay resistance (Shen et al., 2013). Moreover, by identifying genes crucial for cancer cell viability, CRISPR screens aid in mapping key oncogenic pathways.

6.7.7 CRISPR-Cas Technology for Organoid and 3D Culture Systems in Cancer Research

Traditional two-dimensional (2D) cell culture systems have long been used in cancer research, but their inability to fully replicate the complexity of human tumors has been a significant limitation (Jinek et al., 2012). Organoids and three-dimensional (3D) culture systems have emerged as more sophisticated tools, providing a more accurate representation of tumor biology (Nishimasu et al., 2014). The integration of CRISPR-Cas technology into these systems has further expanded their capabilities, allowing for precise genetic modifications that result in models more closely resembling human cancers (Wang et al., 2013). Similarly, 3D culture systems allow cells to grow within a matrix or scaffold that mirrors the extracellular environment, facilitating the formation of complex, multicellular structures that better represent physiological conditions (Mali et al., 2013). CRISPR-Cas has transformed genome editing with its precision and efficiency, enabling targeted genetic alterations in various organisms (Doudna and Charpentier, 2014). When combined with organoid and 3D culture systems, CRISPR-Cas technology allows for the direct engineering of genetic mutations within these complex tissue models. This approach leads to the creation of genetically precise cancer models that more accurately reflect the genetic and phenotypic diversity seen in patient tumors. CRISPR-engineered organoids and 3D culture systems effectively capture the phenotypic and genotypic variety of the tumors in humans (Clevers, 2016). These models are invaluable for studying cell-to-cell interactions, clonal evolution, and responses to therapies in a controlled environment. Also, these advanced culture systems enable comprehensive studies into the roles of specific genes and genetic pathways in cancer progression. Moreover, CRISPR-modified organoids and 3D cultures are used as preclinical models for evaluating drug effectiveness, toxicity, and mechanisms of resistance. Their relevance to clinical conditions makes them powerful tools for guiding personalized treatment strategies.

6.7.8 Gene Modulation Studies

CRISPR interference (CRISPRi) and CRISPR activation (CRISPRa) denote the advanced adaptations of the CRISPR-Cas9 system, designed to regulate gene expression with precision while leaving the DNA sequence unaltered. These techniques allow to selectively suppress or enhance the activity of specific genes, offering powerful tools for exploring gene functions, understanding regulatory pathways, and developing new therapeutic approaches in cancer research. Both CRISPRi and CRISPRa utilize dCas9, which can bind to targeted DNA sites without causing DSBs. The dCas9 is directed to precise genomic locations by sgRNAs, operating in a manner akin to the traditional CRISPR-Cas9 system. In CRISPRi, dCas9 is coupled with a transcriptional repressor domain, such as the Kruppel-associated box (KRAB). The dCas9–KRAB complex, directed to a gene's promoter or regulatory region by an sgRNA, binds to the DNA and recruits repressive chromatin modifiers, leading to transcriptional repression (Komor et al., 2016). The recruitment of KRAB, and its associated repressive machinery, results in the formation of heterochromatin, effectively blocking transcription initiation or elongation and leading to reversible gene knockdown (Kim and Kim, 2014). The applications of CRISPRi and CRISPRa are graphically shown in **Figure 6.1** and **Figure 6.2**, respectively.

6.7.9 CRISPR-Cas for Cancer Model Development

The development of cancer cell lines has long been a cornerstone in cancer research, offering vital platforms for exploring tumor biology and testing therapeutic strategies. Traditionally, cancerous cell lines are obtained from patient tumors through methods such as immortalization and transformation, which often results in cell populations with significant genetic diversity that does not fully represent the intricate mutational landscape of tumors. However, the advent of CRISPR-Cas technology has dramatically enhanced the creation of cancer cell line models, enabling precise genetic modifications that accurately reflect the mutations found in human cancers. The cancerous cell lines are crucial tools in the study of cancer, allowing researchers to investigate key aspects of tumor development, oncogenic processes, and responses to treatment. Historically, these cell lines are generated by culturing tumor cells

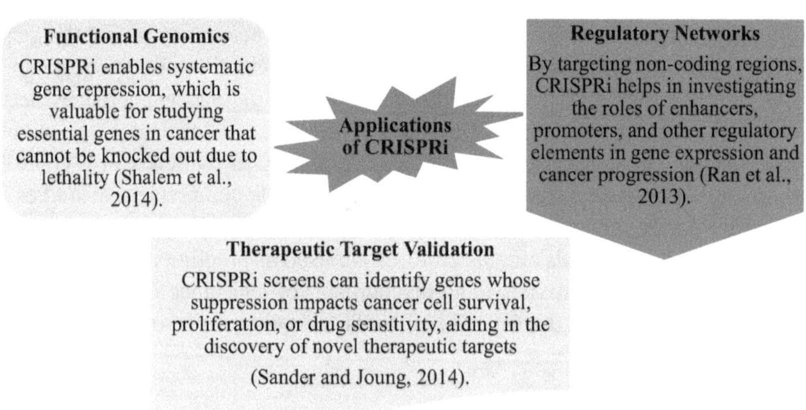

FIGURE 6.1 Applications of CRISPRi in cancer research.

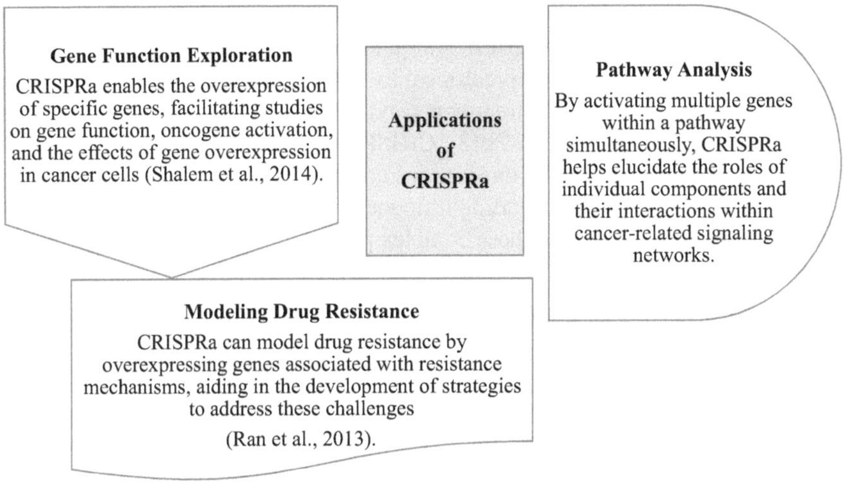

FIGURE 6.2 Applications of CRISPRa in cancer research.

extracted from patients, leading to heterogeneous populations with varying genetic backgrounds. This heterogeneity often reduced the translational relevance of these models, prompting the need for more refined techniques to create *in vitro* models that more accurately represent human cancers. CRISPR-Cas-mediated editing allows for the precise introduction of specific mutations observed in patient tumors into cancer cell lines. This enables the creation of models that more accurately represent tumor development and progression, facilitating the study of oncogenic processes (Doudna and Charpentier, 2014). In addition, engineered cell lines are invaluable for exploring the mechanisms behind drug resistance by introducing mutations that confer resistance to targeted therapies or chemotherapy (Cong et al., 2013). CRISPR-Cas-edited cell lines are powerful tools for functional genomics research, enabling study of the roles of the individual genes or pathways in cancer biology (Mali et al., 2013).

6.7.10 Advancing Cancer Research: Generating Animal Models with CRISPR-Cas Technology

Animal models have been indispensable in cancer research, offering critical insights into tumor biology, therapeutic strategies, and the development of personalized medicine. However, traditional techniques for creating genetically modified animal models were often time-consuming and lacked precision. The advent of CRISPR-Cas technology has transformed this process, providing unmatched accuracy and efficiency in engineering genetic changes in model organisms (Doudna and Charpentier, 2014). CRISPR-generated animal models accurately replicate the genetic mutations associated with hereditary cancer syndromes, offering crucial insights into the biology of initiation and progression of the tumors (Jinek et al., 2012). By introducing specific mutations into genes that drive cancer, researchers can uncover the molecular pathways involved in cancer development and identify potential targets for new therapies (Shalem et al., 2014). Animal models that carry mutations found in human cancers serve as valuable preclinical tools for testing the effectiveness of targeted therapies and combination treatments, aiding in the design of clinical trials and treatment regimens (Wang et al., 2013). CRISPR-generated animal models provide a powerful means to explore the interactions amongst cancerous cells and their surrounding microenvironment, including immune responses, angiogenesis, and metastasis, offering new insights into these complex processes (Nishimasu et al., 2014).

6.7.11 Cancer Diagnostics

The early identification and treatment of cancer play a crucial role in lowering mortality rates and enhancing the quality of life for patient's post-treatment. While various methods for cancer detection are currently in use, there is a pressing need for advancements in their speed, sensitivity, and specificity. Consequently, through genetic diagnostics, pinpointing sensitive genes is essential for cancer prevention (Rahman, 2014). To address these challenges, specific high-sensitivity enzymatic reporter unlocking (SHERLOCK), a CRISPR-mediated diagnostic approach, was developed by Gootenberg et al. (2017). This cutting-edge system utilizes RNA-guided RNase Cas13a, enabling substantial non-specific trans-cleavage of single-stranded DNA (ssDNA) as a collateral effect, in conjunction with a reporter signal that is triggered post–RNA cleavage. This approach has demonstrated outstanding sensitivity in identifying mutations, such as BRAF V600E and EGFR L858R, in mammalian cells (Gootenberg et al., 2017). A comparable system, known as DETECTR (DNA endonuclease-targeted CRISPR trans reporter), incorporates Cas12a along with recombinase polymerase amplification (RPA). It serves as a detection mechanism for identifying viral infections in cancer patients and for amplifying micro-samples. This mechanism is both quick and cost-effective, particularly for the detection of various types of high-risk human papilloma virus (HPV), including HPV16 and HPV18, in infected samples by several HPV strains (Chen et al., 2018; Yang et al., 2019). MicroRNAs (miRNAs) play a significant role in various diseases and hold considerable promise for cancer diagnosis and treatment. Recent advancements have demonstrated a novel isothermal amplification approach utilizing CRISPR-Cas9 that can effectively detect miRNAs (Qiu et al., 2018). This marks the inaugural application of the CRISPR-Cas9 approach for miRNA detection.

6.8 CRISPR-CAS9: PRECLINICAL RESEARCH TO CLINICAL PRACTICE IN ONCOLOGY AND CANCER TREATMENT

6.8.1 Preclinical Applications of CRISPR-Cas9

The dawn of CRISPR-Cas9 methodology, known for its precision in genome editing, has revolutionized cancer research by allowing scientists to explore the genetic foundations of cancer with unprecedented effectiveness. Preclinical studies utilizing CRISPR-Cas9 yield significant insights into cancer biology, facilitate drug discovery, and contribute to the creation of innovative therapies. The following highlights key domains where CRISPR-Cas9 is making a crucial impact in preclinical cancer research.

6.8.1.1 Functional Genomic Screens to Identify Cancer Drivers

In preclinical studies, CRISPR-Cas has become an important means specifically in exploring the roles of oncogenes and tumor suppressor genes. By employing CRISPR-Cas9 to knock out or modify these genes, scientists can investigate their contributions to tumor development and evaluate their potential as targets for therapeutic intervention (Doudna and Charpentier, 2014). CRISPR-Cas9 technology has been quite significant in uncovering key mechanisms of cancer development.

6.8.1.2 Overcoming Drug Resistance

In the treatment of cancer, one of the main concerns is the development of resistance to chemotherapy. The CRISPR-Cas9 technology presents a novel method to tackle this issue by precisely targeting genes that are related with the drug resistance. For instance, in osteosarcoma cell lines, the inclusion of CRISPR-Cas9 to knock down the multidrug resistance (MDR1) gene has been shown to restore sensitivity to chemotherapy drugs, indicating a potential method to tackle drug resistance (Li et al., 2018).

6.8.1.3 Preclinical Models and Therapeutic Delivery

CRISPR-Cas9 has facilitated the creation of sophisticated preclinical models that closely mimic human cancers, serving as valuable tools for understanding tumor biology and testing novel therapies. One innovative technique involves inducing gene knockouts in mouse models by injecting DNA plasmids carrying Cas9 and sgRNA directly into the liver of normal mice. This approach has helped investigate the function of specific cancer genes and evaluate the potential of CRISPR-based treatments. Additionally, new delivery methods like lipid nanoparticles (LNPs) have been used to improve the accuracy and effectiveness of CRISPR-Cas9 delivery to cancer cells. LNPs encapsulate the CRISPR components, ensuring targeted delivery and minimizing off-target effects. Preclinical studies have reported that LNP-mediated CRISPR delivery can effectively disrupt oncogenic pathways and trigger tumor regression.

6.8.1.4 Enhancing Immune Response and CAR-T Cell Engineering

Against the tumors, the CRISPR-Cas9 method has been used quite well to enhance the immune response, particularly through the engineering of CAR-T cells. By editing the T cells via the CRISPR-Cas9 approach, researchers can develop CAR-T cells that are more proficient at identifying and eliminating cancer cells.

6.8.1.5 Exploring Synergistic Therapies

Combining CRISPR-Cas9 with other therapeutic approaches has shown potential in amplifying anti-cancer effects. Preclinical research has investigated the synergistic potential of CRISPR-Cas9 in amalgamation with chemotherapy, radiation therapy, and other targeted treatments (Smith et al., 2020). For example, combining CRISPR-Cas9-driven gene editing with chemotherapy has been found to elevate the sensitivity of the cancerous cells to chemotherapeutic agents, potentially overcoming resistance and improving treatment outcomes. Similarly, integrating CRISPR-Cas9 with radiation therapy can enhance tumor cell death by targeting DNA repair mechanisms.

6.8.2 Clinical Applications of CRISPR-Cas in Oncology

The advancement of CRISPR-Cas9 technology has progressed beyond the laboratory and into clinical settings, particularly within oncology. This section examines how CRISPR-Cas9 is being applied in clinical practice, highlighting its transformative impact on cancer treatment. The integration of CRISPR-Cas9 into clinical practice has been marked by significant strides, especially in oncology. This section explores the advancements and potential of gene-editing therapies and *in vivo* CRISPR-Cas9 applications for cancer treatment.

6.8.2.1 Gene-Editing Therapies

The clinical journey of CRISPR-Cas9 began with its application to blood disorders, which laid the groundwork for its use in oncology. Notable trials, such as those by CRISPR Therapeutics and Vertex Pharmaceuticals, focus on *ex vivo* gene editing. This involves extracting HSCs from patients, editing them with CRISPR-Cas9 to correct genetic anomalies or enhance therapeutic functions, and reintroducing these cells into the patient. This technique has shown promise in treating blood cancers, like leukemia and lymphoma. The process includes isolating HSCs, applying CRISPR-Cas9 to modify genes either by knocking out inhibitory genes or inserting genes that boost cancer-fighting capabilities, and expanding these edited cells before their reinfusion. Clinical trials have shown that these modified HSCs can successfully engraft, produce functional blood cells, and contribute to long-term remission in certain blood cancers. Building on successes with blood cancers, researchers are adapting similar strategies for solid tumors. Initial trials with advanced solid tumors have been encouraging, showing that edited cells persist and demonstrate improved anti-tumor effects. These ex vivo approaches are particularly beneficial for cancers resistant to traditional therapies, offering a personalized treatment strategy that targets specific genetic and immunological profiles, thus potentially reducing side effects by focusing on cancer cells while sparing healthy tissues.

6.8.2.2 CAR-T Cell Therapy

The advent of CRISPR-Cas9 has revolutionized CAR-T cell therapy by boosting its effectiveness against cancer cells. By genetically modifying T cells with CRISPR, scientists can enhance their precision in detecting and destroying tumors. Notably, clinical trials have involved CRISPR-based deletion of the Programmed Death-1 (PD-1) gene in T cells, which has been shown to heighten anti-tumor responses. These studies indicate that CRISPR-engineered CAR-T cells outperform their unmodified counterparts in eliminating cancer cells.

6.8.2.3 Oncolytic Viruses

Oncolytic viruses are specifically designed to target and destroy cancer cells. Although this method is currently undergoing clinical trials, it has demonstrated potential, especially in treating cancers, such as melanoma and glioblastoma. An example of this therapy is talimogene laherparepvec (T-VEC), which has received approval for melanoma treatment and utilizes the herpes virus to attack tumors (Bommareddy et al., 2017).

6.8.2.4 Gene-Editing Approaches

CRISPR-Cas9 and similar gene-editing methods present the opportunity to directly modify genes linked to cancer, which could lead to the creation of more specialized and effective treatment options. Initial research has demonstrated the use of CRISPR-Cas9 to enhance immune cells' ability to identify and combat cancer cells. While still in the experimental stage, this approach shows significant potential. CRISPR-Cas9 enables precise disruption of genes that lead to the development and progression of cancer.

6.8.2.5 Personalized Immunotherapy

CRISPR-Cas9 facilitates the development of personalized cancer treatments by allowing for the modification of genes specific to a patient's tumor. This includes altering cells of the immune system to identify and attack cancer-specific antigens, leading to highly individualized and effective immunotherapy strategies. Clinical trials are actively refining these methods for various cancer types.

6.8.2.6 Tumor Genetic Profiling and Targeted Therapy

CRISPR-Cas9 supports comprehensive genetic profiling of tumors, uncovering mutations and genetic alterations driving cancer progression. This detailed genetic information is then used to create targeted therapies that specifically addresses the genetic makeup of a patient's tumor, potentially enhancing treatment results and reducing side effects. Ongoing clinical trials are showcasing the potential of this precision medicine approach across a range of cancers.

6.8.3 CRISPR-Cas9 in Cancer Prevention

CRISPR-Cas9, a revolutionary gene-editing technology, is gaining attention not only for its potential in cancer treatment but also as a promising method for cancer prevention. The aim of cancer prevention is to stop or diminish the likelihood of cancer occurring, and CRISPR's precise gene-editing capabilities may facilitate this by rectifying precancerous mutations or altering genetic factors that increase cancer risk. The following outlines the exploration of CRISPR-Cas9 in cancer prevention.

6.8.3.1 Genetic Screening and Early Intervention

The significant potential of CRISPR-Cas9 in cancer prevention lies in its ability to revolutionize genetic screening and early intervention strategies. By targeting and correcting mutations in genes such as BRCA1 and BRCA2, which are sturdily related with increased risk of breast and ovarian cancers, CRISPR-Cas9 holds potential for preventing these cancers before they manifest (Ding et al., 2023).

6.8.3.2 Prophylactic Gene Editing

Prophylactic gene editing with CRISPR-Cas9 involves making pre-emptive genetic modifications to lower cancer risk. For example, modifying the Adenomatous

Polyposis Coli (APC gene), frequently mutated in colorectal cancer, can inhibit tumor formation (McCart et al., 2008). Studies in animal models have shown that CRISPR-Cas9 effectively disrupts key oncogenes like KRAS and MYC, preventing tumor initiation in models of lung and pancreatic cancers (Bu et al., 2023).

6.8.3.3 Restoration of Tumor Suppressor Genes

The development of many cancers is linked to the loss or inactivation of tumor suppressor genes that typically serve to protect cells from malignant transformation. Reactivating or restoring these genes prior to the onset of cancer may play a crucial role in preventing tumor formation. The p53 gene, often referred to as the "guardian of the genome," is vital for preventing DNA mutations that can lead to cancer. In numerous cancers, p53 is either mutated or silenced. Employing CRISPR technology to restore or activate p53 in precancerous cells may help maintain genomic stability and thwart cancer progression (Zhu et al., 2020). Additionally, the retinoblastoma (RB) gene is another significant tumor suppressor frequently inactivated in various cancers. The CRISPR-Cas9 system could be utilized to reactivate the RB gene in affected cells, thereby averting the development of malignancies linked to its inactivation (Xiao et al., 2021).

6.8.3.4 Enhancing Immune Surveillance

The CRISPR-Cas9 system has the potential to strengthen the immune response, improving its ability to recognize and destroy cancerous cells. By modifying T cells to produce chimeric antigen receptors (CARs) that are designed to recognize particular cancer markers, researchers have developed CAR-T cell therapies that are effective in treating certain blood cancers (Zhang et al., 2017).

6.8.3.5 Carcinogenesis Prevention

CRISPR technology offers the capability to directly modify or remove precancerous cells, thereby preventing their progression to malignant forms (Gao et al., 2022). Certain conditions, including precancerous lesions and early genetic changes in tissues, are recognized as risk factors for cancer development (Feng et al., 2018). Additionally, stem cells differentiate into various types of cells, which may harbor mutations that contribute to cancer (Tysnes and Bjerkvig, 2007). Employing CRISPR to correct these mutations in stem cells before they evolve into cancerous tissues could serve as a proactive measure against cancer (Ben Jehuda et al., 2018).

6.8.3.6 Prevention of Lifestyle-Induced Cancer

The prevention of cancer induced by environmental factors or lifestyle choices is crucial. While genetic mutations significantly contribute to cancer development, external elements, such as smoking, ultraviolet radiation, and dietary habits, also have a considerable influence on carcinogenesis (Anand et al., 2008). Although CRISPR technology cannot directly address environmental exposures, it has the potential to alleviate the consequences of DNA damage inflicted by mutagens in the environment. By enhancing the body's inherent DNA repair systems, CRISPR could help counteract the harmful effects of environmental carcinogens, such as those found in tobacco smoke and UV radiation, which are known to cause mutations that could lead to cancer (Liu et al., 2024).

6.8.3.7 Epigenetic Editing

The CRISPR-Cas9 system can be utilized alongside other methodologies to target and alter epigenetic markers, facilitating the reversal or prevention of the silencing of tumor suppressor genes or the activation of oncogenes (Khajuria and Sharma, 2021). This strategy involves employing CRISPR to introduce epigenetic modifiers, such as methyl or acetyl groups, to specific genomic regions, effectively reprogramming gene expression.

6.9 COMBINING CRISPR-CAS SYSTEMS WITH OTHER EMERGING THERAPIES FOR SYNERGISTIC EFFECTS

The combination of CRISPR-Cas technology with other innovative therapies presents a capable approach to enhance the effectiveness of cancer therapy. By harnessing the distinct advantages of each therapeutic approach, this strategy aims to create synergistic effects that can address the limitations of using a single therapy.

6.9.1 CRISPR-Cas and Immunotherapy

The integration of CRISPR-Cas technology with cancer immunotherapy offers a novel way to overcome current challenges and improve treatment effectiveness. Immunotherapy, which activates the defense system of the body to identify and destroy cancerous cells, has transformed cancer care. However, issues such as immune system evasion by cancer cells and the progress of treatment resistance continue to be significant hurdles. CRISPR-Cas systems provide a highly accurate and powerful method for genetic manipulation, allowing for the enhancement of immune system responses against cancer cells.

Recent studies have highlighted the crucial role of natural killer (NK) cells in combating cancer and how their function can be influenced by proteins like inhibitor of DNA binding 2 (ID2). By using CRISPR-Cas technology, it is possible to enhance the maturation and functionality of NK cells, boosting their capacity to detect and kill cancerous cells. This method could lead to the creation of more effective immunotherapies capable of overcoming mechanisms that allow cancer cells to resist treatment. The Cluster of Differentiation 155/T cell immunoreceptor with Ig and ITIM domains (CD155/TIGIT) pathway has arisen as a key target in cancer immunotherapy. TIGIT is an inhibitory receptor found on T and NK cells that plays a role in dampening the immune response. Through CRISPR-Cas, it is likely to alter or disrupt the expression of TIGIT, thereby strengthening the immune system's ability to combat tumors. This approach can be combined with current immunotherapies to achieve synergistic effects, potentially resulting in more potent and effective treatments (Wellhausen et al., 2024).

6.9.2 Enhancing Cancer Treatment Efficacy: The Synergistic Potential of CRISPR-Cas Technology and Chemotherapy

Integrating CRISPR-Cas technology with chemotherapy represents a pioneering approach to improving cancer treatment efficiency by addressing the limitations of conventional therapies. Chemotherapy, a mainstay in cancer management, relies on the usage of cytotoxic drugs to target and terminate quickly dividing cancer cells. However, its effectiveness is often hindered by encounters, such as drug resistance, off-target toxicity, and dearth of specificity (Liu et al., 2024). CRISPR-Cas systems provide a powerful tool for specific genetic modification, enabling to make targeted changes to the genome that can increase cancer cells' sensitivity to chemotherapy or directly address genetic mutations responsible for drug resistance. For example, by using CRISPR-Cas to deactivate or alter genes involved in DNA repair mechanisms, it is possible to make cancer cells more vulnerable to chemotherapeutic drugs. This strategy not only enhances the effectiveness of cancer cell eradication but also has the potential to reduce the necessary chemotherapy dosage, thereby lessening the associated side effects (Huang et al., 2020).

6.9.3 Leveraging CRISPR-Cas in Radiotherapy: A Revolutionary Approach to Cancer Treatment

Radiotherapy has long been a cornerstone in the arsenal against cancer, relying on high-energy radiation to damage the DNA of the cancerous cells, ultimately leading to their death. However, the effectiveness of radiotherapy can be significantly compromised by the capability of cancer cells to repair their DNA damage, allowing them to survive and continue proliferating. This challenge is where the revolutionary gene-editing technology CRISPR-Cas comes into play, offering a powerful new tool to enhance the outcomes of radiotherapy. CRISPR-Cas technology, known for its precision in targeting specific genetic sequences, can be employed to manipulate the DNA repair pathways that cancer cells rely on for survival after radiation exposure. By using CRISPR to disrupt or deactivate key genes involved in the repair process, such as those responsible for HR or NHEJ, cancer cells become significantly more susceptible to radiation-induced DNA damage. This targeted approach not only increases the efficacy of radiotherapy in killing cancer cells but also allows for a more selective treatment that minimizes collateral damage to the surrounding healthy tissues.

This innovative combination of CRISPR-Cas with radiotherapy opens up new possibilities, particularly for treating tumors that are traditionally resistant to radiation. By integrating CRISPR-Cas technology to specifically knock out or modify these DNA repair pathways, these resistant tumors could be rendered far more vulnerable to radiation, potentially overcoming their resistance and improving patient outcomes.

6.9.4 Advancing Targeted Cancer Therapies with CRISPR-Cas: A Precision Medicine Approach

Targeted therapies have revolutionized cancer treatment and are designed to inhibit or modulate particular proteins or pathways that are essential for the survival and proliferation of cancer cells. However, despite their success, the effectiveness of targeted therapies is often limited by drug resistance and the emergence of alternative survival pathways in cancer cells. This is where CRISPR-Cas methods could play a transformative part, enhancing the precision and efficacy of targeted treatments. CRISPR-Cas technology, renowned for its ability, can be leveraged to optimize targeted cancer therapies in several ways. First, it allows for the validation and discovery of new molecular targets by enabling researchers to systematically knock out or modify specific genes and observe the resulting effects on cancer cell behavior. This approach can identify critical liabilities in the cancerous cells that could be exploited therapeutically, providing a robust pipeline for the growth of next-generation targeted therapies. CRISPR-Cas can be used to create highly accurate models of human cancers, allowing for more effective preclinical testing of targeted therapies. By editing the genomes of cell lines or animal models to reflect the genetic landscape of specific cancers, researchers can better understand how tumors respond to different treatments and identify potential resistance mechanisms. This capability is invaluable for designing combination therapies or modifying existing drugs to improve their efficacy against resistant cancer cell populations (Shin et al., 2017).

6.9.5 Combining CRISPR-Cas with Nanotechnology: A Novel Strategy for Cancer Therapy

The convergence of CRISPR-Cas and nanotechnology opens up a groundbreaking frontier in the arena of cancer treatment. Nanotechnology, with its ability to create tiny customizable particles, provides an innovative platform for delivering CRISPR components directly to cancer cells with remarkable precision. By utilizing nanoparticles as delivery vehicles, scientists can ensure that CRISPR machinery reaches its intended target within the tumor, lessening off-target effects and enhancing the specificity of genetic amendments (Xu et al., 2017). A significant benefit of utilizing nanoparticles in conjunction with CRISPR-Cas systems is their versatility. These microscopic carriers can be engineered to transport not only CRISPR components, such as Cas9 proteins and guide RNAs, but also a variety of therapeutic agents, including drugs, RNA molecules, and other biologically active compounds.

In preclinical models, the simultaneous delivery of CRISPR-Cas components and chemotherapeutic drugs via nanoparticles has shown great promise. This combination strategy leverages the precise gene-editing ability of CRISPR to disrupt critical cancer cell survival pathways while the chemotherapeutic agents exert their cytotoxic effects. The result is a potent, multifaceted attack that can lead to significant tumor regression, as observed in several experimental studies. The use of nanotechnology in CRISPR-Cas delivery can address some of the key challenges associated with gene-editing therapies, such as stability, efficiency, and safety. Nanoparticles can protect CRISPR components from degradation in the bloodstream, facilitate their uptake into cancer cells, and reduce immune recognition, thereby enhancing the overall efficiency and safety profile of the treatment.

6.10 SAFETY CONSIDERATIONS AND CHALLENGES IN USING CRISPR-CAS SYSTEM FOR CANCER THERAPEUTICS

The advancements in CRISPR-Cas editing approaches represent an ideal switch in the landscape of cancer biology and therapeutics, presenting unparalleled accuracy and flexibility in addressing the genetic underpinnings of malignancies. Despite being mainstays in oncology, these traditional approaches are hampered by issues like non-specificity, systemic toxicity, and limited efficacy against advanced and heterogeneous tumors. The precision-driven CRISPR-Cas9 technology addresses these limitations by enabling highly specific genetic interventions that target oncogenic drivers, modulate immune responses, and reprogram cellular functions critical to tumorigenesis. The utilization of CRISPR-Cas9 technology in cancer therapy has introduced a groundbreaking approach to gene editing, significantly impacting the comprehensive management of cancer. Preclinical and clinical trials underscore its ability to selectively ablate oncogenes, correct driver mutations, and enhance immunotherapies. Notably, the engineering of T cells, particularly in the context of CAR-T therapies, exemplifies how CRISPR-Cas9 can be harnessed to overcome immune evasion by tumors. Furthermore, the system's adaptability facilitates the targeting of cancer-specific vulnerabilities, making it a valuable tool in precision oncology. While these developments are promising, the translation of CRISPR-based interventions from bench to bedside remains complex due to challenges related to various challenges related to delivery, off-targets, efficiency, genomic instability, cancer cell resistance, tumor heterogenicity, and ethical considerations. The threat of off-target effects resulting in unexpected alterations to the genome is still a significant barrier. A vital problem in the process of gene editing is the efficient and safe delivery of CRISPR components—such as Cas proteins, guide RNAs, and DNA templates—to the intended target cells. Particularly in cancer therapy, it is essential that this delivery is highly selective for cancer cells to prevent any adverse effects on surrounding healthy tissues. Targeted CRISPR therapies carry the potential risk of fostering the development of resistant cancer clones that may evade the effects of the therapy, thus compromising its efficacy. Tumors typically exhibit heterogeneity, with cancer cells displaying a range of genetic mutations and traits. This diversity complicates the creation of a universal CRISPR-based therapy that can effectively address all cancer cells present in a tumor. Additionally, ethical considerations, particularly in germline editing and patient consent, are crucial in steering the responsible development and application of this technology. Addressing these challenges is vital for the clinical acceptance and widespread use of CRISPR-based therapies.

6.11 CONCLUSION

The CRISPR-Cas system offers a powerful and versatile platform for cancer therapeutics, with the potential to redefine the standard of care in oncology. Despite the technical and ethical challenges that lie ahead, the rapid progress in preclinical and clinical research underscores the feasibility of translating this technology into safe, effective, and widely accessible treatments. As CRISPR-based therapies continue to evolve, their integration with existing and novel treatment modalities is expected to drive a new era of precision oncology, wherein personalized interventions offer curative outcomes with minimal adverse effects. The journey from gene editing to clinical application is complex but holds the potential to transform cancer from a life-threatening disease into a manageable condition, fundamentally altering the future of cancer therapeutics. The strength of CRISPR in oncology lies in its precise targeting of the genetic causes of cancer, which enhances the specificity, efficacy, and personalization of treatment strategies.

REFERENCES

Adjei, Isaac M., and Sharma Blanka. "Modulation of the tumor microenvironment for cancer treatment: A biomaterials approach." *Journal of Functional Biomaterials* 6, no. 1 (2015): 81–103.

Anand, Preetha, Ajaikumar B. Kunnumakara, Chitra Sundaram, Kuzhuvelil B. Harikumar, Sheeja T. Tharakan, Oiki S. Lai, Bokyung Sung, and Bharat B. Aggarwal. "Cancer is a preventable disease that requires major lifestyle changes." *Pharmaceutical Research* 25, no. 9 (2008): 2097–2116.

Au, Gough G., A. Michael Lindberg, Richard D. Barry, and Darren R. Shafren. "Oncolysis of vascular malignant human melanoma tumors by Coxsackievirus A21." *International Journal of Oncology* 26, no. 6 (2005): 1471–1476.

Barrangou, Rodolphe, C. Fremaux, H. Deveau, M. Richards, P. Boyaval, S. Moineau, D. A. Romero, and P. Horvath. "CRISPR provides acquired resistance against viruses in prokaryotes." *Science* 315, no. 5819 (2007): 1709–1712.

Ben Jehuda, Ronen, Yuval Shemer, and Ofer Binah. "Genome editing in induced pluripotent stem cells using CRISPR/Cas9." *Stem Cell Reviews and Reports* 14 (2018): 323–336.

Bergers, Gabriele, and Laura E. Benjamin. "Tumorigenesis and the angiogenic switch." *Nature Reviews Cancer* 3, no. 6 (2003): 401–410.

Bishayee, Anupam, and Keith Block. "A broad-spectrum integrative design for cancer prevention and therapy: The challenge ahead." In *Seminars in Cancer Biology*, vol. 35, pp. S1–S4. Academic Press, 2015.

Bommareddy, Praveen K., Anand Patel, Saamia Hossain, and Howard L. Kaufman. "Talimogene laherparepvec (T-VEC) and other oncolytic viruses for the treatment of melanoma." *American Journal of Clinical Dermatology* 18, no. 1 (2017): 1–15.

Bu, Wen, Chad J. Creighton, Kelsey S. Heavener, Carolina Gutierrez, Yongchao Dou, Amy T. Ku, Yiqun Zhang et al. "Efficient cancer modeling through CRISPR-Cas9/HDR-based somatic precision gene editing in mice." *Science Advances* 9, no. 19 (2023): eade0059.

Burrell, Rebecca A., Nicholas McGranahan, Jiri Bartek, and Charles Swanton. "The causes and consequences of genetic heterogeneity in cancer evolution." *Nature* 501, no. 7467 (2013): 338–345.

Chen, Janice S., Enbo Ma, Lucas B. Harrington, Maria Da Costa, Xinran Tian, Joel M. Palefsky, and Jennifer A. Doudna. "CRISPR-Cas12a target binding unleashes indiscriminate single-stranded DNase activity." *Science* 360, no. 6387 (2018): 436–439.

Chen, Mei-Ling, Julia Huei-Mei Chang, Kun-Tu Yeh, Ya-Sian Chang, and Jan-Gowth Chang. "Epigenetic changes in tumor suppressor genes, P15, P16, APC-3 and E-cadherin in body fluid." *The Kaohsiung Journal of Medical Sciences* 23, no. 10 (2007): 498–503.

Chu, Dinh-Toi, Tiep Tien Nguyen, Nguyen Le Bao Tien, Dang-Khoa Tran, Jee-Heon Jeong, Pham Gia Anh, Vo Van Thanh, Dang Tien Truong, and Thien Chu Dinh. "Recent progress of stem cell therapy in cancer treatment: Molecular mechanisms and potential applications." *Cells* 9, no. 3 (2020): 563.

Chulpanova, Daria S., Valeriya V. Solovyeva, Kristina V. Kitaeva, Stephen P. Dunham, Svetlana F. Khaiboullina, and Albert A. Rizvanov. "Recombinant viruses for cancer therapy." *Biomedicines* 6, no. 4 (2018): 94.

Clevers, Hans. "Models of development and disease." *Cell* 165, no. 7 (2016): 1586–1597.

Cong, Le, F. Ann Ran, David Cox, Shuailiang Lin, Robert Barretto, Naomi Habib, Patrick D. Hsu et al. "Multiplex genome engineering using CRISPR/Cas systems." *Science* 339, no. 6121 (2013): 819–823.

Cox, David B. T., Jonathan S. Gootenberg, Omar O. Abudayyeh, Mark J. Franklin, Adrian M. Kellner, Julia Joung, and Feng Zhang. "RNA Editing with CRISPR-Cas13." *Science* 358, no. 6366 (2017): 1019–1027.

Delaney, Geoff, Susannah Jacob, Carolyn Featherstone, and Michael Barton. "The role of radiotherapy in cancer treatment: Estimating optimal utilization from a review of evidence-based clinical guidelines." *Cancer: Interdisciplinary International Journal of the American Cancer Society* 104, no. 6 (2005): 1129–1137.

Dimitri, Alexander, Friederike Herbst, and Joseph A. Fraietta. "Engineering the next-generation of CAR T-cells with CRISPR-Cas9 gene editing." *Molecular Cancer* 21, no. 1 (2022): 78.

Ding, Li, Gad Getz, David A. Wheeler, Elaine R. Mardis, Michael D. McLellan, Kristian Cibulskis, Carrie Sougnez et al. "Somatic mutations affect key pathways in lung adenocarcinoma." *Nature* 455, no. 7216 (2008): 1069–1075.

Ding, Shuai, Jinfeng Liu, Xin Han, and Mengfan Tang. "CRISPR/Cas9-Mediated Genome Editing in Cancer Therapy." *International Journal of Molecular Sciences* 24, no. 22 (2023): 16325.

Doudna, Jennifer A., and Emmanuelle Charpentier. "The new frontier of genome engineering with CRISPR-Cas9." *Science* 346, no. 6213 (2014): 1258096.

Elahi, Reza, Amir Hossein Heidary, Kaveh Hadiloo, and Abdolreza Esmaeilzadeh. "Chimeric antigen receptor-engineered natural killer (CAR NK) cells in cancer treatment; recent advances and future prospects." *Stem Cell Reviews and Reports* 17 (2021): 2081–2106.

Feng, Yixiao, Mia Spezia, Shifeng Huang, Chengfu Yuan, Zongyue Zeng, Linghuan Zhang, Xiaojuan Ji et al. "Breast cancer development and progression: Risk factors, cancer stem cells, signaling pathways, genomics, and molecular pathogenesis." *Genes & Diseases* 5, no. 2 (2018): 77–106.

Fidler, Isaiah J. "The pathogenesis of cancer metastasis: The'seed and soil'hypothesis revisited." *Nature Reviews Cancer* 3, no. 6 (2003): 453–458.

Fisher, Bernard. "The changing role of surgery in the treatment of cancer." In *Radiotherapy, Surgery, and Immunotherapy*, pp. 401–421. Boston, MA: Springer US, 1977.

Freeman, Scott M., Camille N. Abboud, Katharine A. Whartenby, Charles H. Packman, David S. Koeplin, Frederick L. Moolten, and George N. Abraham. "The "bystander effect": Tumor regression when a fraction of the tumor mass is genetically modified." *Cancer Research* 53, no. 21 (1993): 5274–5283.

Gao, Chun, Ping Wu, Lan Yu, Liting Liu, Hong Liu, Xiangyu Tan, Liming Wang, Xiaoyuan Huang, and Hui Wang. "The application of CRISPR/Cas9 system in cervical carcinogenesis." *Cancer Gene Therapy* 29, no. 5 (2022): 466–474.

Garraway, Levi A., Jaap Verweij, and Karla V. Ballman. "Precision oncology: An overview." *Journal of Clinical Oncology* 31, no. 15 (2013): 1803–1805.

Gerlinger, Marco, Andrew J. Rowan, Stuart Horswell, James Larkin, David Endesfelder, Eva Gronroos, Pierre Martinez et al. "Intratumor heterogeneity and branched evolution revealed by multiregion sequencing." *New England Journal of Medicine* 366, no. 10 (2012): 883–892.

Gootenberg, Jonathan S., Omar O. Abudayyeh, Jeong Wook Lee, Patrick Essletzbichler, Aaron J. Dy, Julia Joung, Vanessa Verdine et al. "Nucleic acid detection with CRISPR-Cas13a/C2c2." *Science* 356, no. 6336 (2017): 438–442.

Halperin, Edward C., David E. Wazer, Carlos A. Perez, and Luther W. Brady. *Perez & Brady's Principles and Practice of Radiation Oncology*. Lippincott Williams & Wilkins, 2018.

Hanahan, Douglas, and Robert A. Weinberg. "Hallmarks of cancer: The next generation." *Cell* 144, no. 5 (2011): 646–674.

Hanahan, Douglas, and Robert A. Weinberg. "The hallmarks of cancer." *Cell* 100, no. 1 (2000): 57–70.

Hanahan, Douglas. "Hallmarks of cancer: New dimensions." *Cancer Discovery* 12, no. 1 (2022): 31–46.

Hilton, Isaac B., Anthony M. D'ippolito, Christopher M. Vockley, Pratiksha I. Thakore, Gregory E. Crawford, Timothy E. Reddy, and Charles A. Gersbach. "Epigenome editing by a CRISPR-Cas9-based acetyltransferase activates genes from promoters and enhancers." *Nature Biotechnology* 33, no. 5 (2015): 510–517.

Holohan, Caitriona, Sandra Van Schaeybroeck, Daniel B. Longley, and Patrick G. Johnston. "Cancer drug resistance: An evolving paradigm." *Nature Reviews Cancer* 13, no. 10 (2013): 714–726.

Huang, Dennis, Matthew Miller, Bhaargavi Ashok, Samagra Jain, and Nicholas A. Peppas. "CRISPR/Cas systems to overcome challenges in developing the next generation of T cells for cancer therapy." *Advanced Drug Delivery Reviews* 158 (2020): 17–35.

Jinek, Martin, Krzysztof Chylinski, Ines Fonfara, Michael Hauer, Jennifer A. Doudna, and Emmanuelle Charpentier. "A programmable dual-RNA–guided DNA endonuclease in adaptive bacterial immunity." *Science* 337, no. 6096 (2012): 816–821.

Joiner, Michael C., and Albert van der KogeJ, eds. *Basic Clinical Radiobiology*. CRC Press, 2025.

Jones, Peter A., and Stephen B. Baylin. "The epigenomics of cancer." *Cell* 128, no. 4 (2007): 683–692.

Junttila, Melissa R., and Frederic J. De Sauvage. "Influence of tumour micro-environment heterogeneity on therapeutic response." *Nature* 501, no. 7467 (2013): 346–354.

Kanno, Yuzuka, Chang-Yu Chen, Hsin-Lun Lee, Jeng-Fong Chiou, and Yin-Ju Chen. "Molecular mechanisms of chemotherapy resistance in head and neck cancers." *Frontiers in Oncology* 11 (2021): 640392.

Khajuria, Ocean, and Neha Sharma. "Epigenetic targeting for lung cancer treatment via CRISPR/Cas9 technology." *Advances in Cancer Biology-Metastasis* 3 (2021): 100012.

Kim, Hyongbum, and Jin-Soo Kim. "A guide to genome engineering with programmable nucleases." *Nature Reviews Genetics* 15, no. 5 (2014): 321–334.

Kleinstiver, Benjamin P., Vikram Pattanayak, Michelle S. Prew, Shengdar Q. Tsai, Nhu T. Nguyen, Zongli Zheng, and J. Keith Joung. "High-fidelity CRISPR–Cas9 nucleases with no detectable genome-wide off-target effects." *Nature* 529, no. 7587 (2016): 490–495.

Komor, Alexis C., April H. Badran, and David R. Liu. "Programmable editing of a target base in genomic DNA without double-stranded DNA cleavage." *Nature* 533, no. 7603 (2016): 420–424.

Kufe, Donald W., Raphael E. Pollock, Ralph R. Weichselbaum, Robert C. Bast, Ted S. Gansler, James F. Holland, and Emil Frei III, eds. 2003. *Cancer Medicine*. 6th ed. Hamilton, ON: BC Decker.

Li, Jianyuan, Chun Yen, and Yingwei Shi. "The role of p16INK4a in primary, secondary, and tertiary cell immortalization." *Cold Spring Harbor Symposia on Quantitative Biology* 62 (1997a): 279–287.

Li, Jing, Clifford Yen, Danny Liaw, Katrina Podsypanina, Shikha Bose, Steven I. Wang, Janusz Puc et al. "PTEN, a putative protein tyrosine phosphatase gene mutated in human brain, breast, and prostate cancer." *Science* 275, no. 5308 (1997b): 1943–1947.

Li, Rong, Changli Wu, Hongying Liang, Yinghai Zhao, Chunyan Lin, Xiujuan Zhang, and Caiguo Ye. "Knockdown of TWIST Enhances the Cytotoxicity of Chemotherapeutic Drugs in Doxorubicin-Resistant HepG2 Cells by Suppressing MDR1 and EMT." *International Journal of Oncology* 53, no. 4 (2018): 1763–1773.

Lin, Weiping, Linfeng Huang, Ying Li, Bin Fang, Gang Li, Leilei Chen, and Liangliang Xu. "Mesenchymal stem cells and cancer: clinical challenges and opportunities." *BioMed Research International* 2019, no. 1 (2019): 2820853.

Liu, Nian, Jiahui Du, Jiani Ge, and Song-Bai Liu. "DNA damage-inducing endogenous and exogenous factors and research progress." *Nucleosides, Nucleotides & Nucleic Acids* (2024): 1–33.
Mali, Prashant, Le Cong, Shanchita R. Dhanda, Nicholas M. Ying, Qing Ran, Naushad Bajwa, Terrance G. J. Brown et al. "RNA-guided human genome engineering via Cas9." *Science* 339, no. 6121 (2013): 823–826.
Marraffini, Luciano A., and Erik J. Sontheimer. "CRISPR interference limits horizontal gene transfer in Staphylococci by targeting DNA." *Science* 322, no. 5909 (2008): 1843–1845.
Massague, Joan, and Anna C. Obenauf. "Metastatic colonization by circulating tumour cells." *Nature* 529, no. 7586 (2016): 298–306.
McCart, Amy E., Nicola K. Vickaryous, and Andrew Silver. "Apc mice: Models, modifiers and mutants." *Pathology-Research and Practice* 204, no. 7 (2008): 479–490.
McGranahan, Nicholas, and Charles Swanton. "Clonal heterogeneity and tumor evolution: Past, present, and the future." *Cell* 168, no. 4 (2017): 613–628.
Morris, Luc G. T., and Timothy A. Chan. "Therapeutic targeting of tumor suppressor genes." *Cancer* 121, no. 9 (2015): 1357–1368.
Mroz, Edmund A., and James W. Rocco. "The challenges of tumor genetic diversity." *Cancer* 123, no. 6 (2017): 917–927.
Nakamura, Muneaki, Yuchen Gao, Antonia A. Dominguez, and Lei S. Qi. "CRISPR technologies for precise epigenome editing." *Nature Cell Biology* 23, no. 1 (2021): 11–22.
Nishimasu, Hiroshi, F. Ann Ran, Patrick D. Hsu, Silvana Konermann, Soraya I. Shehata, Naoshi Dohmae, Ryuichiro Ishitani, Feng Zhang, and Osamu Nureki. "Crystal structure of Cas9 in complex with guide RNA and target DNA." *Cell* 156, no. 5 (2014): 935–949.
Nowell, Peter C. "The clonal evolution of tumor cell populations: Acquired genetic lability permits stepwise selection of variant sublines and underlies tumor progression." *Science* 194, no. 4260 (1976): 23–28.
Ouyang, Xiaoming, Melinda L. Telli, and Joseph C. Wu. "Induced pluripotent stem cell-based cancer vaccines." *Frontiers in Immunology* 10 (2019): 1510.
Park, H. J., R. J. Griffin, and S. Hui. "Hall EJ. Radiobiology for the radiologist. 6th edn. Philadelphia: Lippincott Williams and Wilkins, 2006, pp." *Cancer* 33 (2006): 1426–1431.
Pei, Wen-Di, Yan Zhang, Tai-Lang Yin, and Yang Yu. "Epigenome editing by CRISPR/Cas9 in clinical settings: Possibilities and challenges." *Briefings in Functional Genomics* 19, no. 3 (2020): 215–228.
Qiu, Xin-Yuan, Lv-Yun Zhu, Chu-Shu Zhu, Jia-Xin Ma, Tao Hou, Xiao-Min Wu, Si-Si Xie et al. "Highly effective and low-cost microRNA detection with CRISPR-Cas9." *ACS Synthetic Biology* 7, no. 3 (2018): 807–813.
Quail, Daniela F., and Johanna A. Joyce. "Microenvironmental regulation of tumor progression and metastasis." *Nature Medicine* 19, no. 11 (2013): 1423–1437.
Rahman, Nazneen. "Mainstreaming genetic testing of cancer predisposition genes." *Clinical Medicine* 14, no. 4 (2014): 436.
Raty, J. K., J. T. Pikkarainen, T. Wirth, and S. Yla-Herttuala. "Gene therapy: The first approved gene-based medicines, molecular mechanisms and clinical indications." *Current Molecular Pharmacology* 1, no. 1 (2008): 13–23.
Richardson, Christopher D., Graham J. Ray, Mark A. DeWitt, Gemma L. Curie, and Jacob E. Corn. "Enhancing homology-directed genome editing by catalytically active and inactive CRISPR-Cas9 using asymmetric donor DNA." *Nature Biotechnology* 34, no. 3 (2016): 339–344.
Robey, Robert W., Suneet Shukla, Elizabeth M. Finley, Robert K. Oldham, Daryl Barnett, Suresh V. Ambudkar, Tito Fojo, and Susan E. Bates. "Inhibition of P-glycoprotein (ABCB1)-and multidrug resistance-associated protein 1 (ABCC1)-mediated transport by the orally administered inhibitor, CBT-1®." *Biochemical Pharmacology* 75, no. 6 (2008): 1302–1312.
Roy, Rohini, Jarin Chun, and Simon N. Powell. "BRCA1 and BRCA2: Different roles in a common pathway of genome protection." *Nature Reviews Cancer* 12, no. 1 (2012): 68–78.

Russell, Stephen J., Kah-Whye Peng, and John C. Bell. "Oncolytic virotherapy." *Nature Biotechnology* 30, no. 7 (2012): 658–670.
Sander, Jeffry D., and J. Keith Joung. "CRISPR-Cas systems for editing, regulating and targeting genomes." *Nature Biotechnology* 32, no. 4 (2014): 347–355.
Sarvari, Pourya, Pouya Sarvari, Ivonne Ramírez-Díaz, Frouzandeh Mahjoubi, and Karla Rubio. "Advances of epigenetic biomarkers and epigenome editing for early diagnosis in breast cancer." *International Journal of Molecular Sciences* 23, no. 17 (2022): 9521.
Savage, David G., and Karen H. Antman. "Imatinib mesylate—a new oral targeted therapy." *New England Journal of Medicine* 346, no. 9 (2002): 683–693.
Shalem, Ophir, Neville E. Sanjana, Ella Hartenian, Xi Shi, David A. Scott, Tarjei S. Mikkelsen, Dirk Heckl et al. "Genome-scale CRISPR-Cas9 knockout screening in human cells." *Science* 343, no. 6166 (2014): 84–87.
Shen, Bing, Jennifer Zhang, Hongwei Wu, Xuening Hou, Jianxin Su, and Yanhong Pan. "Generation of gene-modified mice via Cas9/RNA-mediated gene targeting." *Cell Research* 23 (2013): 720–723.
Shin, Seung Ho, Ann M. Bode, and Zigang Dong. "Precision medicine: The foundation of future cancer therapeutics." *NPJ Precision Oncology* 1, no. 1 (2017): 12.
Smith, Robert A. et al. "Synergistic potential of CRISPR-Cas9 with conventional cancer therapies: A preclinical review." *Molecular Cancer Therapeutics* 19, no. 5 (2020): 1012–1023.
Stratton, Michael R., Peter J. Campbell, and P. Andrew Futreal. "The cancer genome." *Nature* 458, no. 7239 (2009): 719–724.
Sullivan, Genevieve M., Andrew K. Knutsen, Luca Peruzzotti-Jametti, Alexandru Korotcov, Asamoah Bosomtwi, Bernard J. Dardzinski, Joshua D. Bernstock et al. "Transplantation of induced neural stem cells (iNSCs) into chronically demyelinated corpus callosum ameliorates motor deficits." *Acta Neuropathologica Communications* 8 (2020): 1–23.
Takahashi, Kazutoshi, and Shinya Yamanaka. "Induction of pluripotent stem cells from mouse embryonic and adult fibroblast cultures by defined factors." *Cell* 126, no. 4 (2006): 663–676.
Thakore, Pratiksha I., Anthony M. D'ippolito, Lingyun Song, Alexias Safi, Nishkala K. Shivakumar, Ami M. Kabadi, Timothy E. Reddy, Gregory E. Crawford, and Charles A. Gersbach. "Highly specific epigenome editing by CRISPR-Cas9 repressors for silencing of distal regulatory elements." *Nature Methods* 12, no. 12 (2015): 1143–1149.
Trujillo-Alonso, Vicenta, Edwin C. Pratt, Hongliang Zong, Andres Lara-Martinez, Charalambos Kaittanis, Mohamed O. Rabie, Valerie Longo et al. "FDA-approved ferumoxytol displays anti-leukaemia efficacy against cells with low ferroportin levels." *Nature Nanotechnology* 14, no. 6 (2019): 616–622.
Turner, Nicholas C., and Jorge S. Reis-Filho. "Basal-like breast cancer and the BRCA1 phenotype." *Oncogene* 25, no. 43 (2006): 5846–5853.
Tysnes, Berit B., and Rolf Bjerkvig. "Cancer initiation and progression: Involvement of stem cells and the microenvironment." *Biochimica et Biophysica Acta (BBA)-Reviews on Cancer* 1775, no. 2 (2007): 283–297.
Visvader, Jane E. "Cells of origin in cancer." *Nature* 469, no. 7330 (2011): 314–322.
Vogelstein, Bert, and Kenneth W. Kinzler. "Cancer genes and the pathways they control." *Nature Medicine* 10, no. 8 (2004): 789–799.
Vousden, Karen H., and David P. Lane. "p53 in health and disease." *Nature Reviews Molecular Cell Biology* 8, no. 4 (2007): 275–283.
Wang, Haoyi, Hui Yang, Chikdu S. Shivalila, Meelad M. Dawlaty, Albert W. Cheng, Feng Zhang, and Rudolf Jaenisch. "One-step generation of mice carrying mutations in multiple genes by CRISPR/Cas-mediated genome engineering." *Cell* 153, no. 4 (2013): 910–918.
Wang, Tim, Kıvanç Birsoy, Nicholas W. Hughes, Kevin M. Krupczak, Yorick Post, Jenny J. Wei, Eric S. Lander, and David M. Sabatini. "Identification and characterization of essential genes in the human genome." *Science* 350, no. 6264 (2015): 1096–1101.

Wellhausen, Nils, Joanne Baek, Saar I. Gill, and Carl H. June. "Enhancing cellular immunotherapies in cancer by engineering selective therapeutic resistance." *Nature Reviews Cancer* 24, no. 9 (2024): 614–628.

Xiao, Lulu, Jun Li, Yiyu Sheng, Ye Wang, and Xiaoyan Dou. "Synthetic molecular sensors based on CRISPR-Cas9 redirect anticancer signal flows to treat retinoblastomas." *Clinical and Translational Medicine* 11, no. 11 (2021).

Xu, Ke, Guo Chen, Xiaobo Li, Xiaoqin Wu, Zhijie Chang, Jianhua Xu, Yu Zhu, Peihao Yin, Xin Liang, and Lei Dong. "MFN2 suppresses cancer progression through inhibition of mTORC2/Akt signaling." *Scientific Reports* 7, no. 1 (2017): 41718.

Yang, Dongli, Yong Shi, Yemei Tang, Hongyu Yin, Yujia Guo, Shuxin Wen, Binquan Wang, ChAngming An, Yongyan Wu, and Wei Gao. "Effect of HPV infection on the occurrence and development of laryngeal cancer: A review." *Journal of Cancer* 10, no. 19 (2019): 4455.

Yarden, Yosef, and Mark X. Sliwkowski. "Untangling the ErbB signalling network." *Nature Reviews Molecular Cell Biology* 2, no. 2 (2001): 127–137.

Zhang, Cheng, Jun Liu, Jiang F. Zhong, and Xi Zhang. "Engineering car-T cells." *Biomarker Research* 5 (2017): 1–6.

Zheng, Meijun, Jianhan Huang, Aiping Tong, and Hui Yang. "Oncolytic viruses for cancer therapy: Barriers and recent advances." *Molecular Therapy-Oncolytics* 15 (2019): 234–247.

Zhu, Gaoyang, Chaoyun Pan, Jin-Xin Bei, Bo Li, Chen Liang, Yang Xu, and Xuemei Fu. "Mutant p53 in cancer progression and targeted therapies." *Frontiers in Oncology* 10 (2020): 595187.

7 CRISPR-Cas Systems in the Diagnosis of Infectious and Non-Infectious Diseases

Aarif Ali, Javeed Iqbal Ahmad Bhat, and Farheen Shafique

7.1 OVERVIEW OF INFECTIOUS AND NON-INFECTIOUS DISEASES

Infectious diseases pose a persistent and growing risk to human health and welfare, driven by factors such as the rise of drug-resistant strains, population movements, and the emergence of novel pathogens. Despite the presence of antibiotics and vaccines for many of these infectious agents, high mortality rates persist. With 15% of world mortality attributed to infectious diseases, a major impact on public suffering and economic loss, they are among the most threatening agents to human beings (World Health Organization, 2018; Chen et al., 2019), typically including a range of infectious microorganisms that include bacteria, viruses, fungus, and parasites (Li et al., 2021). The emergence of new infectious diseases in the twenty-first century includes swine flu, bird flu, Ebola virus, Zika virus, severe acute respiratory syndrome (SARS), Middle East respiratory syndrome (MERS), as well as the continuing SARS-CoV-2 epidemic (Nii-Trebi, 2017; Morens and Fauci, 2020). Moreover, quite a few older diseases have re-emerged, such as *human papillomavirus* (HPV), *Mycobacterium tuberculosis* (MTB), *human immunodeficiency virus* (HIV), and the parasitic organisms responsible for dengue and malaria (Brejt and Golightly, 2019; Muller et al., 2017). Infections can spread rapidly in a population, making it imperative to quickly identify the causative agent to make an early diagnosis. Bacteria are the main agents of infectious diseases, and their importance depends on the kind and severity of the damage they cause in the body after penetrating a host. These organisms have the ability to enter through body openings, including the mouth, nostrils, genitals, as well as eyes and skin. These infectious agents grow wherever they reside and release toxins found inside cells or in body fluids and damage the tissues as a result. They often produce and release toxins or enzymes that disrupt normal organ and system functions. Additionally, these harmful substances can spread to other organs and systems, potentially leading to further damage or allowing the pathogen to invade additional areas. The host's immune system functions as a complex defense

DOI: 10.1201/9781003537151-7

mechanism that actively targets and eliminates infectious agents. Infectious diseases develop when this immune response is insufficient to clear pathogenic organisms. Thus, these diseases can appear at various points in time within specific populations and settings. Gaining insight into the dynamics of disease and how it spreads allows for the creation of effective methods for fighting, preventing, and managing these conditions (Walker et al., 2014). However, certain pathogens may, after a phase of dormancy post-elimination, gain new properties that facilitate reinfection of either their original or different hosts, often in larger and more concerning quantities.

Non-infectious diseases are those that are not caused by infectious agents; instead, they are generally the result of genetic predispositions or environmental factors unrelated to pathogens, including exposure to harmful substances or unhealthy lifestyle choices. Most of the non-communicable diseases have multiple etiological factors, which mean that their causes are often a product of gene–environment interaction. Examples of non-infectious diseases include cystic fibrosis (CF), various types of cancer, cardiovascular conditions like coronary artery disease (CAD), and diabetes mellitus (DM). Cancer still occupies its position as the foremost non-infectious disease and the second cause of death in the world. As has been pointed out, a number of non-infectious diseases are chronic in nature, but this feature is not unique to the non-infectious diseases only, as there exist certain infectious diseases, like acquired immunodeficiency syndrome (AIDS), which are chronic in nature. However, there are diseases that are non-infectious and are acute, in most cases resulting to death soon. Non-communicable diseases can also present acute symptoms which include some forms of cancer and heart diseases. Yes, non-infectious diseases, also known as disease of affluence or lifestyle diseases, are mainly caused by irresponsible decision-making on health status. These conditions first gained prominence in wealthier nations, while infectious diseases continued to be a significant issue in poorer countries. Currently, non-infectious diseases are the foremost contributors to global mortality. Although infectious diseases still account for a higher number of deaths in low-income regions, the mortality rates from non-infectious diseases are anticipated to increase in these areas as well. By the year 2030, it is projected that non-infectious diseases will result in over 50 million fatalities annually worldwide, with the majority of these deaths occurring in low- and middle-income countries rather than in high-income nations. The economic impact on national economies will be substantial, as non-infectious diseases lead to considerable worker disability and premature mortality.

7.2 PRINCIPLES OF CRISPR-CAS9-BASED DIAGNOSTICS

Diagnostics is an essential part of the medical field, because without it, there is no focus or clear direction. Prompt, precise, and early diagnosis at the point of care is essential for addressing these challenges effectively in every part of the world, especially where resources are limited. The enforcement of these tests results to early identification, good treatment, and check on the disease's rate of spread. Currently, bacterial pathogen identification primarily relies on culture techniques and staining procedures, which are time-consuming and exhibit fluctuating susceptibility (Bhattacharyya et al., 2018). Although diagnosis based on antigen–antibody interactions is fast and effective, sensitivity can vary, and the development of antibodies requires more time (Caliendo et al., 2013). Advancements in nucleic acid detection techniques, including polymerase chain reaction (PCR), quantitative PCR (qPCR), fluorescence in situ hybridization (FISH), and genome sequencing, have transformed infectious disease diagnostics, emphasizing precision and innovation in modern medical science (Ettinger et al., 2018). It should be noted that within the recent period, there has been an impressive development of molecular diagnostic methods; however, most of these technologies require significant infrastructural support, expensive equipment, and highly qualified staff of the laboratory. This makes it imperative to search for new detection methods that are both precise and sensitive for infections which offer inexpensive and straightforward solutions.

Since the development of a number of nucleic acid–binding proteins, the field of genome editing has undergone a fundamental change (Gaj et al., 2013). More and more are such proteins used in practical applications because they are so simple to design and program to be able to bind to a specific DNA/RNA sequence. In addition, a growing body of literature suggests application of the reagents as new reagents for pathogen nucleic acid identification. The clustered regularly interspaced short palindromic repeats (CRISPR) and CRISPR-associated protein (Cas) system does this today, though it is most effective for nucleic acid binding because of its strong nuclease activity and specificity of binding to specific sequences. Since their discovery (Makarova et al., 2011, 2020), the scope of CRISPR-Cas systems has skyrocketed. Currently, these systems are categorized into two primary classes, with various types and subtypes defined by their evolutionary lineage (Makarova et al., 2015). The classification is based on the structure of the ribonucleoprotein effector complex. In the target immunity systems, class 1 systems contain several effector proteins, and in contrast, class 2 systems have a single protein that binds to CRISPR RNAs (crRNAs). The design guidelines for crRNAs targeting the various effector proteins in CRISPR diagnostics are generalizable from other CRISPR applications. Class 2 systems are especially used in diagnostics because they are relatively easy to reconstitute and include enzymes with collateral activities which are crucial to many CRISPR-based diagnostic applications. Further, the components of type I or type III systems, for instance, nuclease Csm6 or Cas10, have been adapted for diagnostic applications, particularly with components of type II systems as part of the native type III complex or in isolation (Gootenberg et al., 2018). Early CRISPR-based diagnostic tools were primarily the naturally occurring molecules that target naked or non-naked double-stranded DNA. Techniques based on original CRISPR-associated protein 9 (Cas9) have been integrated with other concepts, including the guide-directed restoration of broken proteins with non-reactive members of Cas9 family (Zhang et al., 2017), the degradation of protospacer adjacent motif (PAM)–bearing sites by Cas9 (Pardee et al., 2016), and the untwisting of the opposite strand of the DNA with Cas9 to provide isothermal amplification (Zhou et al., 2018). Some of the approaches are mentioned in what follows.

7.2.1 NASBA–CRISPR Cleavage (NASBACC): A CRISPR-Cas9 Tool for Strain-Specific Viral Detection and Genotyping

The Cas9-based technique known as nucleic acid sequence–based amplification (NASBA) CRISPR cleavage (NASBACC) system is an advanced molecular diagnostic platform that integrates three key technologies for the sensitive and specific detection of nucleic acids. It uses NASBA for isothermal target preamplification, enabling rapid and efficient amplification without thermal cycling. Cas9 cleavage provides precise, programmable target recognition, relying on the PAM for high specificity in distinguishing sequences. A toehold sensor completes the system by delivering a sensitive visual readout through a synthetic RNA-based switch that signals successful target cleavage (Pardee et al., 2016). In fact, a toehold trigger is linked to an RNA segment via reverse transcription during NASBA amplification. In the presence of a PAM sequence within the RNA, Cas9 precisely cleaves the trigger sequence. However, when the PAM sequence is absent, the intact RNA engages the toehold switch, unleashing a color change as a clear signal. This innovative approach leverages unique PAM sites to distinguish between viral strains. For example, it enables the detection of Zika virus in monkey plasma even in extremely low quantities. Additionally, the technique CRISPR-typing PCR (ctPCR) has been used to screen for multiple subtypes of HPV16 and HPV18.

7.2.2 LEOPARD: A Multiplex RNA Detection Method with Single-Nucleotide Precision

A cutting-edge breakthrough known as leveraging engineered tracrRNAs and on-target DNAs for parallel RNA detection (LEOPARD) harnesses the power of Cas9 to identify multiple RNA sequences simultaneously with single-nucleotide precision, setting a new standard for RNA detection (Jiao et al., 2021). In this regard, an approach based on non-Cas trans-activating crRNAs (tracrRNAs), similar to those typically present in type II systems for the establishment of Cas9–crRNA complexes, has been developed. They bind cellular RNA to form crRNAs that are "non-canonical." Like any other crRNAs, these reprogrammed non-canonical crRNAs are designed to bind to the selected cellular transcripts while keeping the capacity to form the Cas9–crRNA complex. This lets Cas9 pretty effectively point to the target DNA sequence.

7.2.3 Specific High-Sensitivity Enzymatic Reporter Unlocking (SHERLOCK)

The collateral cleavage activity research of Cas13 and Cas12 has led to the development of a wide span of CRISPR diagnostic tools which are convenient to implement, small in size, and scalable. This was followed by the development of SHERLOCK technology, based on the identification of Cas13's collateral activity (Gootenberg et al., 2017). At that time, the advantage of the RNA-guided and programmable traits of *Leptotrichia wadei* (LwaCas13a) that conferred the most efficient target-activated collateral degradation to date was utilized. However, LwaCas13a alone was not able to robustly discriminate different viral titers in the specimen arising from infected individuals. The method recombinase polymerase amplification (RPA) (Piepenburg et al., 2006) was used to further increase the precision of the assay. The approach here utilizes the synthesis of double-stranded DNA (dsDNA), followed by *in vitro* transcription of RNA using T7 RNA polymerase to create RNA detectable by Cas13.

The detection approach employs a synthetic single-stranded RNA (ssRNA) molecule, wherein a quencher label is attached to the 5' end of the sense strand while a fluorescent dye is attached to the 3' end of the antisense strand. The action of Cas13 leads to the cleavage of the collateral by a fluorescent signal-emitting fluorophore that could be quantified by a fluorescence scanner.

7.2.4 DNA Endonuclease–Targeted CRISPR Trans Reporter (DETECTR)

The CRISPR system's trans-cleavage potential, DETECTR (DNA endonuclease targeted CRISPR trans reporter), which has become a well-known diagnostic innovation (Chen et al., 2018a), uses a reporter with fluorescence-quenched (FQ) label that measures trans-cleavage activity since the Cas12a–guide RNA (gRNA) complex binds to its complementary target DNA sequence. Once the target DNA is interacted with, the Cas12a gRNA complex becomes active for trans-cleavage activity, breaking the FQ-labeled ssDNA reporters. The quencher is consequently released from the breakage of this DNA, and fluorescence signal of the reporter DNA can be detected on the basis of activation after the target DNA protein is added.

Importantly, the DETECTR assay relies on target amplification, which in this case relies on RPA. Attomolar sensitivity is crucial for DNA detection, and this amplification method is crucial for achieving the ability to detect very low concentrations of target nucleic acids via DNA detection. This approach is based on highly sensitive and efficient nucleic acid detection by combining Cas12a's collateral cleavage activity and the RPA amplification step (Gootenberg et al., 2017).

7.2.5 Finding Low-Abundance Sequences by Hybridization (FLASH)

This method is an advanced CRISPR-Cas9 diagnostic platform based on the efficiency, specificity, and versatility of Cas9 to create a predetermined sequence (Quan et al., 2019). At the same time, the FLASH-NGS can enrich for the target genes up to five orders of magnitude and achieves very high sensitivity in the analysis of the genes present at subattomolar levels with low non-target interference. In this situation, DNA is first blocked with phosphatase, and the integrity of the genomic DNA is then digested by a Cas9 complex carrying specific crRNAs to target the wanted genes (Quan et al., 2019). Lastly, cloning, sequencing, and adapter fusion are carried out. The method has been used to quantitatively detect the resistant gram-positive bacteria and *P. falciparum* in respiratory secretions and on dry blood spots.

7.3 CRISPR-CAS SYSTEMS APPLICATIONS IN DETECTING INFECTIOUS DISEASES

Several areas in medicine are revolutionized by CRISPR-based diagnostics. It is a major tool for identifying nucleic acid biomarkers of many diseases, mutations, and deletions common to genetic disorders. The technology has also been adapted to detect proteins and small molecules, but beyond this, it can also detect viruses. CRISPR diagnostics is largely about identifying pathogen DNA or RNA from viruses, bacteria, and parasites in order to identify and monitor infectious disease.

7.3.1 Detection of Viral Infections

CRISPR-Cas revolution brings an amazing opportunity for diagnosis and, most importantly, for the treatment of the generation of infectious bugs. This powerful molecular tool has already been used beyond bacteria to help fight viral diseases, like HIV, coronavirus-19 (COVID-19), and hepatitis, and heralds a new age for precision medicine.

7.3.1.1 COVID-19

The rise of COVID-19 globally has brought with it a whole host of challenges to societies everywhere in diverse sectors. Regardless of the ongoing vaccination process, the threat of the repeated appearance of new subtypes of SARS-CoV-2 persists. Rapid diagnostic tests (RDTs) should then be used in infections, treatment progress, patient outcomes, and disease spread. RT-qPCR, sequencing-based techniques, and immunological assays have been used to diagnose SARS-CoV-2. Significant challenges have hindered the effectiveness of traditional methods, such as RT-PCR and serology tests; they are often inaccurate and insensitive. And these shortcomings usually arise for lack of sample preparation, bad reagents, faulty equipment, or the variabilities of clinical specimens. This emphasizes that more research is needed to develop more affordable, highly sensitive alternatives for SARS-CoV-2 detection (Javalkote et al., 2022; Yoshimi et al., 2020). Recent examples of diagnostic technologies that have become viable sources of rapid testing include CRISPR-Cas as well. Feng Zhang and his team were the first to present an RNA virus detection tool, which they dubbed SHERLOCK. A biotech company has received FDA approval for its COVID-19 test built upon the SHERLOCK platform amidst the new COVID-19 pandemic. Although different from those earlier presented, this technique enables the detection of SARS-CoV-2 genes in about 1 hour (Yoshimi et al., 2020; Joung et al., 2020). The RNase in the SHERLOCK method is Cas13 enzyme, which does not attack DNA but attacks RNA. The enzyme will bind to complementary RNA targets when the enzyme forms a complex with gRNA and execute its function. When triggered, Cas13 cleaves the RNA target sequences plus the surrounding RNA reporters, which can be quantifiable signals (Joung et al., 2020; Yoshimi et al., 2020).

Broughton et al. (2020) revolutionized the detection of COVID-19 with the cutting-edge CRISPR-based system DETECTR using the latest CRISPR-Cas12 technology. Remarkably, this novel tracing technique is based, originally, on a 2018 technique adapted to detect HPV (Chen et al., 2018b). The SHERLOCK system is distinguished from DETECTR in their speed of detection—approximately 30 minutes—and the fact that, unlike SHERLOCK, DETECTR targets dsDNA with Cas12 (Javalkote et al., 2022; Wang et al., 2020). Cas12a-based detection method, or DECTECTR, a method that detects SARS-CoV-2 cDNA, becomes active when it is bound by the target. This activation results in the breaking of reporter probes, including FAM-labeled oligonucleotide sequences as well as biotin-labeled oligonucleotide sequences that, once fixed with streptavidin beads, can verify the presence of a virus.

Another such innovative diagnostic tool is CASdetect, or CRISPR-assisted detection, using CRISPR-based DNA detection (CDetection) to detect SARS-CoV2, the virus that causes the COVID-19 disease (Guo et al., 2020). The study reported that

CASdetect achieves a detection limit of 1×10^4 copies/mL for the SARS-CoV-2 pseudo-virus without reactivity to other human EMC coronaviruses (Guo et al., 2020). CASdetect reflects a new standard in viral nucleic acid detection which combines sample preparation protocols, nucleic acid amplification, and CDetection methods all in one (Guo et al., 2020). The enormous benefit of this innovative system also includes negligible cross-reactivity, reduced false positive results, and terrific precision. Yet its implementation is not free of challenges, including the need for specialized kits and chemicals to extract nucleic acids (Rahimi et al., 2021).

To provide a quicker and more accurate method of SARS-CoV-2 virus detection, the growing techniques in diagnostics are expanded to go with the CRISPR-Cas12a-NER tool. The SARS-CoV-2-specific CRISPR RNAs (crRNAs), their associated Cas12a protein, and a ssDNA reporter tagged with a green fluorescent protein (GFP) are used in this system. When SARS-CoV-2 nucleic acid is present, the Cas12a protein cleaves the reporter molecule, resulting in a visible green fluorescence under the wavelength of 485 nm (Wang et al., 2020). Fifteen crRNAs have been engineered to target different regions of the Wuhan-Hu-1 strain, namely, open reading frames 1a (orf1a), orf1b, nucleocapsid (N), and envelope (E) genes, which allow discrimination from the other SARS-linked viruses by single nucleotide polymorphisms (SNPs). The advantages of this method are as follows: The method is portable, easy to perform, highly sensitive, and specific; does not require sophisticated equipment; and needs relatively short time, and results can be seen with the naked eye. Nonetheless, the requirement for nucleic acid extraction, along with automated extraction tool kits and chemical supplies, presents ongoing challenges (Rahimi et al., 2021)

The variant nucleotide guard (VaNGuard) assay is a precise method for detecting viral mutations, using either purified RNA or nasopharyngeal swab samples (Ooi et al., 2021). It involves three steps: sample preparation, RT-LAMP amplification, and Cas12a-based detection, with results visualized through fluorescence or lateral flow strips. On-the-spot SARS-CoV-2 testing using this technique is very accurate, extremely fast, and low in cost. Although there are many advantages to the method, there are some challenges, including the requirement for specific reagents, or kits, for nucleic acid extraction (Rahimi et al., 2021).

A new and innovative testing method known as STOP (SHERLOCK testing in one pot) has been developed, yielding results in under an hour. The STOP COVID assay is particularly attractive for point-of-care environments because of its rapid results. Unlike conventional RT-qPCR tests for SARS-CoV-2 detection, its sensitivity is unmatched, even detecting truly the smallest amounts, as little as 100 viral genome copies per microliter. This gives results in about 70 minutes by dipstick method and in as little as 40 minutes by fluorescence detection, and the testing process is simpler.

7.3.1.2 Human Immunodeficiency Virus

The prevention and treatment of HIV/AIDS has advanced, and yet the disease has not been solved as a global health challenge. Highly active antiretroviral therapy (HAART) successfully suppresses viral replication but does not eliminate latent viral reservoirs. This approach using the CRISPR-Cas9 gene-editing system to target cellular co-factors or the HIV-1 genome could eliminate infection, eliminate proviruses, and reactivate latent virus for eradication. As a result, this novel gene-editing method has demonstrated preventing and decreasing HIV-1/AIDS in human cells and animal models. In recent years, much progress has been made in developing CRISPR-Cas9-mediated therapies for viral immunodeficiency. This represents a revolutionary step in using the CRISPR-Cas system to investigate the correction of HIV-1 mutations with the CRISPR-SpCas9 gene-editing technology. Elimination of persistent HIV-1 reservoirs, which are maintained in an inactive state in infected cells, is a major hurdle for AIDS therapy. A highly effective strategy for gene manipulation involves utilizing CRISPR-SpCas9 to cut integrated proviral DNA from the host genome, presenting a potential breakthrough in infection treatment. CRISPR's efficacy has been demonstrated as a potent instrument for the direct modification of the HIV-1 genome in infected cells. In a pioneering study, Ebina et al. (2013) made history by demonstrating for the first time that CRISPR could effectively target and edit the HIV-1 genome within infected HEK293T and HeLa cells (Ebina et al., 2013). Their work demonstrated the potential of CRISPR-Cas9 to significantly suppress viral expression, paving the way for its use in developing future treatments. This breakthrough has spurred extensive research into CRISPR-Cas9 as a promising gene therapy for HIV-1/AIDS, highlighting its potential for innovative treatments (Wang et al., 2014; Ye et al., 2014; Zhu et al., 2015; Chen et al., 2018b). Similarly, Hu et al. (2014) investigated the excision of the HIV-1 genome using CRISPR-SpCas9 targeting conserved regions within HIV-1 LTR U3 (Hu et al., 2014). In microglial, promonocytic, and T cells, they found effective suppression of HIV-1 gene expression and replication, with no indication for genotoxicity or other adverse side effects. Zhu et al. (2015) also eliminated latent HIV-1 provirus in JLat10.6 cells via CRISPR-SpCas9 targeting of ten genomic sites. To improve editing efficiency, Liao et al. (2015) targeted multiple HIV-1 genome regions and increased resistance in primary cells. Combining two gRNAs effectively suppressed replication and prevented escape in T cells and in reservoir cells, as shown by Lebbink et al. (2017). Ye et al. (2014) developed C-C chemokine receptor 5, 32 base-pair deletion (CCR5-Δ32) monocytes and macrophages resistant to R5-tropic HIV-1 via combining CRISPR-SpCas9 with piggyBac transposase. Likewise, Li et al. (2015) demonstrated CCR5-silencing CRISPR-SpCas9 delivery using a chimeric adenovirus to HIV-1-resistant CD4+ T cells.

In previous research, Berthoux et al. (2005) utilized a CRISPR-Cas9 genome editing technique to alter tripartite motif-containing protein 5 alpha (TRIM5α), creating a variant that could effectively restrict HIV-1 replication through homology-directed repair (HDR). Their results emphasized that modifying both alleles of TRIM5α was crucial for a significant reduction in HIV-1 replication.

While CRISPR-Cas9 has its limitations—such as imprecision and reliance on PAM alternative delivery systems—like CRISPR-SaCas9, using adeno-associated virus (AAV) presents distinct benefits. AAV delivery methods are recognized for their low toxicity, minimal disease risk, durable gene expression, and ability to effectively target various cell types (Mingozzi and High, 2013). For anti-HIV-1 gene therapy using CRISPR-SaCas9, thorough testing in animal models is essential. Yin et al. (2017) demonstrated the potential of a single AAV vector carrying multiple gRNAs and SaCas9 to target the HIV-1 provirus across three animal models. In experiments with HIV-1 Tg26 mice, which possess a replication-deficient HIV-1 provirus, an injection of AAV-DJ/8 containing four gRNAs and SaCas9 led to a significant decrease in viral replication.

In the fight against HIV-1/AIDS, the "shock and kill" strategy offers a beacon of hope for eradicating latent viral reservoirs. This approach begins by using latency-reversing agents (LRAs) to awaken dormant proviruses from their hidden state during the "shock" phase, making them more vulnerable. Once these viruses are reactivated, they enter the "kill" phase, where they become targets for elimination—either by the harmful effects of HIV-1 itself or by the body's immune system (Chen et al., 2018b). Central to this strategy is the CRISPR activation (CRISPRa) system, a potent LRA that enhances gene expression and stimulates viral replication, which helps make the virus easier to target and eliminate (Deeks, 2012).

The SHERLOCK method involves the RNA-cleaving function of the Cas13a enzyme, and integrated with are RNA amplification techniques, including recombinase polymerase or reverse transcription recombinase polymerase, to enable rapid, user-friendly, and highly sensitive detection of target genes (Gootenberg et al., 2017). Because HIV-1 has two copies of ssRNA, SHERLOCK is ideal in screening individuals with suspected HIV-1 and in measuring the replication rate of active HIV-1 in AIDS patient with a high degree of sensitivity. Much like SHERLOCK, the DNA endonuclease from the Cas12a-derived DETECTR can be used to induce double-strand breaks and non-specific DNA cleavage (Li et al., 2019). As a result, this approach has the ability to cleave the ssDNA reporter, having fluorescence and quenching groups at both termini. By employing specific gRNA/Cas12a, this method can detect HIV-1 proviral DNA in patients with AIDS.

7.3.2 Identification of Bacterial Infections

A bacterial infection occurs when bacteria enter the body, multiply, and provoke a reaction from the immune system. The CRISPR-Cas system serves as a powerful method to identify bacterial infections by detecting nucleic acids and specific biomarkers found in bacteria. Antibiotics usually kill other bacterial species along with the targeted bacteria; CRISPR-Cas precisely targets antibiotic resistance genes (ARGs) and effectively eradicates harmful bacteria while preserving other bacterial species within diverse microbial communities (Greene, 2018). This targeted elimination of pathogenic bacteria is facilitated by the system's capability to focus on genes located on both chromosomes and plasmids (Citorik et al., 2014). CRISPR-Cas systems have been utilized to diagnose various bacterial infections, notably tuberculosis (TB), methicillin-resistant *Staphylococcus aureus* (MRSA), etc.

7.3.2.1 Tuberculosis

TB, a chronic infectious disease caused by MTB, has been among the top ten global causes of death since 2007 and remains the leading fatal infectious disease (Bahlool et al., 2022). Effective and rapid diagnosis of TB is crucial for ensuring that anti-TB treatment begins without delay (Walzl et al., 2018). However, 58.89% of the 9.9 million new TB cases estimated to occur worldwide each year are not diagnosed. Current methods to detect and diagnose TB involve assessing clinical symptoms, radiological changes, and laboratory investigations. Despite its popularity, smear microscopy, a traditional approach of detecting tubercle bacilli in specimens, is less sensitive while detecting tubercle bacilli in specimens from children, in persons with HIV, and in cases of extrapulmonary TB. While mycobacterial culture testing of sputum specimens is recognized as the best approach to diagnosing TB, the test takes a number of weeks to produce results because of the slow growth of *MTB* (Miller et al., 1994). Hence the need to emphasize the development of better diagnostic tools in order to enhance early identification of active TB patients. Although CRISPR-based diagnostics have not yet been employed for clinical diagnosis of TB, they offer several advantages over current diagnostic approaches in practice. The rapid adoption of CRISPR-Cas systems in clinical diagnostics is due to their flexibility, sensitivity, and precision. These methods detect pathogens by converting target nucleic acid segments into fluorescent signals. For instance, the CRISPR-Cas12a system was reported to accurately detect TB and other microbial infections (Jiang et al., 2022; Lei et al., 2022). However, another RNA-guided endonuclease, known as the CRISPR-Cas13a system, has a glorious future both as diagnostic tools. Cas13a still had collateral RNase activity, thus further developing a CRISPR-Cas13a-based platform by integrating Cas13a collateral RNase activity with molecular amplification that can identify viruses with detection limits of attomolar sensitivity and single base resolution (Gootenberg et al., 2017). Thus, Sam et al. (2021) are worthy of being commended for implementing a modern platform for MTB DNA detection called TB-QUICK based on integrating LAMP with CRISPR-Cas12b detection. In 2019, Ai et al. (2019) pioneered a groundbreaking diagnostic technique by merging the power of RPA with the precision of Cas12a-based detection, creating what they called CRISPR-MTB. This new method offers a quick and highly sensitive way to detect both pulmonary and extrapulmonary TB.

The CRISPR interference (CRISPRi) technique was developed to allow for precise gene regulation in microbial organisms. It makes use of a non-reactive version of Cas9, known as dCas9, to manage this regulation. This approach is particularly utilized in mycobacterial species, focusing on reducing gene expression in *M. smegmatis* and *M. tuberculosis* (Rock et al., 2017). CRISPRi holds great promise for aiding in the discovery of new antibiotics against TB, thus expanding the range of targets that are biologically relevant (Rock, 2019). CRISPR-based diagnostic tools have been developed to detect mutant genes or point mutations that lead to drug resistance in MTB, including those linked to resistance against streptomycin and fluoroquinolones. Being capable to discover SNPs early and monitor drug resistance is a major advantage, since it allows clinicians to make better drug choices (Liu et al., 2022).

7.3.2.2 Methicillin-Resistant *Staphylococcus aureus* (MRSA)

Staphylococcus aureus (*S. aureus*) usually does not cause any harmful effects in humans. However, while the immune system is weakened, food poisoning can result from toxin production, and infectious diseases can result, ranging from the mild to the deadly. These diseases include mild skin infections (boils, sores, abscesses) or serious diseases, such as pneumonia, meningitis, and endocarditis (Otto, 2014). The genome of *S. aureus* not only consists of a core set of genes but also includes various mobile genetic elements (MGEs) that can be transferred between cells (Lanza et al., 2015). This genetic structure plays a key role in driving genetic variation, facilitating pathogenic evolution, and enabling the bacteria to adapt to new environments. These factors shed light on why *S. aureus* is such a common pathogen and has the ability to acquire antimicrobial resistance (AMR) genes (Sato'o et al., 2018). Infections caused by this organism are prevalent in both hospital and community settings, with *S. aureus* being a leading pathogen. MRSA infections have a higher mortality rate compared to those caused by methicillin-sensitive strains, making them a significant global health concern. Consequently, it is crucial to have rapid and highly sensitive methods for detecting clinical MRSA infections in patients to implement timely infection control strategies. The CRISPR-Cas system serves as an antimicrobial mechanism, increasing bacterial vulnerability to antibiotics by targeting and eliminating antibiotic resistance genes (ARGs). For the past few years, scientists have been exploring a novel approach using the CRISPR-Cas system to direct antibiotics toward ARGs on plasmids in pathogenic bacteria, aiming to restore the effectiveness of antibiotics against drug-resistant strains. For instance, research has shown that the CRISPR-Cas system effectively restored *S. aureus* sensitivity to kanamycin (Bikard et al., 2014) and to methicillin (Kang et al., 2017). In another study, Yosef et al. (2015) demonstrated that the CRISPR system can excise multiple plasmids that harbor antibiotic resistance genes, which offers a hopeful strategy for tackling multidrug-resistant infections. In summary, the CRISPR-Cas system is vital in preventing the transfer and transformation of plasmids in drug-resistant bacteria. Bikard et al. (2014) studied methicillin-sensitive *S. aureus* and created a phagemid designed to target the plasmid containing the tetracycline resistance gene. The findings indicated that the treated cells showed no tetracycline resistance. Additionally, lysostaphin (Lst), a powerful lytic enzyme against *S. aureus*, was introduced using an electroporation technique (Wu et al., 2019) to transfer CRISPR-dCas9 plasmids into *S. aureus* ATCC 6538 (Monk et al., 2012). Nevertheless, this technique faces challenges due to the potential for cell damage and cytotoxicity, making it more suitable for in vitro applications only.

Ram et al. (2018) developed antibacterial community (ABD) systems that are built from engineered staphylococcal pathogenicity islands (SaPIs) for the treatment of *S. aureus* infections. These kill off ABD function by using SaPICRISPR-Cas9 vectors to target the gene called agr. Evidence indicates that ABDs can directly eradicate contaminated *S. aureus* during *in vivo* testing. The system, however, exhibits limited stability. To address this, Guk et al. (2017) introduced a novel method involving CRISPR-mediated DNA FISH, capable of detecting *MRSA* at concentrations as low as 10 CFU/mL within 0.5 hours.

7.3.3 Detection of Parasitic Infections

Parasitosis, an infection due to parasites, is the biggest health problem around the world (Nordling, 2023). Tropical regions that face poverty, conflict, or poor sanitation are often hotspots for various parasitic diseases. According to the WHO, 14 different parasite diseases cause about 48.4 million cases, claim 59,724 casualties, and a take of 8.78 million disability-adjusted life years (DALYs) annually worldwide (Torgerson et al., 2015). The transmission of parasites through contaminated food constitutes a significant public health challenge in low- and middle-income countries, where it remains a critical concern (Torgerson et al., 2015). Timely therapeutic measures in cases of parasitosis and efficient epidemiological monitoring can only be performed based on the swift and precise detection of the parasite. Therefore, developing a novel, rapid, accurate, and sensitive nucleic acid diagnostic test utilizing nucleic acid amplification techniques (NAATs) in conjunction with isothermal amplification is crucial for clinical applications. A new and attractive strategy is based on the CRISPR-associated (Cas) nuclease to enable the specific detection of nucleic acids, which eliminates shortcomings associated with the use of instruments and complicated procedures. Through careful selection of target genes and the design of specific CRISPR crRNAs, accurate identification of diverse parasites can be assured. CRISPR-Cas systems have also shown promise in detecting parasitic infections, such as malaria and leishmaniasis, among others.

7.3.3.1 Malaria

Malaria has long been one of the greatest global health challenges and has been a leading cause of death in the tropics and the subtropics (Cowman et al., 2016; Monroe et al., 2022). Key intervention tools are becoming ineffective, insecticide-treated nets and anti-malarial medications are becoming increasingly resistant, and there is no vaccine that has proven to be effective till now (Memvanga and Nkanga, 2021). *Plasmodium* is the parasite responsible for causing malaria, living in both human and mosquito hosts. Among the several species of *Plasmodium*, five primarily cause malarial infections in humans worldwide. These include *P. falciparum*, *P. vivax*, *P. malariae*, *P. ovale*, and *P. knowlesi* (Memvanga and Nkanga, 2021; Sato, 2021). Malaria is mainly diagnosed by examining a thick Giemsa-stained peripheral blood smear under a microscope (Fitri et al., 2022). Despite all the aforementioned advantages, there are certain challenges that arise with this method, and these are the technical ability of the microscopist, the quality of the blood sample, parasite density, and most importantly, subjectivity in the interpretation of results (Berzosa et al., 2018; Ospina-Villa et al., 2018).

Cas9/gRNA has been widely applied in *Plasmodium* gene editing, but for identifying malaria nucleic acids, its application is still lacking. On the other hand, other alternative Cas proteins, like Cas12a and Cas13a, have higher targeted DNA cleavage capabilities and are better suited for detecting *Plasmodium*-infected samples. CRISPR-based SHERLOCK assays have been developed as a highly sensitive and specialized molecular technique to identify malaria species (Cunningham et al., 2021). In 2020, Lee and his colleagues replaced the DNA template with *P. vivax* (18s rRNA), *P. falciparum* (Pfr364 gene), *P. malariae* (18s rRNA), and *P. ovale* (18s rRNA) and programmed Cas12a (cpf1) effector proteins to identify and differentiate these dsDNA (Lee et al., 2020). CRISPR-based assays offer innovative methods for detecting drug resistance markers, with advanced tools like FLASH-NGS (finding low-abundance sequences by hybridization next-generation sequencing) enabling high multiplexing capabilities. In contrast, it has been shown to be highly effective at detecting resistance genes related to *P. falciparum* (Quan et al., 2019). With the emergence of CRISPR-Cas9 technology, the *Plasmodium* genome has become an attractive target for manipulation. In *P. falciparum*, CRISPR-Cas9 genome editing has confirmed the role of mutations in the pfcronin (*P. falciparum* coronin-like protein) (Demas et al., 2018) and pfk13 (kelch 13 propeller domain) genes (Stokes et al., 2021; Uwimana et al., 2020) in inducing *in vitro* artemisinin resistance phenotype. Artemisinin resistance is a complex process involving multiple proteins, pathways, and the genetic makeup of the parasite. The CRISPR-Cas9 technology has proven to be an effective method for uncovering and validating the mechanisms behind artemisinin resistance. As resistance to artemisinin-based therapies continues to rise, drug discovery has shifted toward identifying new antimalarial compounds that are effective against *P. falciparum*. Although their mechanisms of action remain unclear, the CRISPR technology provides a powerful tool for uncovering and validating resistance mechanisms, enhancing our understanding of the effects of these drugs.

7.3.3.2 Leishmaniasis

Leishmaniasis is a widespread and pressing public health issue, driven by microscopic *Leishmania* parasites transmitted through the bite of infected sandflies. It is estimated that 12 million people get affected by this disease worldwide and, in 98 countries, disclosing new cases of 1.6 million people every year, which paints a melancholic picture of the world and is continually going on (Alvar et al., 2012). Diagnosing leishmaniasis typically involves examining Giemsa-stained smears from skin lesions to spot Leishmania amastigotes or isolating the parasite through *in vitro* culture. Recent advancements in CRISPR technology have paved the way for its use in Leishmania, greatly improving the efficiency of gene editing (Zhang and Matlashewski, 2015). Unlike traditional gene targeting methods, CRISPR offers a simpler and more effective approach. This advancement has also been crucial in determining the essentiality of certain Leishmania genes, shedding light on their roles and functions (Ishemgulova et al., 2018; Zhang et al., 2015). Bengtson et al. (2022) showcased their technique by successfully detecting kinetoplast DNA (kDNA) from *Leishmania* in blood and urine samples, emphasizing its potential for diagnosis. In another study, Cunningham et al. (2021) used the Cas13a-mediated SHERLOCK method to create a diagnostic tool aimed at identifying *Plasmodium* parasites, which are responsible for malaria.

Despite decades of research, a safe and effective vaccine for leishmaniasis in humans has yet to be developed. The rationale for the creation of its vaccine is based on the finding that most people who get cured from the disease, either by curing or through medicine, become immune to the parasite. This strong immunity is widely regarded as effective, although there have been reports of reinfection, which reinforces the rationale for employing live-attenuated parasites in vaccine strategies. Many promising studies are focused on developing effective vaccine alternatives for leishmaniasis, using various approaches, such as live-attenuated parasites, *Leishmania* antigen fractions, inactivated parasites, DNA vaccines, and recombinant proteins (Yasmin et al., 2022). Vaccines based on attenuated parasites, engineered DNA, and protein formulations have not shown the ability to provide long-term protection (Okwor and Uzonna, 2008). One promising path is inspired by leishmanization, a first-generation vaccine approach that involves injecting live *Leishmania* major parasites into the skin to provide protection against future infections (Pandey et al., 2020). Live-attenuated Leishmania vaccine candidates demonstrate promise as they trigger a robust protective immune response without inducing disease, potentially providing long-lasting immunity against virulent strains (Pandey et al., 2020; Selvapandiyan et al., 2012; Zhang et al., 2020). Infections with live-attenuated *Leishmania* simulate the experience of being infected with a virulent strain, while also offering the advantage of disease prevention. This strategy allows the host's immune system to interact with a wide variety of *Leishmania* antigens, which is essential for promoting the development of an enduring, protective immunity (Selvapandiyan et al., 2012). Since the introduction of CRISPR-Cas gene-editing technology in 2012, along with the growing availability of CRISPR-based techniques for manipulating *Leishmania* parasites, researchers can more effectively choose safer attenuated strains (Jinek et al., 2012).

7.4 APPLICATIONS OF CRISPR-CAS SYSTEMS IN DETECTING NON-INFECTIOUS DISEASES

Some of the CRISPR-based diagnostic tools have progressed even to the identification of non-infectious diseases as well. Non-infectious disease diagnosis employs CRISPR enzymes with cleavage activity. The CRISPR-Cas9 technique has been employed since the identification of oncogenes for drug resistance based on the interaction of the genes, showing changes in genes when subjected to treatment and effects of exposure to drugs, and for determination of genes that cause resistance when exposed to targeted therapies. Some of the non-infectious diseases that can be detected via CRISPR-Cas systems are mentioned in what follows.

7.4.1 Detection of Genetic Disorders

CRISPR-Cas9 is a revolutionary molecular technique used to deliver gene therapy through the editing of genes that are involved in genetic diseases. This technology can modify DNA in various cells or organisms, facilitate the study of DNA mutations linked to hereditary diseases, and aid in the exploration of potential treatments for these conditions. CRISPR-Cas9 holds great promise for advancing research into treatments for various human genetic disorders, such as CF, hemophilia, Parkinson's disease, β-thalassemia, tyrosinemia, Alzheimer's disease, and more.

7.4.1.1 Cystic Fibrosis

CF is a genetic condition that is passed down from both parents and follows an autosomal recessive pattern. This condition is caused by mutations in the CFTR gene, which is responsible for producing a protein that plays a crucial role in managing the movement of chloride ions across cell membranes in various tissues, including the lungs and digestive system. Recent advancements in CRISPR technology, along with promising early results from both preclinical and clinical studies, suggest that this technology could soon be applied to treat various genetic diseases, including CF. Numerous studies demonstrating successful genome editing in different CF models offer hope for its future clinical use. Schwank and his team were among the first to utilize the CRISPR-Cas9 system to repair the CFTR gene in cultured intestinal stem cells from patients who are homozygous for the F508del mutation, representing an important milestone in the potential treatment of CF (Schwank et al., 2013). The gene-corrected stem cells successfully developed into organoids that showed functional responses to forskolin through volume swelling (Schwank et al., 2013). In another study, Firth et al. (2015) employed CRISPR-Cas9 to correct the gene mutation (F508del) in CFTR, using iPSCs generated from skin fibroblasts of homozygous CF patients. The corrected CFTR gene restored the chloride transport function in iPSCs, which then differentiated into mature airway epithelial cells. Similarly, acDNA of the entire CFTR gene was successfully integrated into GGTA1, a safe harbor gene in pig cells that have been cultivated, in another study (Zhou et al., 2019). These findings underscore the feasibility of correcting multiple mutations with a single gene-editing vector (Zhou et al., 2019). Graham and his colleagues utilized the CRISPR-Cas9 system to tackle multiple alleles in CF intestinal organoid cells (Graham and Hart, 2021). They applied base editing to correct the R785X mutation and successfully addressed two other mutations, R553X and W1282X, in clonal cell lines (Graham and Hart, 2021). This advanced technology enables the creation of sophisticated cellular and animal models that closely mimic respiratory diseases, including CF, chronic obstructive pulmonary disease (COPD), and asthma. The CRISPR-Cas system has been utilized to create new model organisms for CF. Using CRISPR-Cas9 technology, scientists developed a knockout sheep by removing the CFTR gene through a process known as non-homologous end joining (NHEJ) (Fan et al., 2018). These CFTR-/-sheep exhibited severe CF-like symptoms, including pancreatic fibrosis, intestinal obstructions, and the absence of the vas deferens in their testes (Fan et al., 2018). To investigate the pathophysiology of the ΔF508 mutation, researchers generated rat models using CRISPR-Cas9 to create a break at ΔF508 (phenylalanine) of the CFTR gene (Dreano et al., 2019). Furthermore, CRISPR nucleases were used to create a mutant CFTR mouse model carrying the G542X nonsense mutation, which displayed typical CF symptoms due to the loss of the CFTR ion channel (McHugh et al., 2018). In summary, CRISPR genome editing represents a highly promising approach for developing new therapeutic strategies and valuable experimental models to assess treatments for various CF-related mutations.

7.4.1.2 Sickle Cell Anemia

Sickle cell disease (SCD) is a prevalent genetic condition that impacts hemoglobin. It arises from a single-point mutation in the HBB gene, which changes the amino acid at codon 6 from glutamic acid to valine (Odame, 2014). This genetic alteration occurs in both alleles—meaning, individuals with this mutation are homozygous for sickle hemoglobin, or HbS. The condition is marked by vaso-occlusive crises, where sickled hemoglobin obstructs small blood vessels, causing severe pain and potential organ failure (Kato et al., 2018). Although SCD was the first molecular condition identified with a known genetic basis over 60 years ago, treatment alternatives for SCD continue to be quite restricted, and the average life expectancy for patients has not seen significant improvement in recent decades (Lanzkron et al., 2013). The CRISPR-Cas9 technique has been applied to treat SCD through two primary approaches: first, by repairing the mutation to restore the wild-type allele, which leads to increased expression of HbA and alleviates the SCD phenotype; second, by promoting the production of HbF through the disruption of gene regulatory elements, thereby reducing sickling and enhancing the overall quality of life. A 2016 cross-sectional study assessed the safety and durability of Cas9 RNPs, including Cas9 protein, guide RNA, HDR donor, and templates delivered via adeno-associated viral vectors in CD34+ hematopoietic stem and progenitor cells (Dever et al., 2016). These frequencies of HDR at the HBB locus using the TALENs were further demonstrated in this strategy (Dever et al., 2016). Furthermore, several other studies have narrated the effects of the ex vivo correction of SCD mutation, the mutation rate of which ranges between 18% and 94% in CD34+ HSPCs (Lattanzi et al., 2021). Khosravi et al. (2019) found that some experiments indicate enhancement of both efficacy and safety insofar as the abolition of the BCL11A gene results in augmented levels of γ-globin. Eventually, in a significant breakthrough, researchers utilized the CRISPR-Cas9 system to edit a 13.6 kb segment that includes the δ- and β-globin genes, as well as a potential γ-δ intergenic HbF repressor site in erythroblasts. This approach boosts HbF synthesis and improves SCD pathology by inhibiting repressor proteins that target the HGB1 and HGB2 promoters, increasing γ-globin expression (Weber et al., 2020). Comparing various CRISPR-Cas9 strategies to increase HbF, both gene knockouts (e.g., KLF1, BCL11A) and non-coding element insertions (e.g., HBG1, HBG2) achieved therapeutically active HbF levels (Lamsfus-Calle et al., 2020). Further, the approval of Casgevy demonstrates the progressive nature of genetic interventions, especially the CRISPR-Cas9, to treat SCD and other hereditary diseases, which mark a giant leap in genetic therapies (Dever et al., 2016).

7.4.2 Diagnosis of Autoimmune Diseases

Autoimmune diseases are disorders in which the immune system attacks and damages the body's own tissues (Marrack et al., 2001). Generally, the immune system functions with a level of immunologic tolerance that prevents it from harming the body. However, in autoimmune diseases, the body's tolerance is disrupted, causing the immune system to attack its own organs and tissues. These conditions affect approximately 3–5% of the general population (Marrack et al., 2001). Managing autoimmune diseases typically involves controlling the immune response, which often requires the use of immunosuppressive medications. Our increasing knowledge of the immunological and molecular mechanisms underlying autoimmune diseases has highlighted gene therapy as a promising treatment option (Shu et al., 2015). A central aim of gene therapy is to modulate cytokine levels and prevent lymphocytes from moving to affected areas, thereby helping reduce inflammation (Shu et al., 2015). As the need for innovative treatment approaches grows, CRISPR-Cas9 has emerged as a powerful tool in gene editing. This advanced technology shows potential in treating various autoimmune diseases, such as rheumatoid arthritis (RA) and multiple sclerosis (MS), providing new hope for individuals facing these chronic illnesses.

7.4.2.1 Rheumatoid Arthritis

Rheumatoid arthritis (RA) is an autoimmune disease where the immune system mistakenly attacks the joints, leading to inflammation. This condition is more prevalent among females and affects about 0.5–1.0% of adults in developed countries (Scott et al., 2010). One potential strategy for managing RA is the use of

CRISPR-Cas9 technology within immunotherapy. Unlike traditional immunotherapies that generally suppress the immune system, CRISPR-Cas9 enables precise targeting of specific immune components without weakening the overall immune response. Furthermore, B cells can be effectively targeted using CRISPR-Cas9 as a possible treatment for rheumatoid arthritis. A promising and cost-effective approach could involve delivering CRISPR-Cas9 systems directly into peripheral blood to specifically target T and B cells. Future delivery methods might investigate the use of nanoparticles as a safe and efficient way to transport CRISPR components into human cells. Several genes could be potential targets for future CRISPR-based editing in RA treatment, depending on experimental validation of their functional significance. According to Yang et al. (2020), there is a link between the MYC and FOXO1 genes and RA. It has been suggested that CD4+ T cells in RA patients show increased levels of autophagy, with MYC acting as a regulatory factor in this process (Yang et al., 2020). Additionally, FOXO1 is believed to play a role in RA activity (Yang et al., 2020). In a significant study, Yu et al. (2016) identified an intergenic SNP, rs6927172, in the 6q23 region, which was significantly associated with the progression of RA. This genome-wide association study highlighted several genes, including IL20RA, IL22RA2, IFNGR1, OLIG3, and TNFAIP3, in proximity to the SNP region (Yu et al., 2016). MicroRNAs (miRNAs), particularly miR155, play a vital role in the development of inflammatory diseases, like rheumatoid arthritis (Su et al., 2017). Targeting miR155 may provide potential therapeutic benefits. A recent study employing CRISPR-Cas9 to eliminate miR155 in the RAW264.7 macrophage cell line demonstrated a reduction in the production of pro-inflammatory cytokines (Jing et al., 2015). Furthermore, this technology has the potential to enhance personalized medicine by correcting mutations that lead to autoinflammatory diseases and restoring the functionality of the affected gene for individual patients (Vicente et al., 2016). In addition, this method can be used to modify cell therapy approaches to increase regulatory T cell–suppressive activities in RA, or to augment immunomodulatory and regenerative characteristics of MSCs that are in the process of phase II trial for knee osteoarthritis (Pers et al., 2016).

7.4.2.2 Multiple Sclerosis

Multiple sclerosis (MS) is a long-term inflammatory disease that causes damage to the myelin sheath within the central nervous system (CNS) and is characterized by the damage of axons. Despite, the exact cause of the ailment not being definitive, MS progression is downstream of immune responses, both inherent and acquired, that target antigens in the myelin sheath. As will be shown, there are many different therapies and treatments for schizophrenia that aim to mitigate symptomatology and improve the quality of life. Another approach involves the use the CRISPR-Cas9 system using a neuronal-specific AAV9 with dead Cas9 linked with an activation domain (Carlstrom et al., 2020). This AAV is designed to generate a gRNA to knock down GSTA4, but with the purpose of increasing GSTA4 transcription with dCas9 fused with an activation domain. Research has shown that the upregulation of GSTA4 can prevent apoptosis of oligodendrocytes when they are differentiating and thus enhance signal conduction in the nerves together with a decrease in symptoms of MS (Carlstrom et al., 2020).

Recent research has identified several genes that may be targeted through CRISPR-Cas9 technology to aid in the treatment of various diseases. One such gene is IR7R, which is involved in regulating the immune system. Another gene of interest is RNA helicase DEAD box polypeptide 39B, which plays a role in RNA processing and the body's stress response (Galarza-Munoz et al., 2017). The IL2RA gene, essential for signaling in immune cells and the activation of T cells, also shows promise as a target for enhancing immune function (Maier et al., 2009). Furthermore, the TNFRSF1A gene, which helps regulate inflammation and cell death, could potentially be targeted to treat conditions characterized by chronic inflammation (Gregory et al., 2012). These findings bring us closer to the possibility of using CRISPR-Cas9 for the precise and effective treatment of genetic disorders and immune-related diseases.

7.5 CONCLUSION

The CRISPR system found in bacteria has been successfully adapted to edit the genomes of different organisms quickly and efficiently. CRISPR-Cas methods are simple, are cost-effective, and do not require advanced equipment or extensive expertise. Custom synthetic single guide RNAs are created to enable CRISPR systems to recognize nucleic acids associated with both infectious and non-infectious diseases, which enhances diagnostic tools for on-the-go detection and improves the diagnosis and prevention of diseases. This CRISPR-based diagnostic system provides infection diagnoses with accuracy on par with traditional methods, while being more affordable. The initial focus of research on CRISPR-based treatments for infectious diseases has been on the detection and management of conditions caused by drug-resistant bacteria and chronic viral infections. These diseases significantly impact global health, necessitating the development of modern tools to investigate their underlying mechanisms and to accurately detect and treat infections in various contexts. As CRISPR research and gene therapy continue to advance, expanding from rare genetic disorders to more common issues, the expectation of gene therapy being available for a few thousand dollars is increasingly seen as feasible and has been a recurring notion for many years.

REFERENCES

Ai, Jing-Wen, Xian Zhou, Teng Xu, Minling Yang, Yuanyuan Chen, Gui-Qing He, Ningp Pan et al. "CRISPR-based rapid and ultra-sensitive diagnostic test for Mycobacterium tuberculosis." *Emerging Microbes & Infections* 8, no. 1 (2019): 1361–1369.

Alvar, Jorge, Iván D. Vélez, Caryn Bern, Mercé Herrero, Philippe Desjeux, Jorge Cano, Jean Jannin, Margriet den Boer, and WHO Leishmaniasis Control Team. "Leishmaniasis worldwide and global estimates of its incidence." *PloS ONE* 7, no. 5 (2012): e35671.

Bahlool, Ahmad Z., Conor Grant, Sally-Ann Cryan, Joseph Keane, and Mary P. O'Sullivan. "All trans retinoic acid as a host-directed immunotherapy for tuberculosis." *Current Research in Immunology* 3 (2022): 54–72.

Bengtson, Michel, Mitasha Bharadwaj, Oskar Franch, Jaco van der Torre, Veronique Meerdink, Henk Schallig, and Cees Dekker. "CRISPR-dCas9 based DNA detection scheme for diagnostics in resource-limited settings." *Nanoscale* 14, no. 5 (2022): 1885–1895.

Berthoux, Lionel, Sarah Sebastian, Elena Sokolskaja, and Jeremy Luban. "Cyclophilin A is required for TRIM5α-mediated resistance to HIV-1 in Old World monkey cells." *Proceedings of the National Academy of Sciences* 102, no. 41 (2005): 14849–14853.

Berzosa, P., A. de Lucio, M. Romay-Barja, Z. Herrador, V. González, L. García, A. Fernández-Martínez, M. Santana-Morales, P. Ncogo, and B. Valladares. "Comparison of three diagnostic methods (microscopy, RDT, and PCR) for the detection of malaria parasites in representative samples from Equatorial Guinea." *Malaria Journal* 17(2018): 1–12.

Bhattacharyya, Roby P., Sri Gowtham Thakku, and Deborah T. Hung. "Harnessing CRISPR effectors for infectious disease diagnostics." *ACS Infectious Diseases* 4, no. 9 (2018): 1278–1282.

Bikard, David, Chad W. Euler, Wenyan Jiang, Philip M. Nussenzweig, Gregory W. Goldberg, Xavier Duportet, Vincent A. Fischetti, and Luciano A. Marraffini. "Exploiting CRISPR-Cas nucleases to produce sequence-specific antimicrobials." *Nature Biotechnology* 32, no. 11 (2014): 1146–1150.

Brejt, Josef A., and Linnie M. Golightly. "Severe malaria: Update on pathophysiology and treatment." *Current Opinion in Infectious Diseases* 32, no. 5 (2019): 413–418.

Broughton, James P., Xianding Deng, Guixia Yu, Clare L. Fasching, Venice Servellita, Jasmeet Singh, Xin Miao et al. "CRISPR–Cas12-based detection of SARS-CoV–2." *Nature Biotechnology* 38, no. 7 (2020): 870–874.

Caliendo, Angela M., David N. Gilbert, Christine C. Ginocchio, Kimberly E. Hanson, Larissa May, Thomas C. Quinn, Fred C. Tenover et al. "Better tests, better care: Improved diagnostics for infectious diseases." *Clinical Infectious Diseases* 57, no. suppl_3 (2013): S139–S170.

Carlstrom, Karl E., Keying Zhu, Ewoud Ewing, Inge E. Krabbendam, Robert A. Harris, Ana Mendanha Falcão, Maja Jagodic, Gonçalo Castelo-Branco, and Fredrik Piehl. "Gsta4 controls apoptosis of differentiating adult oligodendrocytes during homeostasis and remyelination via the mitochondria-associated Fas-Casp8-Bid-axis." *Nature Communications* 11, no. 1 (2020): 4071.

Chen, Hui, Kengku Liu, Zhao Li, and Ping Wang. "Point of care testing for infectious diseases." *Clinica Chimica Acta* 493 (2019): 138–147.

Chen, Janice S., Enbo Ma, Lucas B. Harrington, Maria Da Costa, Xinran Tian, Joel M. Palefsky, and Jennifer A. Doudna. "CRISPR-Cas12a target binding unleashes indiscriminate single-stranded DNase activity." *Science* 360, no. 6387 (2018a): 436–439.

Chen, Shuliang, Xiao Yu, and Deyin Guo. "CRISPR-Cas targeting of host genes as an antiviral strategy." *Viruses* 10, no. 1 (2018b): 40.

Citorik, Robert J., Mark Mimee, and Timothy K. Lu. "Sequence-specific antimicrobials using efficiently delivered RNA-guided nucleases." *Nature Biotechnology* 32, no. 11 (2014): 1141–1145.

Cowman, Alan F., Julie Healer, Danushka Marapana, and Kevin Marsh. "Malaria: Biology and disease." *Cell* 167, no. 3 (2016): 610–624.

Cunningham, Clark H., Christopher M. Hennelly, Jessica T. Lin, Ratawan Ubalee, Ross M. Boyce, Edgar M. Mulogo, Nicholas Hathaway et al. "A novel CRISPR-based malaria diagnostic capable of Plasmodium detection, species differentiation, and drug-resistance genotyping." *EBioMedicine* 68 (2021).

Deeks, Steven G. "Shock and kill." *Nature* 487, no. 7408 (2012): 439–440.

Demas, Allison R., Aabha I. Sharma, Wesley Wong, Angela M. Early, Seth Redmond, Selina Bopp, Daniel E. Neafsey, Sarah K. Volkman, Daniel L. Hartl, and Dyann F. Wirth. "Mutations in Plasmodium falciparum actin-binding protein coronin confer reduced artemisinin susceptibility." *Proceedings of the National Academy of Sciences* 115, no. 50 (2018): 12799–12804.

Dever, Daniel P., Rasmus O. Bak, Andreas Reinisch, Joab Camarena, Gabriel Washington, Carmencita E. Nicolas, Mara Pavel-Dinu et al. "CRISPR/Cas9 β-globin gene targeting in human haematopoietic stem cells." *Nature* 539, no. 7629 (2016): 384–389.

Dreano, Elise, Marc Bacchetta, Juliette Simonin, Louise Galmiche, Claire Usal, Lotfi Slimani, Jérémy Sadoine et al. "Characterization of two rat models of cystic fibrosis—KO and F508del CFTR—Generated by CRISPR-Cas9." *Animal Models and Experimental Medicine* 2, no. 4 (2019): 297–311.

Ebina, Hirotaka, Naoko Misawa, Yuka Kanemura, and Yoshio Koyanagi. "Harnessing the CRISPR/Cas9 system to disrupt latent HIV-1 provirus." *Scientific Reports* 3, no. 1 (2013): 2510.

Ettinger, David S., Dara L. Aisner, Douglas E. Wood, Wallace Akerley, Jessica Bauman, Joe Y. Chang, Lucian R. Chirieac et al. "NCCN guidelines insights: Non–small cell lung cancer, version 5.2018." *Journal of the National Comprehensive Cancer Network* 16, no. 7 (2018): 807–821.

Fan, Zhiqiang, Iuri Viotti Perisse, Calvin U. Cotton, Misha Regouski, Qinggang Meng, Chaim Domb, Arnaud J. Van Wettere et al. "A sheep model of cystic fibrosis generated by CRISPR/Cas9 disruption of the CFTR gene." *JCI Insight* 3, no. 19 (2018).

Firth, Amy L., Tushar Menon, Gregory S. Parker, Susan J. Qualls, Benjamin M. Lewis, Eugene Ke, Carl T. Dargitz et al. "Functional gene correction for cystic fibrosis in lung epithelial cells generated from patient iPSCs." *Cell Reports* 12, no. 9 (2015): 1385–1390.

Fitri, Loeki Enggar, Tarina Widaningrum, Agustina Tri Endharti, Muhammad Hatta Prabowo, Nuning Winaris, and Rivo Yudhinata Brian Nugraha. "Malaria diagnostic update: From conventional to advanced method." *Journal of Clinical Laboratory Analysis* 36, no. 4 (2022): e24314.

Gaj, Thomas, Charles A. Gersbach, and Carlos F. Barbas. "ZFN, TALEN, and CRISPR/Cas-based methods for genome engineering." *Trends in Biotechnology* 31, no. 7 (2013): 397–405.

Galarza-Munoz, Gaddiel, Farren B. S. Briggs, Irina Evsyukova, Geraldine Schott-Lerner, Edward M. Kennedy, Tinashe Nyanhete, Liuyang Wang et al. "Human epistatic interaction controls IL7R splicing and increases multiple sclerosis risk." *Cell* 169, no. 1 (2017): 72–84.

Gootenberg, Jonathan S., Omar O. Abudayyeh, Max J. Kellner, Julia Joung, James J. Collins, and Feng Zhang. "Multiplexed and portable nucleic acid detection platform with Cas13, Cas12a, and Csm6." *Science* 360, no. 6387 (2018): 439–444.

Gootenberg, Jonathan S., Omar O. Abudayyeh, Jeong Wook Lee, Patrick Essletzbichler, Aaron J. Dy, Julia Joung, Vanessa Verdine et al. "Nucleic acid detection with CRISPR-Cas13a/C2c2." *Science* 356, no. 6336 (2017): 438–442.

Graham, Carina, and Stephen Hart. "CRISPR/Cas9 gene editing therapies for cystic fibrosis." *Expert Opinion on Biological Therapy* 21, no. 6 (2021): 767–780.

Gregory, Adam P., Calliope A. Dendrou, Kathrine E. Attfield, Aiden Haghikia, Dionysia K. Xifara, Falk Butter, Gereon Poschmann et al. "TNF receptor 1 genetic risk mirrors outcome of anti-TNF therapy in multiple sclerosis." *Nature* 488, no. 7412 (2012): 508–511.

Greene, Adrienne C. "CRISPR-based antibacterials: Transforming bacterial defense into offense." *Trends in Biotechnology* 36, no. 2 (2018): 127–130.

Guk, Kyeonghye, Joo Oak Keem, Seul Gee Hwang, Hyeran Kim, Taejoon Kang, Eun-Kyung Lim, and Juyeon Jung. "A facile, rapid and sensitive detection of MRSA using a CRISPR-mediated DNA FISH method, antibody-like dCas9/sgRNA complex." *Biosensors and Bioelectronics* 95 (2017): 67–71.

Guo, Lu, Xuehan Sun, Xinge Wang, Chen Liang, Haiping Jiang, Qingqin Gao, Moyu Dai et al. "SARS-CoV-2 detection with CRISPR diagnostics." *Cell Discovery* 6, no. 1 (2020): 34.

Hu, Wenhui, Rafal Kaminski, Fan Yang, Yonggang Zhang, Laura Cosentino, Fang Li, Biao Luo et al. "RNA-directed gene editing specifically eradicates latent and prevents new HIV-1 infection." *Proceedings of the National Academy of Sciences* 111, no. 31 (2014): 11461–11466.

Ishemgulova, Aygul, Jana Hlaváčová, Karolina Majerová, Anzhelika Butenko, Julius Lukeš, Jan Votýpka, Petr Volf, and Vyacheslav Yurchenko. "CRISPR/Cas9 in Leishmania mexicana: A case study of LmxBTN1." *PLoS One* 13, no. 2 (2018): e0192723.

Javalkote, Vivek S., Nagesh Kancharla, Bhaskar Bhadra, Manish Shukla, Badrish Soni, Ajit Sapre, Michael Goodin, Anindya Bandyopadhyay, and Santanu Dasgupta. "CRISPR-based assays for rapid detection of SARS-CoV–2." *Methods* 203 (2022): 594–603.

Jiang, Chuanwen, Dagang Tao, Yuanchen Geng, Hao Yang, Bingrong Xu, Yingyu Chen, Changmin Hu, Huanchun Chen, Shengsong Xie, and Aizhen Guo. "Sensitive and specific detection of lumpy skin disease virus in cattle by CRISPR-Cas12a fluorescent assay coupled with recombinase polymerase amplification." *Genes* 13, no. 5 (2022): 734.

Jiao, Chunlei, Sahil Sharma, Gaurav Dugar, Natalia L. Peeck, Thorsten Bischler, Franziska Wimmer, Yanying Yu et al. "Noncanonical crRNAs derived from host transcripts enable multiplexable RNA detection by Cas9." *Science* 372, no. 6545 (2021): 941–948.

Jinek, Martin, Krzysztof Chylinski, Ines Fonfara, Michael Hauer, Jennifer A. Doudna, and Emmanuelle Charpentier. "A programmable dual-RNA–guided DNA endonuclease in adaptive bacterial immunity." *Science* 337, no. 6096 (2012): 816–821.

Jing, Weixia, Xuewu Zhang, Wenyan Sun, Xiujuan Hou, Zhongqiang Yao, and Yuelan Zhu. "CRISPR/Cas9-mediated genome editing of miRNA-155 inhibits proinflammatory cytokine production by RAW264. 7 cells." *BioMed Research International* 2015, no. 1 (2015): 326042.

Joung, Julia, Alim Ladha, Makoto Saito, Nam-Gyun Kim, Ann E. Woolley, Michael Segel, Robert P. J. Barretto et al. "Detection of SARS-CoV-2 with SHERLOCK one-pot testing." *New England Journal of Medicine* 383, no. 15 (2020): 1492–1494.

Kang, Yoo Kyung, Kyu Kwon, Jea Sung Ryu, Ha Neul Lee, Chankyu Park, and Hyun Jung Chung. "Nonviral genome editing based on a polymer-derivatized CRISPR nanocomplex for targeting bacterial pathogens and antibiotic resistance." *Bioconjugate Chemistry* 28, no. 4 (2017): 957–967.

Kato, Gregory J., Frédéric B. Piel, Clarice D. Reid, Marilyn H. Gaston, Kwaku Ohene-Frempong, Lakshmanan Krishnamurti, Wally R. Smith et al. "Sickle cell disease." *Nature Reviews Disease Primers* 4, no. 1 (2018): 1–22.

Khosravi, Mohammad Ali, Maryam Abbasalipour, Jean-Paul Concordet, Johannes Vom Berg, Sirous Zeinali, Arash Arashkia, Kayhan Azadmanesh, Thorsten Buch, and Morteza Karimipoor. "Targeted deletion of BCL11A gene by CRISPR-Cas9 system for fetal hemoglobin reactivation: A promising approach for gene therapy of beta thalassemia disease." *European Journal of Pharmacology* 854 (2019): 398–405.

Lamsfus-Calle, Andrés, Alberto Daniel-Moreno, Justin S. Antony, Thomas Epting, Lukas Heumos, Praveen Baskaran, Jakob Admard et al. "Comparative targeting analysis of KLF1, BCL11A, and HBG1/2 in CD34+ HSPCs by CRISPR/Cas9 for the induction of fetal hemoglobin." *Scientific Reports* 10, no. 1 (2020): 10133.

Lanza, Val Fernández, Ana P. Tedim, José Luís Martínez, Fernando Baquero, and Teresa M. Coque. "The plasmidome of Firmicutes: Impact on the emergence and the spread of resistance to antimicrobials." *Plasmids: Biology and Impact in Biotechnology and Discovery* (2015): 379–419.

Lanzkron, Sophie, C. Patrick Carroll, and Carlton Haywood Jr. "Mortality rates and age at death from sickle cell disease: US, 1979–2005." *Public Health Reports* 128, no. 2 (2013): 110–116.

Lattanzi, Annalisa, Joab Camarena, Premanjali Lahiri, Helen Segal, Waracharee Srifa, Christopher A. Vakulskas, Richard L. Frock et al. "Development of β-globin gene correction in human hematopoietic stem cells as a potential durable treatment for sickle cell disease." *Science Translational Medicine* 13, no. 598 (2021): eabf2444.

Lebbink, Robert Jan, Dorien C. M. de Jong, Femke Wolters, Elisabeth M. Kruse, Petra M. van Ham, Emmanuel J. H. J. Wiertz, and Monique Nijhuis. "A combinational CRISPR/Cas9 gene-editing approach can halt HIV replication and prevent viral escape." *Scientific Reports* 7, no. 1 (2017): 41968.

Lee, Rose A., Helena De Puig, Peter Q. Nguyen, Nicolaas M. Angenent-Mari, Nina M. Donghia, James P. McGee, Jeffrey D. Dvorin, Catherine M. Klapperich, Nira R. Pollock, and James J. Collins. "Ultrasensitive CRISPR-based diagnostic for field-applicable detection of Plasmodium species in symptomatic and asymptomatic malaria." *Proceedings of the National Academy of Sciences* 117, no. 41 (2020): 25722–25731.

Lei, Rong, Limei Li, Pinshan Wu, Xinyu Fei, Yuting Zhang, Jingyi Wang, Di Zhang, Qingfang Zhang, Na Yang, and Xinyi Wang. "RPA/CRISPR/Cas12a-based on-site and rapid nucleic acid detection of toxoplasma gondii in the environment." *ACS Synthetic Biology* 11, no. 5 (2022): 1772–1781.

Li, Chang, Xinmeng Guan, Tao Du, Wei Jin, Biao Wu, Yalan Liu, Ping Wang et al. "Inhibition of HIV-1 infection of primary CD4+ T-cells by gene editing of CCR5 using adenovirus-delivered CRISPR/Cas9." *Journal of General Virology* 96, no. 8 (2015): 2381–2393.

Li, Peipei, Li Wang, Junning Yang, Li-Jun Di, and Jingjing Li. "Applications of the CRISPR-Cas system for infectious disease diagnostics." *Expert Review of Molecular Diagnostics* 21, no. 7 (2021): 723–732.

Li, Zichong, Jun Wu, Leonard Chavez, Rebecca Hoh, Steven G. Deeks, Satish K. Pillai, and Qiang Zhou. "Reiterative Enrichment and Authentication of CRISPRi Targets (REACT) identifies the proteasome as a key contributor to HIV-1 latency." *PLoS Pathogens* 15, no. 1 (2019): e1007498.

Liao, Hsin-Kai, Ying Gu, Arturo Diaz, John Marlett, Yuta Takahashi, Mo Li, Keiichiro Suzuki et al. "Use of the CRISPR/Cas9 system as an intracellular defense against HIV-1 infection in human cells." *Nature Communications* 6, no. 1 (2015): 6413.

Liu, Peng, Xinjie Wang, Juan Liang, Qian Dong, Jinping Zhang, Dongxin Liu, Shuai Wang et al. "A recombinase polymerase amplification-coupled Cas12a mutant-based module for efficient detection of streptomycin-resistant mutations in mycobacterium tuberculosis." *Frontiers in Microbiology* 12 (2022): 796916.

Maier, Lisa M., Christopher E. Lowe, Jason Cooper, Kate Downes, David E. Anderson, Christopher Severson, Pamela M. Clark et al. "IL2RA genetic heterogeneity in multiple sclerosis and type 1 diabetes susceptibility and soluble interleukin-2 receptor production." *PLoS Genetics* 5, no. 1 (2009): e1000322.

Makarova, Kira S., Daniel H. Haft, Rodolphe Barrangou, Stan J. J. Brouns, Emmanuelle Charpentier, Philippe Horvath, Sylvain Moineau et al. "Evolution and classification of the CRISPR–Cas systems." *Nature Reviews Microbiology* 9, no. 6 (2011): 467–477.

Makarova, Kira S., Yuri I. Wolf, Omer S. Alkhnbashi, Fabrizio Costa, Shiraz A. Shah, Sita J. Saunders, Rodolphe Barrangou et al. "An updated evolutionary classification of CRISPR–Cas systems." *Nature Reviews Microbiology* 13, no. 11 (2015): 722–736.

Makarova, Kira S., Yuri I. Wolf, Jaime Iranzo, Sergey A. Shmakov, Omer S. Alkhnbashi, Stan J. J. Brouns, Emmanuelle Charpentier et al. "Evolutionary classification of CRISPR–Cas systems: A burst of class 2 and derived variants." *Nature Reviews Microbiology* 18, no. 2 (2020): 67–83.

Marrack, Philippa, John Kappler, and Brian L. Kotzin. "Autoimmune disease: Why and where it occurs." *Nature Medicine* 7, no. 8 (2001): 899–905.

McHugh, Daniel R., Miarasa S. Steele, Dana M. Valerio, Alexander Miron, Rachel J. Mann, David F. LePage, Ronald A. Conlon, Calvin U. Cotton, Mitchell L. Drumm, and Craig A. Hodges. "A G542X cystic fibrosis mouse model for examining nonsense mutation directed therapies." *PLoS One* 13, no. 6 (2018): e0199573.

Memvanga, Patrick B., and Christian I. Nkanga. "Liposomes for malaria management: The evolution from 1980 to 2020." *Malaria Journal* 20, no. 1 (2021): 327.

Miller, Nancimae, Sandra G. Hernandez, and Timothy J. Cleary. "Evaluation of gen-probe amplified Mycobacterium tuberculosis direct test and PCR for direct detection of Mycobacterium tuberculosis in clinical specimens." *Journal of Clinical Microbiology* 32, no. 2 (1994): 393–397.

Mingozzi, Federico, and Katherine A. High. "Immune responses to AAV vectors: Overcoming barriers to successful gene therapy." *Blood, The Journal of the American Society of Hematology* 122, no. 1 (2013): 23–36.

Monk, Ian R., Ishita M. Shah, Min Xu, Man-Wah Tan, and Timothy J. Foster. "Transforming the untransformable: Application of direct transformation to manipulate genetically Staphylococcus aureus and Staphylococcus epidermidis." *MBio* 3, no. 2 (2012): 10–1128.

Monroe, April, Nana Aba Williams, Sheila Ogoma, Corine Karema, and Fredros Okumu. "Reflections on the 2021 World Malaria Report and the future of malaria control." *Malaria Journal* 21, no. 1 (2022): 154.

Morens, David M., and Anthony S. Fauci. "Emerging pandemic diseases: How we got to COVID–19." *Cell* 182, no. 5 (2020): 1077–1092.

Muller, David A., Alexandra C. I. Depelsenaire, and Paul R. Young. "Clinical and laboratory diagnosis of dengue virus infection." *The Journal of Infectious Diseases* 215, no. suppl_2 (2017): S89–S95.

Nii-Trebi, Nicholas Israel. "Emerging and neglected infectious diseases: Insights, advances, and challenges." *BioMed Research International* 2017, no. 1 (2017): 5245021.

Nordling, Linda. "Malaria's modelling problem." *Nature* (2023): S34–S35.

Odame, Isaac. "Perspective: We need a global solution." *Nature* 515, no. 7526 (2014): S10–S10.

Okwor, Ifeoma, and Jude Uzonna. "Persistent parasites and immunologic memory in cutaneous leishmaniasis: Implications for vaccine designs and vaccination strategies." *Immunologic Research* 41 (2008): 123–136.

Ooi, Kean Hean, Mengying Mandy Liu, Jie Wen Douglas Tay, Seok Yee Teo, Pornchai Kaewsapsak, Shengyang Jin, Chun Kiat Lee et al. "An engineered CRISPR-Cas12a variant and DNA-RNA hybrid guides enable robust and rapid COVID-19 testing." *Nature Communications* 12, no. 1 (2021): 1739.

Ospina-Villa, Juan David, César López-Camarillo, Carlos A. Castañón-Sánchez, Jacqueline Soto-Sánchez, Esther Ramírez-Moreno, and Laurence A. Marchat. "Advances on aptamers against protozoan parasites." *Genes* 9, no. 12 (2018): 584.

Otto, M. (2014). Staphylococcus aureus toxins. *Current Opinion in Microbiology*, 17, 32–37.

Pandey, Satish Chandra, Awanish Kumar, and Mukesh Samant. "Genetically modified live attenuated vaccine: A potential strategy to combat visceral leishmaniasis." *Parasite Immunology* 42, no. 9 (2020): e12732.

Pardee, Keith, Alexander A. Green, Melissa K. Takahashi, Dana Braff, Guillaume Lambert, Jeong Wook Lee, Tom Ferrante et al. "Rapid, low-cost detection of Zika virus using programmable biomolecular components." *Cell* 165, no. 5 (2016): 1255–1266.

Pers, Yves-Marie, Lars Rackwitz, Rosanna Ferreira, Oliver Pullig, Christophe Delfour, Frank Barry, Luc Sensebe et al. "Adipose mesenchymal stromal cell-based therapy for severe osteoarthritis of the knee: A phase I dose-escalation trial." *Stem Cells Translational Medicine* 5, no. 7 (2016): 847–856.

Piepenburg, Olaf, Colin H. Williams, Derek L. Stemple, and Niall A. Armes. "DNA detection using recombination proteins." *PLoS Biology* 4, no. 7 (2006): e204.

Quan, Jenai, Charles Langelier, Alison Kuchta, Joshua Batson, Noam Teyssier, Amy Lyden, Saharai Caldera et al. "FLASH: A next-generation CRISPR diagnostic for multiplexed detection of antimicrobial resistance sequences." *Nucleic Acids Research* 47, no. 14 (2019): e83.

Rahimi, Hossein, Marziyeh Salehiabar, Murat Barsbay, Mohammadreza Ghaffarlou, Taras Kavetskyy, Ali Sharafi, Soodabeh Davaran et al. "CRISPR systems for COVID-19 diagnosis." *ACS Sensors* 6, no. 4 (2021): 1430–1445.

Ram, Geeta, Hope F. Ross, Richard P. Novick, Ivelisse Rodriguez-Pagan, and Dunrong Jiang. "Conversion of staphylococcal pathogenicity islands to CRISPR-carrying antibacterial agents that cure infections in mice." *Nature Biotechnology* 36, no. 10 (2018): 971–976.

Rock, Jeremy. "Tuberculosis drug discovery in the CRISPR era." *PLoS Pathogens* 15, no. 9 (2019): e1007975.

Rock, Jeremy M., Forrest F. Hopkins, Alejandro Chavez, Marieme Diallo, Michael R. Chase, Elias R. Gerrick, Justin R. Pritchard et al. "Programmable transcriptional repression in mycobacteria using an orthogonal CRISPR interference platform." *Nature Microbiology* 2, no. 4 (2017): 1–9.

Sam, I. Kuan, Ying-Ying Chen, Jun Ma, Shi-Yuan Li, Ruo-Yan Ying, Lin-Xian Li, Ping Ji et al. "TB-QUICK: CRISPR-Cas12b-assisted rapid and sensitive detection of Mycobacterium tuberculosis." *Journal of Infection* 83, no. 1 (2021): 54–60.

Sato, Shigeharu. "Plasmodium—a brief introduction to the parasites causing human malaria and their basic biology." *Journal of Physiological Anthropology* 40, no. 1 (2021): 1.

Sato'o, Yusuke, Junzo Hisatsune, Liansheng Yu, Tetsushi Sakuma, Takashi Yamamoto, and Motoyuki Sugai. "Tailor-made gene silencing of Staphylococcus aureus clinical isolates by CRISPR interference." *PLoS One* 13, no. 1 (2018): e0185987.

Schwank, Gerald, Bon-Kyoung Koo, Valentina Sasselli, Johanna F. Dekkers, Inha Heo, Turan Demircan, Nobuo Sasaki et al. "Functional repair of CFTR by CRISPR/Cas9 in intestinal stem cell organoids of cystic fibrosis patients." *Cell Stem Cell* 13, no. 6 (2013): 653–658.

Scott, David L., Frederick Wolfe, and Tom W. J. Huizinga. "Rheumatoid Effusions." *Lancet* 376 (2010): 1094.

Selvapandiyan, Angamuthu, Ranadhir Dey, Sreenivas Gannavaram, Ines Lakhal-Naouar, Robert Duncan, Poonam Salotra, and Hira L. Nakhasi. "Immunity to visceral leishmaniasis using genetically defined live-attenuated parasites." *Journal of Tropical Medicine* 2012, no. 1 (2012): 631460.

Shu, Shang-An, Jinjun Wang, Mi-Hua Tao, and Patrick S. C. Leung. "Gene therapy for autoimmune disease." *Clinical Reviews in Allergy & Immunology* 49 (2015): 163–176.

Stokes, Barbara H., Satish K. Dhingra, Kelly Rubiano, Sachel Mok, Judith Straimer, Nina F. Gnädig, Ioanna Deni et al. "Plasmodium falciparum K13 mutations in Africa and Asia impact artemisinin resistance and parasite fitness." *Elife* 10 (2021): e66277.

Su, Lin-Chong, An-Fang Huang, Hong Jia, Yi Liu, and Wang-Dong Xu. "Role of micro RNA-155 in rheumatoid arthritis." *International Journal of Rheumatic Diseases* 20, no. 11 (2017): 1631–1637.

Torgerson, Paul R., Brecht Devleesschauwer, Nicolas Praet, Niko Speybroeck, Arve Lee Willingham, Fumiko Kasuga, Mohammad B. Rokni et al. "World Health Organization estimates of the global and regional disease burden of 11 foodborne parasitic diseases, 2010: A data synthesis." *PLoS Medicine* 12, no. 12 (2015): e1001920.

Uwimana, Aline, Eric Legrand, Barbara H. Stokes, Jean-Louis Mangala Ndikumana, Marian Warsame, Noella Umulisa, Daniel Ngamije et al. "Emergence and clonal expansion of in vitro artemisinin-resistant Plasmodium falciparum kelch13 R561H mutant parasites in Rwanda." *Nature Medicine* 26, no. 10 (2020): 1602–1608.

Vicente, Rita, Julie Quentin, Anne-Laure Mausset-Bonnefont, Paul Chuchana, Delphine Martire, Maïlys Cren, Christian Jorgensen, and Pascale Louis-Plence. "Nonclassical CD4+ CD49b+ regulatory T cells as a better alternative to conventional CD4+ CD25+ T cells to dampen arthritis severity." *The Journal of Immunology* 196, no. 1 (2016): 298–309.

Walker, Mark J., Timothy C. Barnett, Jason D. McArthur, Jason N. Cole, Christine M. Gillen, Anna Henningham, K. S. Sriprakash, Martina L. Sanderson-Smith, and Victor Nizet. "Disease manifestations and pathogenic mechanisms of group A Streptococcus." *Clinical Microbiology Reviews* 27, no. 2 (2014): 264–301.

Walzl, Gerhard, Ruth McNerney, Nelita du Plessis, Matthew Bates, Timothy D. McHugh, Novel N. Chegou, and Alimuddin Zumla. "Tuberculosis: Advances and challenges in development of new diagnostics and biomarkers." *The Lancet Infectious Diseases* 18, no. 7 (2018): e199–e210.

Wang, Meng, Rui Zhang, and Jinming Li. "CRISPR/cas systems redefine nucleic acid detection: principles and methods." *Biosensors and Bioelectronics* 165 (2020): 112430.

Wang, Weiming, Chaobaihui Ye, Jingjing Liu, Di Zhang, Jason T. Kimata, and Paul Zhou. "CCR5 gene disruption via lentiviral vectors expressing Cas9 and single guided RNA renders cells resistant to HIV-1 infection." *PLoS One* 9, no. 12 (2014): e115987.

Weber, Leslie, Giacomo Frati, Tristan Felix, Giulia Hardouin, Antonio Casini, Clara Wollenschlaeger, Vasco Meneghini et al. "Editing a γ-globin repressor binding site restores fetal hemoglobin synthesis and corrects the sickle cell disease phenotype." *Science Advances* 6, no. 7 (2020): eaay9392.

World Health Organization. *Global Status Report on Alcohol and Health 2018.* World Health Organization, 2018.

Wu, Xia, Jian Zha, Mattheos A. G. Koffas, and Jonathan S. Dordick. "Reducing Staphylococcus aureus resistance to lysostaphin using CRISPR-dCas9." *Biotechnology and Bioengineering* 116, no. 12 (2019): 3149–3159.

Yang, Jing, Amanda McGovern, Paul Martin, Kate Duffus, Xiangyu Ge, Peyman Zarrineh, Andrew P. Morris et al. "Analysis of chromatin organization and gene expression in T cells identifies functional genes for rheumatoid arthritis." *Nature Communications* 11, no. 1 (2020): 4402.

Yasmin, Hadida, Anureeta Adhikary, Mohammed N. Al-Ahdal, Syamal Roy, and Uday Kishore. "Host–pathogen interaction in leishmaniasis: Immune response and vaccination strategies." *Immuno* 2, no. 1 (2022): 218–254.

Ye, Lin, Jiaming Wang, Ashley I. Beyer, Fernando Teque, Thomas J. Cradick, Zhongxia Qi, Judy C. Chang et al. "Seamless modification of wild-type induced pluripotent stem cells to the natural CCR5Δ32 mutation confers resistance to HIV infection." *Proceedings of the National Academy of Sciences* 111, no. 26 (2014): 9591–9596.

Yin, Chaoran, Ting Zhang, Xiying Qu, Yonggang Zhang, Raj Putatunda, Xiao Xiao, Fang Li et al. "In vivo excision of HIV-1 provirus by saCas9 and multiplex single-guide RNAs in animal models." *Molecular Therapy* 25, no. 5 (2017): 1168–1186.

Yosef, Ido, Miriam Manor, Ruth Kiro, and Udi Qimron. "Temperate and lytic bacteriophages programmed to sensitize and kill antibiotic-resistant bacteria." *Proceedings of the National Academy of Sciences* 112, no. 23 (2015): 7267–7272.

Yoshimi, Kazuto, Kohei Takeshita, Seiya Yamayoshi, Satomi Shibumura, Yuko Yamauchi, Masaki Yamamoto, Hiroshi Yotsuyanagi, Yoshihiro Kawaoka, and Tomoji Mashimo. "Rapid and accurate detection of novel coronavirus SARS-CoV-2 using CRISPR-Cas3." *MedRxiv* (2020). https://doi.org/10.1101/2020.06.02.20119875

Yu, D., Z. Jin, X. Liu, W. Gao, X. Gao, S. Zhang, S. Ma, J. Yu, and S. Wang. "Manipulation of an ra-associated intergenic snp by CRISPR/Cas9 system alters expression of multiple genes at the tnfaip3 locus, poster presentation abstracts." *International Journal of Rheumatic Diseases* 19 (2016).

Zhang, Wen-Wei, Subir Karmakar, Sreenivas Gannavaram, Ranadhir Dey, Patrick Lypaczewski, Nevien Ismail, Abid Siddiqui et al. "A second generation leishmanization vaccine with a markerless attenuated Leishmania major strain using CRISPR gene editing." *Nature Communications* 11, no. 1 (2020): 3461.

Zhang, Wen-Wei, and Greg Matlashewski. "CRISPR-Cas9-mediated genome editing in Leishmania donovani." *MBio* 6, no. 4 (2015): 10–1128.

Zhang, Yihao, Long Qian, Weijia Wei, Yu Wang, Beining Wang, Pingping Lin, Wenchao Liu et al. "Paired design of dCas9 as a systematic platform for the detection of featured nucleic acid sequences in pathogenic strains." *ACS Synthetic Biology* 6, no. 2 (2017): 211–216.

Zhou, Wenhua, Li Hu, Liming Ying, Zhen Zhao, Paul K. Chu, and Xue-Feng Yu. "A CRISPR–Cas9-triggered strand displacement amplification method for ultrasensitive DNA detection." *Nature Communications* 9, no. 1 (2018): 5012.

Zhou, Zhichang Peter, Liang Leo Yang, Huibi Cao, Ziyan Rachel Chen, Yiqian Zhang, Xiao-Yan Wen, and Jim Hu. "In vitro validation of a CRISPR-mediated CFTR correction strategy for preclinical translation in pigs." *Human Gene Therapy* 30, no. 9 (2019): 1101–1116.

Zhu, Weijun, Rongyue Lei, Yann Le Duff, Jian Li, Fei Guo, Mark A. Wainberg, and Chen Liang. "The CRISPR/Cas9 system inactivates latent HIV-1 proviral DNA." *Retrovirology* 12 (2015): 1–7.

8 CRISPR-Cas Systems in Hemolytic Diseases

Huma Hameed, Mahtab Ahmad Khan, Saleha Faheem, Anam Hameed, Muhammad Zaman, Ali Irfan, Tooba Mehboob, Tehseen Riaz, Naila Tabassam, and Nelofer Ereej

8.1 INTRODUCTION

Hemolytic diseases refer to medical conditions where red blood cells (RBCs) are destroyed at a rate that surpasses their production by the bone marrow, a process known as hemolysis. This leads to a deficiency in red blood cells, resulting in anemia and various associated symptoms. *Anemia* is the term for a low red blood cell count (RBC count) that is unable to meet the body's physiological needs (Bhadra & Deb, 2020). Blood loss, reduced or impaired erythropoiesis, or an accelerated rate of RBC deterioration can all be contributing factors (Paul et al., 2021). Anemia is characterized by either excessive RBC loss or insufficient oxygen delivery to the tissues (Hsu et al., 2022). Decreased erythrocytes impair oxygen flow, resulting in fatigue, weakness, impaired eyesight, and poor performance. Iron overload can potentially result in heart failure from cirrhosis and chronic liver diseases (Farooq et al., 2020). Iron deficiency is a frequent cause of anemia globally, but inherited diseases, such hemolytic anemias, are further exacerbated by increased RBC destruction.

The term *hemolytic disease* includes a variety of disorders that can be either inherited or acquired, each with distinct causes and severity levels. These diseases can affect individuals of all ages, but certain forms are primarily observed in newborns or during pregnancy. Hemolytic diseases are further classified into the following types: autoimmune hemolytic anemia (AIHA), hemolytic disease of the newborn (HDN), hereditary hemolytic anemia, and hemolytic uremic syndrome (HUS), as illustrated in **Figure 8.1**, respectively.

8.1.1 Autoimmune Hemolytic Anemia

In the case of AIHA, the immune system incorrectly attacks and eliminates the body's RBCs. Antibodies are produced that mistakenly recognize these cells as foreign, resulting in their premature destruction. This condition can be triggered by infections, certain drugs, or autoimmune disorders, including lupus. Symptoms of AIHA often include fatigue, pallor, shortness of breath, and jaundice.

DOI: 10.1201/9781003537151-8

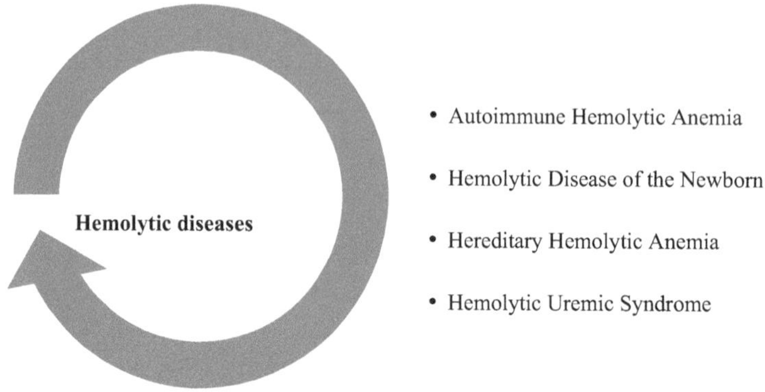

FIGURE 8.1 Classification of hemolytic diseases.

8.1.2 Hemolytic Disease of the Newborn

The HDN condition arises from a mismatch between the blood types of the mother and her baby, commonly related to as the rhesus (Rh) factor, or ABO blood group incompatibilities. It frequently occurs when a Rh-negative mother is pregnant with a Rh-positive child, resulting in complications such as anemia, jaundice, and in severe instances, hydrops fetalis, which is characterized by fluid accumulation in various body cavities of the fetus.

8.1.3 Hereditary Hemolytic Anemia

This refers to a collection of genetic disorders that result in the premature breakdown of RBCs. Notable hereditary hemolytic anemias include the following.

8.1.3.1 Sickle Cell Disease (SCD)

A hereditary condition characterized by the abnormal shape of RBCs (sickle-shaped), which can obstruct small blood vessels, causing pain, anemia, and potential organ damage. The most common hemoglobinopathy is sickle cell disease (SCD) (Basso et al., 2020). SCD is a recessive genetic condition characterized by low RBC count, recurrent infections, and severe veno-occlusive episodes. It is caused by a mutation in the β-globin gene. Allogeneic hematopoietic stem cell transplantation (HSCT) is the only available treatment option (Basso et al., 2020).

8.1.3.2 Thalassemia

A genetic disorder that leads to the production of an abnormal type of hemoglobin, causing the destruction of RBCs and resulting in anemia.

8.1.3.3 Hereditary Spherocytosis (HS)

A disorder in which RBCs take on a spherical shape instead of the typical disc shape, making them more susceptible to rupture. These conditions are typically passed down in an autosomal recessive manner, which requires both parents to possess a copy of the faulty gene for the disease to become apparent. The specific disorders can lead to chronic anemia, pain, organ damage, and various other complications. For severe cases of hereditary anemias, which carry a significant danger to health, this HSCT therapy is offered.

8.1.4 Hemolytic Uremic Syndrome

HUS is a medical condition that predominantly impacts the kidneys and is marked by hemolytic anemia, a reduced platelet count (thrombocytopenia), and acute renal failure. It is frequently triggered by infections from specific strains of *Escherichia coli* (*E. coli*) that generate toxins. These toxins inflict damage on blood vessels and facilitate hemolysis, resulting in kidney impairment, anemia, and bleeding issues.

Due to the large number of afflicted individuals and the extensive understanding of their genetic origins, hemoglobinopathies are considered as the first hereditary-based red blood illnesses and reported to be treated with the help of the gene of interest via gene therapy. RBCs and other hematopoietic cells can easily be produced by corrected hematopoietic stem cells (HSPCs) to preserve the patient's hematopoietic system (Demirci et al., 2020). Two approaches to treating hematopoietic inherited disorders include addition and gene-editing treatment. For hereditary RBC diseases, this chapter will concentrate on addition gene therapy and gene-editing treatment, including everything from fundamental research to clinical application (Anurogo et al., 2021). New advances in diagnostic and therapeutic methods have expanded our knowledge of inherited anemias and spurred the development of innovative gene therapy approaches. Next-generation sequencing and gene therapy have enabled the identification of specific genetic alterations responsible for a range of hereditary anemias, paving the way for the development of innovative genomic treatment–based strategies (Quintana-Bustamante et al., 2022). When autologous HSPCs are utilized as gene correctors in gene therapy, monogenic blood diseases can be effectively treated. Allogeneic HSCT utilization is hindered by the minimum number of donors that have compatibility to human leukocyte antigen (HLA) and the possibility of graft-versus-host disease (GvHD) (Yesilipek, 2020). Prior to being reinfused into the patients, the HSPCs are extracted from them, purified, and then *ex vivo* gene-corrected. In order to reconstruct a corrected hematopoiesis, autologous HSPCs having corrected genes will be engrafted into the bone marrow of the treated patient, where they will begin to proliferate and differentiate (Ferrari et al., 2021).

Overcoming the drawbacks of allogeneic HSCT, autologous HSCT is a viable therapy for hereditary anemias (Oikonomopoulou & Goussetis, 2021). This technique allows for the repair of transduced cells and their offspring by delivering therapeutic genetic elements using viral vectors derived from the family retroviridae. This is especially crucial in highly proliferative organs, like the hematopoietic system, where a limited number of stem cells are in charge of sustaining the entire system (Mann et al., 2022). All distinct, functioning mature cells can work well through corrected stem cells that would happen through gene correction. Seminal articles have shown that this method employing gamma-retroviral vectors is feasible, and clinical trials have shown that they are clinically applicable. Gene therapy has been a major advancement in the profession, especially in persons having X-linked severe combined immunodeficiency (X-SCID) (Allenspach et al., 2021). One significant obstacle was the management of leukemia brought on by gamma-retroviral insertional mutagenesis. The option of gene therapy permanent repairment is considered an advancement because of studies on regulatory sequences and lentiviral vectors. Presently, lentiviral gene therapy is the favored method for genetically modifying HSPCs, managing more than 350 patients with diverse pathologies, such as hereditary hematological disorders, like β-thalassemia (De Simone et al., 2022).

HSPCs with lentivirus correction have been authorized for clinical usage, and β-thalassemia patients will be considered, having problem of transfusions (Grech et al., 2022). Hemoglobin subunit beta (HBB) gene–linked mutation causes β-thalassemia, a condition characterized by abnormal hemoglobin production with varying effects (Satthakarn et al., 2020). The global incidence is 1 in 100,000. Polycystic kidney disease (PKD) is very important and is an autosomal recessive disorder linked to pyruvate kinase, liver and red blood cell (PKLR) gene mutation, responsible for encoding of the erythroid pyruvate kinase protein (RPK) (Quintana-Bustamante et al., 2022). A lentiviral vector has been designed to genetically correct PKD. When protein activity in erythrocytes drops below 25% of normal levels, PKD becomes clinically significant. Chronic kidney disease, with an estimated frequency of 1:20,000, is a major reason of chronic non-spherocytic hemolytic anemia (CNSHA) (Iskander et al., 2024).

A phase I clinical trial was conducted to assess the safety and efficacy of lentiviral gene therapy (Fumagalli et al., 2022). The first two patients with PKD who received autologous HSPCs corrected with lentiviral showed rapid recovery in hemolytic disease due to recovery in erythroid parameters after transplant over 18 months later, indicating that the potential linked to this approach is high for hemolytic anemia treatment. Unfortunately, people receiving lentivirally corrected T cells and HSPC transplants have recently been shown to have uncontrolled cell proliferation, which leads to pre-leukemic or leukemic events (Yannaki et al., 2021). This finding deserves further examination in patients treated with lentiviral gene therapy for hemoglobinopathies, notably SCD. Offering the safest and most effective gene therapy still confronts considerable barriers, notwithstanding the exceptionally fewer side effects. The two most crucial points are the (1) regulation of the therapeutic gene, endogenously, and the (2) elimination of insertional oncogenesis, among others (Bushman, 2020).

8.2 IDENTIFICATION OF TARGET GENES RESPONSIBLE FOR HEMOLYTIC DISEASE

Restrictions enzymes were used to create targeted DNA cuts throughout the 1960s and early 1970s, giving rise to directed gene modification (GDM) (Caradus, 2023). Capecchi showed that homologous recombination (HR) may introduce new DNA sequences into the genome and transmit alterations into live cells. In 1985, Smithies developed targeted genetic modifications to address deleterious mutations in thalassemia and SCD patients (Lal & Prajapati, 2023). In eukaryotic cells, HR efficiency was constrained. Double-strand break (DSB) induction effectively increased HR at a particular genetic target in the late 1980s and early 1990s. The place where the cut was produced is where DSB activates the cell's natural DNA repair processes

(Aleksandrov et al., 2020). Various processes are used, such as microhomology-mediated end joining (MMEJ) and nonhomologous end joining (NHEJ), which include changing two damaged ends in order to make them suitable for rejoining without the need for a template. Without utilizing a model, NHEJ ligates the broken ends directly (Stinson & Loparo, 2021). However, end resection-derived microhomology sequences are aligned by MMEJ to repair the damaged ends. The damaged HR strand requires template-directed repair. A plethora of endonuclease-based techniques has been developed to produce very accurate DSBs. Different types of endonucleases may precisely modify genomic DNA by removing, adding, or even swapping out certain DNA segments in a very focused and selective way (Anzalone et al., 2020).

8.2.1 Zinc Finger Nucleases (ZFNs)

Zinc finger nucleases (ZFNs) are genetic instruments that may be modified to selectively target particular DNA sequences (Paris, 2022). Zinc finger protein motifs were fused to the FokI endonuclease's cleavage domain in 1996 to form the first ZFN. ZFN improved gene targeting in human cell lines, as shown by Baltimore and Porteus in 2003 (Lu et al., 2022). The goal of the first clinical experiment was to interfere with the human immunodeficiency virus (HIV) coreceptor C-C chemokine receptor type 5 (CCR5). In HSPCs, ZFNs have been employed as gene therapy for SCD patients in order to disrupt the erythroid enhancer of the B-cell lymphoma/leukemia 11A (BCL11A) gene, increasing fetal globulin production and decreasing sickling in erythroid cells. Two phase 1/2a clinical studies are now being conducted to evaluate this method for SCD and transfusion-dependent β-thalassemia (Gamage et al., 2023).

8.2.2 Activator-Like Effector Nucleases in Transcription (TALENs)

A novel class of nucleases known as transcription activator–like effector nucleases (TALENs) was created in order to get around ZFNs' drawbacks (Bhardwaj & Nain, 2021). Like ZFN plus, TALENs are restriction enzymes developed synthetically linked to the FokI endonuclease's catalytic domains, which are demonstrated to work as dimers for target spot cleavage in DNA. It has been demonstrated that this approach effectively targets the genomes of human pluripotent stem cells as well as human somatic cells (Guan et al., 2022). TALEN technology was further checked by the outcomes of a clinical investigation to check therapeutic potential that treated B cell acute lymphoblastic leukemia. It was reported that TALEN are better than chimeric antigen receptor T cells (CART) technology. Hemoglobinopathies and PKD are examples of RBC hereditary illnesses for which TALEN technology has recently been used (Fañanas-Baquero et al., 2021).

8.2.3 Clustered Regularly Interspaced Short Palindromic Repeats and Cas (CRISPR-Cas)

The clustered regularly interspaced short palindromic repeats (CRISPR) approach is a genome editing method that was initially employed in 2012 by a team of University of Berkeley academics (Wang & Doudna, 2023). Additionally, there are other, comparable recombinant DNA editing methods, notably ZFN and TALEN; however, CRISPR offers a superior compromise of three important aspects: affordability, accessibility, and accuracy (Bisht et al., 2024; Panigrahi et al., 2023). In addition, it is simpler to utilize than the alternatives, because the latter need for more workers with specialized training and more time (Mindell & Reynolds, 2023). It is a process that begins with bacterial DNA, which acts as an immune system to fight viruses (Bernheim & Sorek, 2020). These sequences are able to identify viruses that infiltrate bacteria and "activate" an enzyme that degrades the bacterium (Grabowski et al., 2021). The enzyme uses the pieces that are produced as a result to immunize the bacteria against the virus (Seeberger, 2021). This procedure is made feasible by CRISPR (the RNA molecule that transfers the biological information found in DNA for protein synthesis) and CRISPR associated protein 9 (Cas9), a specialized bacterial enzyme that is capable of precisely repairing DNA breaks (Schwartz et al., 2021). Prokaryotic organisms have an innate defensive mechanism called CRISPR-Cas9 that protects them against invading genetic material, like plasmids and bacteriophages. Short virus-related DNA molecules can be added to the CRISPR spot by CRISPR-Cas9 when a plasmid/virus infects a bacterium cell (Abavisani et al., 2023; Balcha & Neja, 2023).

8.3 CRISPR-CAS SYSTEM FOR HEMOLYTIC DISEASES

Short DNA repeats with plasmid or viral origin, or CRISPR sequences, are a type of repeats found in bacterial genomes that are consistently spaced and clustered together (Pinilla-Redondo et al., 2022). CRISPR-Cas9 are the genes that encode helicase plus nuclease proteins that are connected to CRISPR repeat sequences that cut or disintegrate DNA (Mohamadi et al., 2020). Cas9 is an endonuclease that belongs to the Cas gene family and has the ability to cut DNA from two active cut sites on either end of the DNA double helix (Kantor et al., 2020; Khanzadi & Khan, 2020). The source was found to be *Streptococcus pyogenes* (Efstratiou & Lamagni, 2022). The CRISPR-Cas system recognizes a virus's or bacteria's DNA when it enters the body, and the Cas protein ultimately destroys it. It was later demonstrated that genome editing may be accomplished by instructing the CRISPR-Cas system to find as well as cut specific target DNA regions (Li et al., 2023). The discovery that CRISPR-Cas9 can modify the human genome, which can be utilized for therapeutic and diagnostic applications, signaled the start of genetic engineering's new era (Cao et al., 2021; Zaib et al., 2022). After 2020, an excess of CRISPR-Cas9-based tools for DNA and RNA gene editing had been developed due to the technology's fast progress (Batool et al., 2021).

8.4 DELIVERY STRATEGY OF EDITED GENES IN HEMOLYTIC DISEASES

The ideal carrier/vector should target components for gene editing to a particular tissue/cell type by enhanced delivery, be non-immunogenic, allow for foreign genes of the right size, and produce enough transgenic expression to solve the issue (Rosanwo & Bauer, 2021). The primary obstacle to using gene editing as a therapeutic strategy is getting the editing instruments to the right cells and tissues in an efficient manner. It is necessary for edited gene components to deliver either *in vivo* or *ex vivo* to the target cell or nucleus. The significance of this constraint increases when one considers HSPC editing. Such HSPCs are resistant to viral vector transduction and challenging to transfect (Bloomer et al., 2021). Physical barriers (such as cell/nucleus membranes), gene-editing instruments, and their stability within target cells, and the possibility of host immune system rejection, are all factors to take into consideration. Gene delivery systems fall into three categories, namely, physicochemical, viral and non-viral carriers. Gene delivery to target cells is more achievable via methods that include microinjection and electroporation. Microinjection involves using a microscope and a needle with a diameter of 0.5–5.0 µm (Munir et al., 2024).

Cargoes are transported to a specific target spot within the cell via piercing its membrane. This mechanism bypasses delivery obstacles, such as extracellular matrix, cell membranes, and cytoplasmic components (Cao et al., 2020). Individual cells can be injected with plasmid DNA, mRNA, or Cas9 protein using single guide RNA (sgRNA). This method can only handle a limited number of cells concurrently, requires a lot of training, and is time-consuming. High-voltage currents are used in electroporation to make nanometer-sized pores in cell membranes that let components into the cell (Kaladharan et al., 2021). A flexible delivery method that may be used to introduce cargo into sometimes-difficult-to-handle cells is electroporation. Mammalian cell electroporation is a commonly used technology that has several described methods. Electroporation, however, has the potential to kill cells. To reduce toxicity and increase delivery efficiency, new electroporation platforms have been developed (Dong & Chang, 2021). The preservation of engraftment and multipotency by electroporation of HSPCs makes gene editing more feasible in clinical settings. The second method of dissemination relies on viruses. Effectively infecting cells, viruses transfer genetic information without causing an immune response to the host (Shirley et al., 2020). By this methodology, viruses become a promising carrier/vector for gene therapy delivery. Engineered viral carriers/vectors can convey the desired genetic material while remaining safe. Viral vectors cannot replicate themselves. Currently, one gene-editing technology used for nuclease entrance is electroporation, and for donor template distribution, the use of viral carriers/vectors is common (Roma-Rodrigues et al., 2020).

The best way to distribute donor templates is by adeno-associated viral vectors, or AAV. Single-stranded DNA viruses, or AAVs, are often employed in gene therapy. There is no proof that AAVs cause diseases in people. Different recognized serotypes have different levels of specificity when it comes to infecting cells (Wells et al., 2024). Effective transduction of HSPCs is achieved when electroporation and AAV serotype 6 are combined. Transduction can occur in up to 70% of HSPCs without compromising their ability to engraft. Hereditary anemia gene therapy can be effectively achieved by electroporation of HSPCs with nucleases and rAAV6. Hemoglobinopathies, PKD, X-linked chronic granulomatous disease (X-CGD), and Wiskott–Aldrich syndrome are a few examples of hematological diseases (Odiba et al., 2021).

Adenoviral vectors (HDAdV) are one of the other viral vector systems that have been studied. In mice with a human β-globin locus, a study was done. HDAdVs in yeast, being an artificial chromosome, can re-express γ-globin. For HSPCs, many non-viral delivery techniques have been investigated (Molaei et al., 2024). Despite the limited effectiveness of targeting HSPCs, the ease of fabrication and replication of these systems may eventually enable them to surpass viral vector-based delivery platforms. These fixes could also be less-expensive. Many options are now being investigated. It has been shown that nanoparticles may revive fetal globins and transmit ribonucleoproteins (RNPs) (Sii-Felice et al., 2020). In a humanized mouse model, Intellia, therapeutics-modified HSPCs using bone marrow tropic lipid nanoparticles (LNPs) were used, and it was proven that this method was effective in treating SCD (Murugesan et al., 2023).

8.5 POTENTIAL ADVANTAGES OF CRISPR-CAS SYSTEMS FOR IMMUNOLOGICAL DISORDERS MANAGEMENT

Since its ability to edit DNA was demonstrated in 2012, systems based on editing, like CRISPR, have emerged in molecular biology labs (Ansari et al., 2020). They have accomplished a great deal in the fields of organ xenotransplantation issue resolution, pathogenic mutation correction, and gene discovery for cancer immunotherapy. Regretfully, there are still some issues with CRISPR-Cas systems that need to be resolved. These issues include less specificity due to higher off-target effects and poor efficiency, leading to a limited scope of genome targeting (Naeem et al., 2020). Further, this system is constrained by protospacer adjacent motif (PAM) sequences. As a result, several research groups are involved to enhance its efficiency by reducing the possible off-target effects due to high specificity (Manghwar et al., 2020).

8.5.1 Dead Cas9 System

By inserting mutations at two points, such as H840A and D10A, into the His-Asn-His endonuclease (HNH) and RuvC nuclease domains, scientists have produced a dead Cas9 nuclease system (dCas9) (Losito, 2022). As a result, nuclease-dead Cas9 (dCas9) exhibits no DNA-binding activity but lacks DNA cleavage activity. The target gene transcription can then be inhibited (CRISPRi) or activated (CRISPRa) using the CRISPR-dCas9 technology. Moreover, epigenetic modifiers along with fluorescent proteins may be recruited in particular sequences for epigenetic modification in addition to genome imaging via dCas9 (Cai et al., 2023). dCas9 also has the ability for fusion of various effector domains. Additionally, this technology becomes easy to use by allowing concurrent regulation of many genes in a single cell (Baysoy et al., 2023).

8.5.2 Base Editing System

Site-directed mutagenesis can be improved with various base editing methods, such as dCas9 in combination with adenosine deaminase (ABE: adenine base editor), in addition to cytosine deaminase (CBE: cytidine base editor) (Shelake et al., 2023). dCas9, along with these enzymes, can insert point mutations from CGG to TAA or AAT to GCC into sgRNA, which acts as the editing window for the target sites without breaking the double strands of DNA. Such base editing techniques mostly prohibit the development of arbitrary insertions or deletions of genes (Jiang et al., 2020). By this way, gene mutation results are more predictable. Nevertheless, base editing methods are not suitable for every target sequence in the genome due to the small base editing window. Further, C-rich sequences could be responsible for the initiation of a lot of off-target mutations. In an effort to circumvent this issue, scientists have been creating and refining innovative base editing technologies (Neugebauer et al., 2023; Wang & Doudna, 2023). Base editing–linked systems are currently widely utilized to achieve efficient site-directed mutagenesis in a wide range of microbes, plants, animals, cell lines, and human embryos. This technique may have several uses in gene therapy, industry, and fundamental research.

8.5.3 Cas9-Linked System

An NGG PAM has to be present at the target DNA site's 30 ends in order for the Cas9 protein to detect and cleave the target gene (Collias & Beisel, 2021). There are more PAM sites outside the typical NGG PAM sites, such as NAG and NGA; however, their efficacy in genome editing is rather low. The human genome has just around 7% of these PAM sites, so severely restricting the genetic areas that may be targeted. Several Cas9 variants, due to this reason, have been developed to improve PAM stability and compatibility. Engineered cas9 (xCas9) is able to recognize several PAMs (GAT, GAA, NG, etc.) (Kondrateva et al., 2021). This opportunity was provided in 2018 by David Liu et al. using phage-assisted continuous evolution (PACE) linked approach. Streptococcus pyogenes Cas9-NG (SpCas9-NG) was developed by Nishimasu and associates during the latter part of that year; It can recognize NG PAMs that are relaxed. Miller et al. created three novel SpCas9 variants in 2020 that have the ability to identify non-G PAMs, such as NRCH, NRTH, and NRRH PAMs (Li et al., 2021a). Walton et al. created SpG, an SpCas9 version, by which such a huge range of NGN PAMs can be targeted, later that year. This SpG system further improved to create SpRY, mainly PAMless variation responsible for modifying all forms of PAMs, including NRN plus NYN. In literature, many studies have linked to the correction of genetic defects associated with sickness using Cas9 variations; however, greater advancements in molecular engineering will be needed for widespread genome editing applications due to Cas9's limited efficiency and limits on cleavage activity (Sledzinski et al., 2021).

8.5.4 RNA Editing System

CRISPR-Cas systems are applied for the modification of both DNA and RNA. Cas13 protein, which is a single RNA-guided gene, along with ribonuclease activity, has the ability to be used in class 2 type VI CRISPR-Cas13 systems for cleavage of single-stranded RNA (ssRNA) selectively (Perčulija et al., 2021). There are now four known Cas13 proteins: Cas13a, which is also known as C2c2, and Cas13b, Cas13c, and Cas13d. All these four variants are considered useful in the detection of viruses, RNA knockdown, transcript tagging, and splicing regulation. Two RNA base editing systems were developed by Feng Zhang et al. by coupling inactivated Cas13 catalytically (dCas13), along with a domain of adenosine deaminase acting on RNA type 2 (ADAR2), by using adenine-cytidine deaminase (Xu & Li, 2020): RESPAIR system for replacement of A to I (G) and RNA Editing for Specific C to U Exchange (RESCUE system for replacement of C-to-U). Compared to DNA editing, RNA editing is more precise and efficient. It also minimizes the risks and moral dilemmas related to ongoing genome editing by enabling reversible genetic changes to the genome. Preclinical research on a wide range of diseases has already routinely used RNA editing, opening the door to new RNA-level study, diagnosis, and treatment possibilities (Bhokisham et al., 2023).

8.5.5 Prime Editing System

Prime editing, a method of genome editing developed by Anzalone et al. (2020) enables targeted insertions, deletions, and all 12 types of base substitutions without the need for donor DNA templates or double-strand breaks (Matsoukas, 2020). Reverse transcriptase and a main editing guide RNA (pegRNA), which designates the target site and encodes the desired alteration, are coupled to a catalytically inhibited Cas9 in this approach. Reverse transcriptase inserts new genetic material into the target location using pegRNA as a template after Cas9 cleaves it. Prime editing increases the accuracy and efficiency of genome editing, increasing its potential for biological and therapeutic uses (Zhao et al., 2023). Up to 89% of disease-causing gene mutations may be fixed in theory. Gene-editing instruments are increasingly being transferred into the cytoplasm of HSPCs via non-viral techniques. This means that they are appropriate strategies linked to editing of genes based on the use of nucleases, prime/base editors. Numerous techniques, including chemical modifications, alternative untranslated region (UTR) additions, and increasing nuclease nuclear localization via nuclear localization signal (NLS), attempted to improve the stability-linked tools used in the editing of the gene within tissue/cells (Pia, 2024).

8.6 CRISPR-CAS9 APPROACHES TO RESTORE HEMOGLOBIN PRODUCTION

8.6.1 Gene Editing in Blood Cells at *In Vitro* Level

Because hematopoietic cell lines are simple to work with, expand, and enrich with modified cells, they provide a gRNA specificity model for testing and CRISPR-Cas9-based systems design for experiments (Bowling et al., 2020). In order to study gene activity following NHEJ disruption, insert point mutations or DNA fragments, repair point mutations, and create chromosomal translocations, cell lines have been employed. Editing success is not always transferable to difficult primary cells, even if it might be helpful as a proof-of-concept. Numerous investigations have demonstrated that hematopoietic cell lines may effectively modify genes using CRISPR-Cas9. NHEJ-mediated DSB repair has the ability to introduce indels that result in nonsense or frameshift mutations, which can shorten proteins and alter their functionality. Gene knockouts can be achieved more quickly with CRISPR-Cas9 as compared to homologous recombination-based targeting of genes used conventionally (Rainha et al., 2020).

Because of its simplicity, CRISPR-CAS9 becomes an optimal method for knockout investigations. When it comes to figuring out gene function, CRISPR-Cas9 is on the upper end, as that of RNA interference technology for editing of genes, since the latter usually produces hypomorphic phenotypes that do not fully represent the loss of function brought on by genetic changes. Point mutations or gene segments have been inserted into specific loci using CRISPR-Cas9-mediated cleavage and HDR (Sansbury et al., 2022). The sequences of 3' and 5' end that are homologous to the target area encircle the donor template DNA. Homologous recombination using CRISPR-Cas9 or embryonic stem cells to create a knock-in allele are equally important considering their cost and time employed in editing (Leal et al., 2024).

When CRISPR-Cas9 is used for the editing of genes to create cellular models, the effects of patient mutations can be better understood. Splicing gene mutations are commonly found in patients with myelodysplastic syndromes. One such mutation is the proline (P) at position 95 is substituted with histidine (P95H) mutation in Serine/Arginine-rich Splicing Factor 2 (SRSF2) responsible for controlling splicing of pre-mRNA. In literature, CRISPR-Cas9-mediated HDR technology was used to generate an SRSF2/P95H cell line (Tatwavedi et al., 2023). Splicing mis-regulation followed, resulting in function phenotype enhancement, and RNA-binding preferences decreased. This demonstrates, in myelodysplastic syndromes, how gene mutations can alter splicing patterns, potentially influencing disease and providing a therapeutic avenue. Alleles involving the driver gene may be involved in acute myeloid leukemia (AML). Creating novel, customized treatments needs precise AML models. Isocitrate dehydrogenase 2 (IDH2), a Krebs cycle enzyme, has the arginine (R) replaced with a glutamine (Q) at position 140 (R140Q) mutation, which results in neomorphic enzyme activity in cells and an oncometabolite that alters with epigenetic regulation and fuels the development of cancer (Carosi et al., 2024).

The IDH2/R140Q was introduced into K562 cells through genome editing. The cells created by this mutation are a good model for evaluating drugs because they mimic the genetic, epigenetic, and functional problems seen in IDH2 (Mehrjardi et al., 2020). The use of single-stranded donornucleotides as HDR DNA templates has been used to correct disease-associated genes. In a chronic myeloid leukemia cell line, a loss of function in the Additional sex combs-like 1 (ASXL1) gene was effectively corrected. The cell function was returned to wild-type levels after the IDH2R140Q was fixed. The results show that it is possible to do gene repair in primary hematopoietic cells. Human-induced pluripotent stem cells can have disease-related mutations corrected using CRISPR-Cas9 genome editing, but it can also correct mutations linked to illness in human-induced pluripotent stem cells (iPSCs) (Paolini Sguazzi et al., 2021).

By using CRISPR-Cas9-mediated HDR to correct the HCLS1-associated protein X-1 (HAX1) gene, severe congenital neutropenia was successfully treated in patient-specific iPSCs (Dannenmann, 2022). A promising platform for disease modeling and customized regenerative medicine is iPSC. A patient's mutated hematopoietic cells can be removed, transformed into iPSCs, altered, grown into HSCP, and then autologously transplanted back into the body. One obstacle to the therapeutic translation of iPSC-derived HSCs is the long-term restoration of the blood system (Demirci et al., 2020).

8.6.2 β-Thalassemia and SCD: Sickle Cell Disease

Single-gene-linked genetic illnesses, such as SCD and β-thalassemia dependent on transfusion, are very important to consider for treatment at the genetic level (Kavanagh et al., 2022). Hemolytic anemia and hemoglobin chain imbalance, which are hallmarks of sickle cell disease, are frequently managed with iron chelation therapy and blood transfusions. Individuals with β-thalassemia often require blood transfusions, hydroxyurea, and painkillers in the clinic due to their RBCs having a sickle shape, responsible for less oxygen-carrying capacity and causing discomfort. Both diseases have also been treated with bone marrow transplantation, albeit matching is difficult (Ciurea et al., 2020).

New scientific discoveries have led to a rapid expansion of the field of gene editing. Because gene editing is safer and more effective, its use has risen quickly; the first clinical trials are currently being conducted. CRISPR-Cas9 gene-editing studies are among the more than 30 authorized clinical gene-editing research. Treating β-hemoglobinopathies is the goal of at least 13 of these trials, which represents a noteworthy advancement in biotechnology and medicine (Cavazzana et al., 2017). Since 1980, a number of groups have employed lentiviral vector gene therapy to treat illnesses. Because globin expression to the erythroid lineage is very restricted, there has been difficulty in managing hemoglobinopathies. Clinical studies are now investigating lineage-specific promoters, which are necessary for lentiviral constructs intended for β-thalassemia and SCD treatment. In light of such major gap, editing of gene is suggested as a suitable gene therapy advancement to satisfy the need for transgenic control at the physiological level (Ereej et al., 2024; Pena et al., 2020). The role of CRISPR-Cas9-based gene editing is shown in **Figure 8.2** for the treatment of hemolytic disorders.

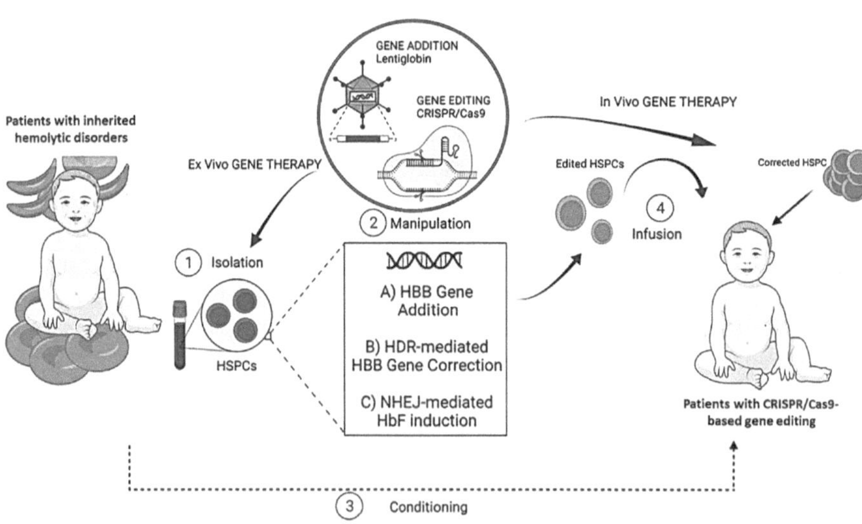

FIGURE 8.2 Role of CRISPR-Cas9-based gene editing for the restoration of hemoglobin production.

A number of institutions are investigating gene-editing methods, such as NHEJ and HDR, in an effort to find a treatment plan (Guitart et al., 2016). After myeloablative conditioning, a number of clinical studies are available to evaluate the feasibility of treating SCD and β-thalassemia with genetically modified HSPCs. The previously published method of utilizing ZFN, CRISPR-Cas9, or Cas12a to damage the erythroid enhancer of the BCL11A gene has been applied in nine clinical investigations. This sequence, as previously shown, stops adult fetal globin activity. Those who have TDT β-thalassemia or SCD may find this approach helpful. There are six open-phase 1/2 multicenter clinical trials to evaluate gene-edited products based on autologous HSPC (Crees et al., 2023). The subject of two trials (NCT03745287 plus NCT03655678) is being evaluated.

8.7 CHALLENGES IN CURRENT TREATMENT OPTIONS

Funded by CRISPR Therapeutics and Vertex Pharmaceuticals Inc., CTX001 is a CRISPR-Cas9-based therapeutic treatment for TDT β-thalassemia and sickle cell disease. Enrolling 45 patients with untreated SCD, having two severe vaso-occlusive crises at least, and annually showing eligibility for use of autologous HSCT are part of the phases 1–3 clinical study of NCT03745287. TDT β-thalassemia patients receiving transfusion of packed red blood cells of at least 100 mL/kg/year or more than 10 units/year in the last two years are being enrolled in the phase 1/2/3 research of NCT03655678. Clinical research to treat SCD is being conducted by Novartis Pharmaceuticals and Intellia Therapeutics (NCT04443907). This investigation attempts to evaluate the efficacy and safety of two autologous HSPCs from SCD patients (OTQ923 and HIX763) that have had their CRISPR-Cas9 genes altered (Germino-Watnick et al., 2022). Thirty individuals with a confirmed diagnosis of SCD and showing one severe SCD side effect, mainly acute chest syndrome, recurrent priapism, or vaso-occlusive pain crisis, are sought for this phase 1/2 clinical research. Participants who have had RBC alloimmunization or chronic transfusions are also eligible. ZFN technology is being used in two further clinical studies (NCT03432364 plus NCT03653247), where the same genetic editing techniques were used. Six β-thalassemia-linked TDT patients who have undergone eight or more RBC transfusions are participating in a phase 1–2 clinical research (NCT03432364) by Sangamo Therapeutics to evaluate the safety, tolerability, and effectiveness of ST-400 in the last two years. Bioverativ is evaluating BIVV003's efficacy and the safety of HSPCT (autologous) in eight different patients suffering from severe sickle cell disease (Quintana-Bustamante et al., 2022).

Preliminary results have come from recent clinical investigations. The first findings of NCT03432364, which investigated ST-400 for TDT β-thalassemia, did not provide any encouraging results because of the low expression of HbF and low genome editing efficiency (Grech et al., 2022). Two patients with SCD and beta-thalassemia (TDT β-thalassemia) received CTX001 infusion. The patients' bone marrow and blood showed allelic editing (up to 80% higher level and then 60%, respectively), and their total hemoglobin level was approximately 14 g/dL. Additionally, the SCD patients' fetal hemoglobin levels increased; they were transfusion-independent, and their vaso-occlusive episodes stopped more than a year after treatment. These encouraging outcomes show that gene editing is a successful treatment for β-thalassemia and SCD.

A phase 1/2 clinical research (NCT04819841) by Graphite Bio aims to cure SCD by employing CRISPR-Cas9 and rAAV6 to fix the E6V mutation by gene editing (Hardouin et al., 2023). In a reported study, to evaluate the safety and effectiveness of the medication GPH101, first human trial was conducted. Fifteen patients with recurrent severe vaso-occlusive episodes and severe SCD will be examined for this product. This study aims to restore the detrimental sickle mutation in the HBB gene while maintaining normal gene expression, which sets it apart from other previous SCD investigations. By eliminating pathologic HbS expression, this gene repair lessens the pathogenic protein's competitiveness. Apart from hemoglobinopathies, our group is producing preclinical data that will probably result in clinical trials in the near future (Karamperis et al., 2021). Clinical investigations for hemolytic disorders using gene editing are displayed in **Table 8.1**.

8.8 ESSENTIAL TO ADVANCE THE HEMATOLOGY FIELD?

Decades later, CRISPR-Cas has developed into a broad family-linked single–base gene editing, control at the transcription level, and strand cutting–linked RNA tools rather from being limited to cleaving DNA strands. Therefore, most human ailments, including blood problems and hereditary diseases resulting from a single gene, may be treated using these systems (Cring & Sheffield, 2022). Cells have displayed extraordinary gene-editing capabilities using systems such as CRISPR-Cas9. On the other hand, creating a reliable, stable, and safe method of gene-editing tools delivery at the target site at *in vivo* level are diseased cells, considered a major obstacle to their use in therapeutics. Even though mRNA, RNPs, and plasmids are all immunologically extracted from blood, CRISPR is frequently administered as any of these three (Mohammadian Gol et al., 2023). Even without loading capacity, introducing AAV-based gene therapy delivery into the human genome can result in illness and will be problematic for people's acceptance. AAV is considered as a more promising carrier/vector for edited gene delivery. Clinical studies have demonstrated the superiority of non-viral vectors, especially LNPs, for delivery because of their greater loading capacities and improved safety profiles. Not derived from living things, vectors like AuNPs and LNPs often induce immunogenicity *in vivo* (Ferreira et al., 2020).

These vectors are also prone to incorrect absorption by the gastrointestinal tract, yet scientists have added peptides or changes to prevent immune clearance. Exosomes derived from organisms or biofilms used as masks for non-viral vectors are incredibly effective in evading immune clearance, and the concentrated proteins and peptides within the biofilm may help transfer gene therapy to the target cells. The microenvironment around sick cells alters as a disease develops, setting it apart from healthy tissue and fostering the development of delivery vectors (Mata et al., 2021). The overexpression of inflammatory factors in cells or other membrane proteins linked to the disease occurs at the same time as the illness. Peptides that target these membrane proteins selectively enable nanomaterials to identify sick cells and introduce CRISPR-Cas to them, with the intention of altering their genomes.

TABLE 8.1
Clinical Trials Data for Challenges in Hemolytic Diseases Based on Gene-Editing Technology

Conditions	Gene-Editing Approach	Title of Study	NCT Number	Sponsor/Collaborators
Sickle cell disease	ZFN repression of γ-globin disrupting BCL11A gene enhancer	"An investigation to evaluate BIVV003's safety, tolerability, and effectiveness in autologous hematopoietic stem cell transplantation for patients with severe sickle cell disease (PRECIZN-1)"	NCT03653247	Bioverativ, a Sanofi company/ Sangamo
Sickle cell disease	CRISPR-Cas9 repression of γ-globin that targets the erythroid-specific enhancer BCL11A	"An Assessment of CTX001's Safety and Effectiveness in Patients with Severe Sickle Cell Disease"	NCT03745287	Vertex Pharmaceuticals Incorporated
β-thalassemia Thalassemia SCD: sickle cell disease	CRISPR-Cas9 γ-globin re-expression that targets the erythroid-specific enhancer BCL11A	"An Extended Monitoring Study in Participants Receiving CTX001"	NCT04208529	Vertex Pharmaceuticals Incorporated CRISPR Therapeutics
β-thalassemia major	CRISPR-Cas9 γ-globin re-expression by targeting globin regulators	"Autologous HSC in Major β-thalassemia Patients Was Restored by β-globin"	NCT04205435	Bioray Laboratories PLA 923 Hospital
Sickle cell disease	Targeting CRISPR-Cas9-ssODN mutations that cause SCD	"Patients with Severe Sickle Cell Disease Can Receive Transplantation of CRISPR_SCD001 Modified Hematopoietic Progenitor Stem Cells: Clustered Regularly Interspaced Short Palindromic Repeats"	NCT04774536	University of California
Thalassemia	Before iPSCs differentiated into iHSCs, HDR of CRISPR-Cas9-piggyBac transposons corrected mutation causing β-thalassemia	"Gene-corrected iHSCs with HBB mutations prevent subjects with β-thalassemia mutations"	NCT03728322	Allife Medical Science and Technology Co. Ltd.

(*Continued*)

TABLE 8.1 (*Continued*)
Clinical Trials Data for Challenges in Hemolytic Diseases Based on Gene-Editing Technology

Conditions	Gene-Editing Approach	Title of Study	NCT Number	Sponsor/Collaborators
β-thalassemia	ZFN linked to γ-globin re-expression disrupting a BCL11A gene enhancer	"An investigation to evaluate ST-400's safety, tolerability, and effectiveness in treating t in patients suffering from -Dependent β-Thalassemia Transfusion"	NCT03432364	Sangamo Therapeutics
β-thalassemia	CRISPR-Cas9 re-expression of γ-globin that targets the erythroid-specific enhancer BCL11A	"An Investigation of the Safety and Effectiveness of CTX001 in patients suffering from -Dependent β-Thalassemia Transfusion"	NCT03655678	Vertex Pharmaceuticals Incorporated
Transfusion-dependent β-thalassemia	CRISPR-Cas9 re-expression of γ-globin that targets the erythroid-specific enhancer BCL11A	"An Investigation of the Safety and Effectiveness of ET-01 in patients suffering from -Dependent β-Thalassemia Transfusion"	NCT04925206	EdiGene Inc.
β-thalassemia	CRISPR-Cas9 re-expression of γ-globin that targets the erythroid-specific enhancer BCL11A	"An Investigation of the Safety and Effectiveness of CTX001 in Patients with Transfusion-Dependent β-Thalassemia"	NCT03655678	Vertex Pharmaceuticals Incorporated
Sickle cell disease	CRISPR-Cas9 re-expression of γ-globin that targets the BCL11A gene	"Investigation of the Safety and Effectiveness of Hematopoietic Stem and Progenitor Cells with Edited Genomes in Sickle Cell Disease (SCD)"	NCT04443907	Novartis/Intellia
Sickle cell disease Hemoglobinopathies	Cas12a RNP re-expression of γ-globin that targets the HBG1/2 promoters	"EDIT-301 for Autologous Hemopoietic Stem Cell Therapy in Patients having Severe (SCD) i.e., Sickle Cell Disease"	NCT04853576	Editas Medicine Inc.
Sickle cell disease	HDR repair of the CRISPR-Cas9-rAAV6 mutation causing SCD	"To Treat Severe Sickle Cell Disease, Autologous CD34+ Hematopoietic Stem Cells with Gene Correction (HbS to HbA)"	NCT04819841	Graphite Bio Inc.

Furthermore, the concentration of nanomaterials at the target location may be enhanced by using nanoparticles that respond to circumstances specific to a disease (Yetisgin et al., 2020). Therefore, a detailed examination of the process of illness onset, together with the numerous pathways involved, changes in protein expression, and the microenvironment in which the disease cells are situated, may considerably benefit from the creation of vectors for the targeted delivery of gene therapies. In addition, the efficacy and safety of CRISPR-Cas9 evaluation are important to consider in therapeutic contexts. Because CRISPR-Cas9 effects on the off-target level might have unbearable consequences, it is important to identify and address these cases prior to clinical use (Piergentili et al., 2021). Researchers reported the cleavage mechanism of CRISPR-Cas9 in great detail, by which production of a large number of Cas9 variants can be possible that are both efficient and less likely to cause off-target effects. Modifying the sgRNA effectively prevents off-target effects. Ultimately, it is imperative to carry out comprehensive investigations into the etiology of the illness, create more effective and targeted delivery vectors, and create superior Cas9 variants with more extensive and secure modifications. These pillars will enable the complete clinical application of CRISPR-Cas technology to treat human diseases (Nidhi et al., 2021).

Safety and effectiveness are crucial when implementing novel treatment methods. New vector methods should be developed to optimize delivery to target cells while minimizing adverse effects (Han et al., 2022). It would be ideal to have platforms for both systemic and local CRISPR-Cas9 delivery, such as ribonucleoproteins, nanoparticle technologies, non-integrative gene therapy vectors, and other cutting-edge methods. Over the past ten years, molecular diagnosis for hematologic malignancies has been greatly enhanced by next-generation sequencing. Numerous diseases have been linked to somatic mutations, which may be highly relevant in terms of prognosis. The functional impact of these mutations on tumor formation and maintenance will be effectively ascertained in the coming years using CRISPR-Cas9-based screenings performed in patient-derived cells (Zhang et al., 2021). Mutations in epigenetic modifiers, which alter DNA at cytosine residues or cause post-translational histone changes, have been seen in several hematologic malignancies. Treatments for AML and myelodysplastic syndrome involve the use of hypomethylating medications, such as decitabine and azacytidine, as well as epigenetic targets, like DNA methyltransferase 3A (DNMT3A) and Histone deacetylases (HDAC) (Stomper et al., 2021).

Hematology research on targeted epigenome editing, which alters the epigenome at certain locations rather than the entire genome, may be very promising. This may assist in regulating DNA methylation or identify the purpose of certain locations, which may help fine-tune gene expression (Willbanks et al., 2021). Epigenome-linked editing does not include alterations at the genetic level or the possibility of unintended repercussions; it may be a safer procedure. Providing editing tools for epigenomes at *in vivo* level, accomplishing modifications at reversible level in epigenomes at particular locations, plus guaranteeing heritability throughout cell division are among the challenges. Significant advancements in CAR-T cell multiplex editing have been made possible by CRISPR-Cas9, resulting in safer and more efficient therapies (Al Saber et al., 2021). With the use of this technique, personalized T cell treatments for cancer immunotherapy may progress. In order to increase public accessibility to

CAR-T cells, academic institutions and pharmaceutical corporations should collaborate on their production. This would enable each patient to receive therapy that is unique to them. To enable a larger scale-up of these products, drug regulatory bodies have to fund university research and provide resources to streamline and automate the production processes for CAR-T cells (van den Bos et al., 2024).

8.9 CONCLUSION

The novel genome editing technique known as CRISPR-Cas9 has several uses in various fields of molecular biology and medicines. It provides the opportunity to research the functioning of genes, model diseases, and alter human DNA for medical reasons (Claussnitzer et al., 2020). It is currently employed in biotechnology and research facilities more frequently than more costly and time-consuming techniques, like zinc finger nucleases or TALEN. CRISPR-Cas9 can be used in hematology to represent illnesses in organisms or cultured cells. It can also rectify *ex vivo* mutations and chromosomal abnormalities in patient cells with blood disorders by autologous HSC transplantation. Several challenges must be overcome before CRISPR-Cas9 gene editing may be applied in medical settings (Jacinto et al., 2020). We are unsure if gene editing might be employed to fix the minimal number of altered cells required to fix a genetic flaw or to treat multigenic illnesses. More research is needed to deploy CRISPR-Cas9 in therapeutic settings and give patients access to genome editing drugs. Ultimately, the CRISPR-Cas9 revolution provides a flexible, precise, and effective instrument for modifying the genome. Advances in technology have made it possible for scientists to alter DNA in nearly any manner (Fan et al., 2020).

8.10 FUTURE PERSPECTIVE

Ex vivo and *in vivo* techniques can successfully treat hereditary hematological and nonhematopoietic disorders, as evidenced by preclinical and early clinical outcomes. Innovative therapeutics for long-term HSPC illnesses have three major challenges: boosting effectiveness, monitoring safety, and improving patient access (Koniali et al., 2021). Hemoglobinopathies are disproportionately impacted, and gene editing for these disorders should be made more accessible globally. Newly redesigned HDR components might improve efficiency and apply to a broader variety of genetic illnesses. Various methods are being studied for every gene mutation (Koh et al., 2021). To more successfully correct mutant sequences, prime editing uses longer gRNAs, including both the genetic material and the genomic recognition site.

It has been demonstrated that co-localization of enhancer proteins, donor templates, and nucleases enhances effectiveness and ease of usage. To enable subcellular colocalization and raise the likelihood of therapeutic gene editing, further developments are required. Successful gene editing depends on getting the necessary components to the cell nucleus and avoiding their elimination by natural cellular processes (Leibowitz et al., 2021). Preclinical development is ongoing. Treating inherited hematopoietic diseases with HSCT employing *ex vivo* gene-modified cells is still limited. This is particularly crucial in conditions when a high proportion of long-term corrected HSPCs have engrafted and where corrected cells do not have a selection advantage over uncorrected cells. To get rid of unhealthy cells that could compete with corrected cells injected after *ex vivo* editing, a conditioning strategy is required. These conditioning programs can result in tissue damage, immunosuppression, and even sterility. They are also highly genotoxic (Jensen et al., 2021).

In order to create regimens based on non-genotoxic conditioning that support healthy cells engrafting for therapeutic reasons, new strategies will be required. Technologies for *in vivo* administration may be used to treat genetic disorders impacting the hematological system as well as other organs, including the liver, kidney, lung, and brain, that cannot be fixed *ex vivo* (Li et al., 2021b). It might be possible to do gene editing without the requirement for complex cell manipulation methods if recombinant virus-based systems or nanoparticles were developed that target certain tissues or cell types, such as HSPCs. Strict control of the influence on non-target tissues is required to reduce the negative consequences of introducing gene-editing instruments into patient tissues. Gene editing is still in its infancy, and new instruments and methods are constantly being developed (Mushtaq et al., 2021). For those suffering from inherited RBC abnormalities, gene editing offers novel therapeutic alternatives. Making this technology available to a wide audience is still difficult, though.

REFERENCES

Abavisani, Mohammad, Reza Khayami, Melika Hoseinzadeh, Mansoor Kodori, Prashant Kesharwani, and Amirhossein Sahebkar. 2023. "CRISPR-Cas system as a promising player against bacterial infection and antibiotic resistance." *Drug Resistance Updates* 68: 100948.

Al Saber, Md, Partha Biswas, Dipta Dey, Md Abu Kaium, Md Aminul Islam, Miss Ismoth Ara Tripty, MD Hasanur Rahman, Tanjim Ishraq Rahaman, Md Yeaman Biswas, and Priyanka Paul. 2021. "A comprehensive review of recent advancements in cancer immunotherapy and generation of CAR T cell by CRISPR-Cas9." *Processes* 10 (1): 16.

Aleksandrov, Radoslav, Rossitsa Hristova, Stoyno Stoynov, and Anastas Gospodinov. 2020. "The chromatin response to double-strand DNA breaks and their repair." *Cells* 9 (8): 1853.

Allenspach, Eric J, David J Rawlings, Aleksandra Petrovic, and Karin Chen. 2021. "X-linked severe combined immunodeficiency." Europe PMC.

Ansari, Waquar A, Sonali U Chandanshive, Vacha Bhatt, Altafhusain B Nadaf, Sanskriti Vats, Jawahar L Katara, Humira Sonah, and Rupesh Deshmukh. 2020. "Genome editing in cereals: Approaches, applications and challenges." *International Journal of Molecular Sciences* 21 (11): 4040.

Anurogo, Dito, Nova Yuli Prasetyo Budi, Mai-Huong Thi Ngo, Yen-Hua Huang, and Jeanne Adiwinata Pawitan. 2021. "Cell and gene therapy for anemia: Hematopoietic stem cells and gene editing." *International Journal of Molecular Sciences* 22 (12): 6275.

Anzalone, Andrew V, Luke W Koblan, and David R Liu. 2020. "Genome editing with CRISPR–Cas nucleases, base editors, transposases and prime editors." *Nature Biotechnology* 38 (7): 824–844.

Balcha, Fikre Birhanu, and Sultan Abda Neja. 2023. "CRISPR-Cas9 mediated phage therapy as an alternative to antibiotics." *Animal Diseases* 3 (1): 4.

Basso, Sabrina, Francesca Compagno, Paola Zelini, Giovanna Giorgiani, Stella Boghen, Elena Bergami, Jessica Bagnarino, Mariangela Siciliano, Claudia Del Fante, and Mario Luppi. 2020. "Harnessing T cells to control infections after allogeneic hematopoietic stem cell transplantation." *Frontiers in Immunology* 11: 567531.

Batool, Asiya, Fayaz Malik, and Khurshid Iqbal Andrabi. 2021. "Expansion of the CRISPR/Cas genome-sculpting toolbox: Innovations, applications and challenges." *Molecular Diagnosis & Therapy* 25: 41–57.

Baysoy, Alev, Zhiliang Bai, Rahul Satija, and Rong Fan. 2023. "The technological landscape and applications of single-cell multi-omics." *Nature Reviews Molecular Cell Biology* 24 (10): 695–713.

Bernheim, Aude, and Rotem Sorek. 2020. "The pan-immune system of bacteria: Antiviral defence as a community resource." *Nature Reviews Microbiology* 18 (2): 113–119.

Bhadra, Preetha, and A Deb. 2020. "A review on nutritional anemia." *Indian Journal of Natural Sciences* 10 (59): 18466–18474.

Bhardwaj, Anuradha, and Vikrant Nain. 2021. "TALENs—an indispensable tool in the era of CRISPR: A mini review." *Journal of Genetic Engineering and Biotechnology* 19 (1): 125.

Bhokisham, Narendranath, Ethan Laudermilch, Lindsay L Traeger, Tonya D Bonilla, Mercedes Ruiz-Estevez, and Jordan R Becker. 2023. "CRISPR-Cas system: The current and emerging translational landscape." *Cells* 12 (8): 1103.

Bisht, Deepanker, Sagar Salave, Nimeet Desai, Purnima Gogoi, Dhwani Rana, Prachurya Biswal, Gautami Sarma, Derajram Benival, Nagavendra Kommineni, and Dhruv Desai. 2024. "Genome editing and its role in vaccine, diagnosis, and therapeutic advancement." *International Journal of Biological Macromolecules* 131802.

Bloomer, Hanan, Richard H Smith, Waleed Hakami, and Andre Larochelle. 2021. "Genome editing in human hematopoietic stem and progenitor cells via CRISPR-Cas9-mediated homology-independent targeted integration." *Molecular Therapy* 29 (4): 1611–1624.

Bowling, Sarah, Duluxan Sritharan, Fernando G Osorio, Maximilian Nguyen, Priscilla Cheung, Alejo Rodriguez-Fraticelli, Sachin Patel, Wei-Chien Yuan, Yuko Fujiwara, and Bin E Li. 2020. "An engineered CRISPR-Cas9 mouse line for simultaneous readout of lineage histories and gene expression profiles in single cells." *Cell* 181 (6): 1410–1422. e27.

Bushman, Frederic D. 2020. "Retroviral insertional mutagenesis in humans: Evidence for four genetic mechanisms promoting expansion of cell clones." *Molecular Therapy* 28 (2): 352–356.

Cai, Ruijie, Runyu Lv, Xin'e Shi, Gongshe Yang, and Jianjun Jin. 2023. "CRISPR/dCas9 tools: Epigenetic mechanism and application in gene transcriptional regulation." *International Journal of Molecular Sciences* 24 (19): 14865.

Cao, Genmao, Xuezhen Xuan, Ruijing Zhang, Jie Hu, and Honglin Dong. 2021. "Gene therapy for cardiovascular disease: Basic research and clinical prospects." *Frontiers in Cardiovascular Medicine* 8: 760140.

Cao, Jun, Dennis Huang, and Nicholas A Peppas. 2020. "Advanced engineered nanoparticulate platforms to address key biological barriers for delivering chemotherapeutic agents to target sites." *Advanced Drug Delivery Reviews* 167: 170–188.

Caradus, John R. 2023. "Perceptions of plant breeding methods–from 'phenotypic selection'to 'genetic modification'and 'new breeding technologies'." *New Zealand Journal of Agricultural Research* 1–49.

Carosi, Francesca, Elisabetta Broseghini, Laura Fabbri, Giacomo Corradi, Riccardo Gili, Valentina Forte, Roberta Roncarati, Daria Maria Filippini, and Manuela Ferracin. 2024. "Targeting isocitrate dehydrogenase (IDH) in solid tumors: Current evidence and future perspectives." *Cancers* 16 (15): 2752.

Cavazzana, Marina, Chiara Antoniani, and Annarita Miccio. 2017. "Gene therapy for β-hemoglobinopathies." *Molecular Therapy* 25 (5): 1142–1154.

Ciurea, Stefan O, Monzr M Al Malki, Piyanuch Kongtim, Ephraim J Fuchs, Leo Luznik, Xiao-Jun Huang, Fabio Ciceri, Franco Locatelli, Franco Aversa, and Luca Castagna. 2020. "The European Society for Blood and Marrow Transplantation (EBMT) consensus recommendations for donor selection in haploidentical hematopoietic cell transplantation." *Bone Marrow Transplantation* 55 (1): 12–24.

Claussnitzer, Melina, Judy H Cho, Rory Collins, Nancy J Cox, Emmanouil T Dermitzakis, Matthew E Hurles, Sekar Kathiresan, Eimear E Kenny, Cecilia M Lindgren, and Daniel G MacArthur. 2020. "A brief history of human disease genetics." *Nature* 577 (7789): 179–189.

Collias, Daphne, and Chase L Beisel. 2021. "CRISPR technologies and the search for the PAM-free nuclease." *Nature Communications* 12 (1): 555.

Crees, Zachary D, Michael P Rettig, and John F DiPersio. 2023. "Innovations in hematopoietic stem-cell mobilization: A review of the novel CXCR4 inhibitor motixafortide." *Therapeutic Advances in Hematology* 14: 20406207231174304.

Cring, Matthew R, and Val C Sheffield. 2022. "Gene therapy and gene correction: Targets, progress, and challenges for treating human diseases." *Gene Therapy* 29 (1): 3–12.

Dannenmann, Benjamin. 2022. "Establishment of a patient-specific iPSC model of stepwise leukemogenesis in severe congenital neutropenia." Universität Tübingen.

De Simone, Giovanna, Alberto Quattrocchi, Benedetta Mancini, Alessandra di Masi, Clara Nervi, and Paolo Ascenzi. 2022. "Thalassemias: From gene to therapy." *Molecular Aspects of Medicine* 84: 101028.

Demirci, Selami, Alexis Leonard, and John F Tisdale. 2020. "Hematopoietic stem cells from pluripotent stem cells: Clinical potential, challenges, and future perspectives." *Stem Cells Translational Medicine* 9 (12): 1549–1557.

Dong, Zaizai, and Lingqian Chang. 2021. "Recent electroporation-based systems for intracellular molecule delivery." *Nanotechnology and Precision Engineering* 4 (4).

Efstratiou, Androulla, and Theresa Lamagni. 2022. "Epidemiology of Streptococcus pyogenes." In *Streptococcus Pyogenes: Basic Biology to Clinical Manifestations [Internet]*. 2nd edition. University of Oklahoma Health Sciences Center; Chapter 19.

Ereej, Nelofer, Huma Hameed, Mahtab Ahmad Khan, Saleha Faheem, and Anam Hameed. 2024. "Nanoparticle-based Gene Therapy for Neurodegenerative Disorders." *Mini Reviews in Medicinal Chemistry* 24 (19): 1723–1745.

Fan, Daoqing, Juan Wang, Erkang Wang, and Shaojun Dong. 2020. "Propelling DNA computing with materials' power: Recent advancements in innovative DNA logic computing systems and smart bio-applications." *Advanced Science* 7 (24): 2001766.

Fañanas-Baquero, Sara, Oscar Quintana-Bustamante, Daniel P Dever, Omaira Alberquilla, Rebeca Sanchez-Dominguez, Joab Camarena, Isabel Ojeda-Perez, Mercedes Dessy-Rodriguez, Rolf Turk, and Mollie S Schubert. 2021. "Clinically relevant gene editing in hematopoietic stem cells for the treatment of pyruvate kinase deficiency." *Molecular Therapy Methods & Clinical Development* 22: 237–248.

Farooq, Faheem, Peter J Mogayzel, Sophie Lanzkron, Carlton Haywood, and John J Strouse. 2020. "Comparison of US federal and foundation funding of research for sickle cell disease and cystic fibrosis and factors associated with research productivity." *JAMA Network Open* 3 (3): e201737.

Ferrari, Samuele, Valentina Vavassori, Daniele Canarutto, Aurelien Jacob, Maria Carmina Castiello, Attya Omer Javed, and Pietro Genovese. 2021. "Gene editing of hematopoietic stem cells: Hopes and hurdles toward clinical translation." *Frontiers in Genome Editing* 3: 618378.

Ferreira, Daniela, David Fontinha, Catarina Martins, David Pires, Alexandra R Fernandes, and Pedro V Baptista. 2020. "Gold nanoparticles for vectorization of nucleic acids for cancer therapeutics." *Molecules* 25 (15): 3489.

Fumagalli, Francesca, Valeria Calbi, Maria Grazia Natali Sora, Maria Sessa, Cristina Baldoli, Paola Maria V Rancoita, Francesca Ciotti, Marina Sarzana, Maddalena Fraschini, and Alberto Andrea Zambon. 2022. "Lentiviral haematopoietic stem-cell gene therapy for early-onset metachromatic leukodystrophy: Long-term results from a non-randomised, open-label, phase 1/2 trial and expanded access." *The Lancet* 399 (10322): 372–383.

Gamage, Udani, Kesari Warnakulasuriya, Sonali Hansika, and Gayathri N Silva. 2023. "CRISPR Gene therapy: A promising one-time therapeutic approach for transfusion-dependent β-Thalassemia—CRISPR-Cas9 gene editing for β-thalassemia." *Thalassemia Reports* 13 (1): 51–69.

Germino-Watnick, Paula, Malikiya Hinds, Anh Le, Rebecca Chu, Xiong Liu, and Naoya Uchida. 2022. "Hematopoietic stem cell gene-addition/editing therapy in sickle cell disease." *Cells* 11 (11): 1843.

Grabowski, Łukasz, Krzysztof Łepek, Małgorzata Stasiłojć, Katarzyna Kosznik-Kwaśnicka, Karolina Zdrojewska, Monika Maciąg-Dorszyńska, Grzegorz Węgrzyn, and Alicja Węgrzyn. 2021. "Bacteriophage-encoded enzymes destroying bacterial cell membranes and walls, and their potential use as antimicrobial agents." *Microbiological Research* 248: 126746.

Grech, Laura, Karen Borg, and Joseph Borg. 2022. "Novel therapies in β-thalassaemia." *British Journal of Clinical Pharmacology* 88 (6): 2509–2524.

Guan, Jingyang, Guan Wang, Jinlin Wang, Zhengyuan Zhang, Yao Fu, Lin Cheng, Gaofan Meng, Yulin Lyu, Jialiang Zhu, and Yanqin Li. 2022. "Chemical reprogramming of human somatic cells to pluripotent stem cells." *Nature* 605 (7909): 325–331.

Guitart Jr, Joan Ramon, Jodi L Johnson, and Wade W Chien. 2016. "Research techniques made simple: The application of CRISPR-Cas9 and genome editing in investigative dermatology." *Journal of Investigative Dermatology* 136 (9): e87–e93.

Han, Xuejiao, Aqu Alu, Hongmei Liu, Yi Shi, Xiawei Wei, Lulu Cai, and Yuquan Wei. 2022. "Biomaterial-assisted biotherapy: A brief review of biomaterials used in drug delivery, vaccine development, gene therapy, and stem cell therapy." *Bioactive Materials* 17: 29–48.

Hardouin, Giulia, Elisa Magrin, Alice Corsia, Marina Cavazzana, Annarita Miccio, and Michaela Semeraro. 2023. "Sickle cell disease: From genetics to curative approaches." *Annual Review of Genomics and Human Genetics* 24 (1): 255–275.

Hsu, Christine C, Nizar H Senussi, Kleber Y Fertrin, and Kris V Kowdley. 2022. "Iron overload disorders." *Hepatology Communications* 6 (8): 1842–1854.

Iskander, Deena, Lucio Luzzatto, and Anastasios Karadimitris. 2024. "The molecular basis of bone marrow failure syndromes and red cell enzymopathies." *Molecular Hematology* 145–168.

Jacinto, Filipe V, Wolfgang Link, and Bibiana I Ferreira. 2020. "CRISPR/Cas9-mediated genome editing: From basic research to translational medicine." *Journal of Cellular and Molecular Medicine* 24 (7): 3766–3778.

Jensen, Nicolai B, Signe D Justesen, Agnete Larsen, Erik Ernst, and Lars H Pedersen. 2021. "A systematic overview of the spermatotoxic and genotoxic effects of methotrexate, ganciclovir and mycophenolate mofetil." *Acta Obstetricia et Gynecologica Scandinavica* 100 (9): 1557–1580.

Jiang, Zhengzheng, Xulin Hong, Shun Zhang, Ruilian Yao, and YI Xiao. 2020. "CRISPR base editing and prime editing: DSB and template-free editing systems for bacteria and plants." *Synthetic and Systems Biotechnology* 5 (4): 277–292.

Kaladharan, Kiran, Ashish Kumar, Pallavi Gupta, Kavitha Illath, Tuhin Subhra Santra, and Fan-Gang Tseng. 2021. "Microfluidic based physical approaches towards single-cell intracellular delivery and analysis." *Micromachines* 12 (6): 631.

Kantor, Ariel, Michelle E McClements, and Robert E MacLaren. 2020. "CRISPR-Cas9 DNA base-editing and prime-editing." *International Journal of Molecular Sciences* 21 (17): 6240.

Karamperis, Kariofyllis, Maria T Tsoumpeli, Fotios Kounelis, Maria Koromina, Christina Mitropoulou, Catia Moutinho, and George P Patrinos. 2021. "Genome-based therapeutic interventions for β-type hemoglobinopathies." *Human Genomics* 15 (1): 32.

Kavanagh, Patricia L, Titilope A Fasipe, and Ted Wun. 2022. "Sickle cell disease: A review." *JAMA* 328 (1): 57–68.

Khanzadi, Manzoor N, and Abid A Khan. 2020. "CRISPR/Cas9: Nature's gift to prokaryotes and an auspicious tool in genome editing." *Journal of Basic Microbiology* 60 (2): 91–102.

Koh, Gene, Andrea Degasperi, Xueqing Zou, Sophie Momen, and Serena Nik-Zainal. 2021. "Mutational signatures: Emerging concepts, caveats and clinical applications." *Nature Reviews Cancer* 21 (10): 619–637.

Kondrateva, Ekaterina, Anna Demchenko, Alexander Lavrov, and Svetlana Smirnikhina. 2021. "An overview of currently available molecular Cas-tools for precise genome modification." *Gene* 769: 145225.

Koniali, Lola, Carsten W Lederer, and Marina Kleanthous. 2021. "Therapy development by genome editing of hematopoietic stem cells." *Cells* 10 (6): 1492.

Lal, Nand, and Priya Prajapati. 2023. "Curative Therapy of Sickle Cell Disease Using Gene Editing Technologies." *Asian Journal of Biotechnology and Genetic Engineering* 6 (2): 211–235.

Leal, Andrés Felipe, Angelica María Herreno-Pachón, Eliana Benincore-Flórez, Amali Karunathilaka, and Shunji Tomatsu. 2024. "Current Strategies for Increasing Knock-In Efficiency in CRISPR/Cas9-Based Approaches." *International Journal of Molecular Sciences* 25 (5): 2456.

Leibowitz, Mitchell L, Stamatis Papathanasiou, Phillip A Doerfler, Logan J Blaine, Lili Sun, Yu Yao, Cheng-Zhong Zhang, Mitchell J Weiss, and David Pellman. 2021. "Chromothripsis as an on-target consequence of CRISPR–Cas9 genome editing." *Nature Genetics* 53 (6): 895–905.

Li, Chao, Wen Chu, Rafaqat Ali Gill, Shifei Sang, Yuqin Shi, Xuezhi Hu, Yuting Yang, Qamar U Zaman, and Baohong Zhang. 2023. "Computational tools and resources for CRISPR/Cas genome editing." *Genomics, Proteomics & Bioinformatics* 21 (1): 108–126.

Li, Juan, Rongfang Xu, Ruiying Qin, Xiaoshuang Liu, Fanna Kong, and Pengcheng Wei. 2021a. "Genome editing mediated by SpCas9 variants with broad non-canonical PAM compatibility in plants." *Molecular Plant* 14 (2): 352–360.

Li, Yichen, Faisal Raza, Yuhao Liu, Yiqi Wei, Ruonan Rong, Mengyuan Zheng, Weien Yuan, Jing Su, Mingfeng Qiu, and Y Li. 2021b. "Clinical progress and advanced research of red blood cells based drug delivery system." *Biomaterials* 279: 121202.

Losito, Marialucrezia. 2022. "Functional characterization of CRISPR-Cas interactions with DNA." Durham University.

Lu, Yaoyao, Cedric Happi Mbakam, Bo Song, Eli Bendavid, and Jacques-P Tremblay. 2022. "Improvements of nuclease and nickase gene modification techniques for the treatment of genetic diseases." *Frontiers in Genome Editing* 4: 892769.

Manghwar, Hakim, Bo Li, Xiao Ding, Amjad Hussain, Keith Lindsey, Xianlong Zhang, and Shuangxia Jin. 2020. "CRISPR/Cas systems in genome editing: Methodologies and tools for sgRNA design, off-target evaluation, and strategies to mitigate off-target effects." *Advanced Science* 7 (6): 1902312.

Mann, Zoya, Manisha Sengar, Yogesh Kumar Verma, Raja Rajalingam, and Pawan Kumar Raghav. 2022. "Hematopoietic stem cell factors: Their functional role in self-renewal and clinical aspects." *Frontiers in Cell and Developmental Biology* 10: 664261.

Mata, Rani, Yuejun Yao, Wangbei Cao, Jie Ding, Tong Zhou, Zihe Zhai, and Changyou Gao. 2021. "The dynamic inflammatory tissue microenvironment: Signality and disease therapy by biomaterials." *Research* 2021: 1–31.

Matsoukas, Ianis G. 2020. "Prime editing: Genome editing for rare genetic diseases without double-strand breaks or donor DNA." *Frontiers in Genetics* 11: 528.

Mehrjardi, Narges Zare, Daniel Hänggi, and Ulf Dietrich Kahlert. 2020. "Current biomarker-associated procedures of cancer modeling-a reference in the context of IDH1 mutant glioma." *Cell Death & Disease* 11 (11): 998.

Mindell, David A, and Elisabeth Reynolds. 2023. *The Work of the Future: Building Better Jobs in an Age of Intelligent Machines*. MIT Press.

Mohamadi, Sahar, Saeed Zaker Bostanabad, and Reza Mirnejad. 2020. "CRISPR arrays: A review on its mechanism." *Journal of Applied Biotechnology Reports* 7 (2): 81–86.

Mohammadian Gol, Tahereh, Guillermo Ureña-Bailén, Yujuan Hou, Ralph Sinn, Justin S Antony, Rupert Handgretinger, and Markus Mezger. 2023. "CRISPR medicine for blood disorders: Progress and challenges in delivery." *Frontiers in Genome Editing* 4: 1037290.

Molaei, Zahra, Zahra Jabbarpour, Azadeh Omidkhoda, and Naser Ahmadbeigi. 2024. "Exploring non-viral methods for the delivery of CRISPR-Cas ribonucleoprotein to hematopoietic stem cells." *Stem Cell Research & Therapy* 15 (1): 233.

Munir, Ahmad, Muhammad Ali, Sameer H Qari, Nayla Munawar, Muhammad Sulyman Saleem, and Aftab Ahmad. 2024. "CRISPR workflow solutions: Cargos and versatile delivery platforms in genome editing." In *CRISPRized Horticulture Crops*, 67–90. Elsevier.

Murugesan, Ramya, Karthik V Karuppusamy, Srujan Marepally, and Saravanabhavan Thangavel. 2023. "Current approaches and potential challenges in the delivery of gene editing cargos into hematopoietic stem and progenitor cells." *Frontiers in Genome Editing* 5: 1148693.

Mushtaq, Muntazir, Aejaz Ahmad Dar, Milan Skalicky, Anshika Tyagi, Nancy Bhagat, Umer Basu, Basharat Ahmad Bhat, Abbu Zaid, Sajad Ali, and Tanvir-Ul-Hassan Dar. 2021. "CRISPR-based genome editing tools: Insights into technological breakthroughs and future challenges." *Genes* 12 (6): 797.

Naeem, Muhammad, Saman Majeed, Mubasher Zahir Hoque, and Irshad Ahmad. 2020. "Latest developed strategies to minimize the off-target effects in CRISPR-Cas-mediated genome editing." *Cells* 9 (7): 1608.

Neugebauer, Monica E, Alvin Hsu, Mandana Arbab, Nicholas A Krasnow, Amber N McElroy, Smriti Pandey, Jordan L Doman, Tony P Huang, Aditya Raguram, and Samagya Banskota. 2023. "Evolution of an adenine base editor into a small, efficient cytosine base editor with low off-target activity." *Nature Biotechnology* 41 (5): 673–685.

Nidhi, Sweta, Uttpal Anand, Patrik Oleksak, Pooja Tripathi, Jonathan A Lal, George Thomas, Kamil Kuca, and Vijay Tripathi. 2021. "Novel CRISPR–Cas systems: An updated review of the current achievements, applications, and future research perspectives." *International Journal of Molecular Sciences* 22 (7): 3327.

Odiba, Arome Solomon, Nkwachukwu Oziamara Okoro, Olanrewaju Ayodeji Durojaye, and Yanjun Wu. 2021. "Gene therapy in PIDs, hemoglobin, ocular, neurodegenerative, and hemophilia B disorders." *Open Life Sciences* 16 (1): 431–441.

Oikonomopoulou, Christina, and Evgenios Goussetis. 2021. "HSCT remains the only cure for patients with transfusion-dependent thalassemia until gene therapy strategies are proven to be safe." *Bone Marrow Transplantation* 56 (12): 2882–2888.

Panigrahi, Kaushik Kumar, Ayesha Mohanty, Smruti Ranjan Padhan, Priya brata Bhoi, Syed Mohammad Bashir Ali, and Purandar Mandal. 2023. "Recent advances in genome editing towards sustainable agriculture." *Genome Editing and Global Food Security* 155–193.

Paolini Sguazzi, Giulia, Valentina Muto, Marco Tartaglia, Enrico Bertini, and Claudia Compagnucci. 2021. "Induced pluripotent stem cells (iPSCs) and gene therapy: A new era for the treatment of neurological diseases." *International Journal of Molecular Sciences* 22 (24): 13674.

Paris, Katherine. 2022. "Background: Genome editing with programmable nucleases." In *Genome Editing and Biological Weapons: Assessing the Risk of Misuse*, 17–50. Springer.

Paul, Prolay, Dipankar Maiti, and Triyasha Adhikary. 2021. "Iron deficiency anemia-An updated overview." *International Journal of Research in Pharmacy and Pharmaceutical Sciences* 6 (4): 28–33.

Pena, Stefanie A, Rahul Iyengar, Rebecca S Eshraghi, Nicole Bencie, Jeenu Mittal, Abdulrahman Aljohani, Rahul Mittal, and Adrien A Eshraghi. 2020. "Gene therapy for neurological disorders: Challenges and recent advancements." *Journal of Drug Targeting* 28 (2): 111–128.

Perčulija, Vanja, Jinying Lin, Bo Zhang, and Songying Ouyang. 2021. "Functional features and current applications of the RNA-targeting type VI CRISPR-Cas systems." *Advanced Science* 8 (13): 2004685.

Pia, Luisanna V. 2024. "Developing a spatial functional genomics approach for probing ovarian cancer biology." Icahn School of Medicine at Mount Sinai.

Piergentili, Roberto, Alessandro Del Rio, Fabrizio Signore, Federica Umani Ronchi, Enrico Marinelli, and Simona Zaami. 2021. "CRISPR-Cas and its wide-ranging applications: From human genome editing to environmental implications, technical limitations, hazards and bioethical issues." *Cells* 10 (5): 969.

Pinilla-Redondo, Rafael, Jakob Russel, David Mayo-Muñoz, Shiraz A Shah, Roger A Garrett, Joseph Nesme, Jonas S Madsen, Peter C Fineran, and Søren J Sørensen. 2022. "CRISPR-Cas systems are widespread accessory elements across bacterial and archaeal plasmids." *Nucleic Acids Research* 50 (8): 4315–4328.

Quintana-Bustamante, Oscar, Sara Fañanas-Baquero, Mercedes Dessy-Rodriguez, Isabel Ojeda-Pérez, and Jose-Carlos Segovia. 2022. "Gene editing for inherited red blood cell diseases." *Frontiers in Physiology* 13: 848261.

Rainha, João, Joana L Rodrigues, and Lígia R Rodrigues. 2020. "CRISPR-Cas9: A powerful tool to efficiently engineer Saccharomyces cerevisiae." *Life* 11 (1): 13.

Roma-Rodrigues, Catarina, Lorenzo Rivas-García, Pedro V Baptista, and Alexandra R Fernandes. 2020. "Gene therapy in cancer treatment: Why go nano?" *Pharmaceutics* 12 (3): 233.

Rosanwo, Tolulope O, and Daniel E Bauer. 2021. "Editing outside the body: Ex vivo gene-modification for β-hemoglobinopathy cellular therapy." *Molecular Therapy* 29 (11): 3163–3178.

Sansbury, Brett M, Amanda M Hewes, Olivia M Tharp, Sophia B Masciarelli, Salma Kaouser, and Eric B Kmiec. 2022. "Homology directed correction, a new pathway model for point mutation repair catalyzed by CRISPR-Cas." *Scientific Reports* 12 (1): 8132.

Satthakarn, Surada, Sitthichai Panyasai, and Sakorn Pornprasert. 2020. "Molecular characterization of β-and α-globin gene mutations in individuals with borderline hb a2 levels." *Hemoglobin* 44 (5): 349–353.

Schwartz, Matthew L, M Wayne Davis, Matthew S Rich, and Erik M Jorgensen. 2021. "High-efficiency CRISPR gene editing in C. elegans using Cas9 integrated into the genome." *PLoS Genetics* 17 (11): E1009755.

Seeberger, Peter H. 2021. "Discovery of semi-and fully-synthetic carbohydrate vaccines against bacterial infections using a medicinal chemistry approach: Focus review." *Chemical Reviews* 121 (7): 3598–3626.

Shelake, Rahul Mahadev, Dibyajyoti Pramanik, and Jae-Yean Kim. 2023. "Improved dual base editor systems (iACBEs) for simultaneous conversion of adenine and cytosine in the bacterium Escherichia coli." *MBio* 14 (1): e02296-22.

Shirley, Jamie L, Ype P de Jong, Cox Terhorst, and Roland W Herzog. 2020. "Immune responses to viral gene therapy vectors." *Molecular Therapy* 28 (3): 709–722.

Sii-Felice, Karine, Olivier Negre, Christian Brendel, Alisa Tubsuwan, Eglantine Morel-à-l'Huissier, Camille Filardo, and Emmanuel Payen. 2020. "Innovative therapies for hemoglobin disorders." *BioDrugs* 34 (5): 625–647.

Sledzinski, Pawel, Magdalena Dabrowska, Mateusz Nowaczyk, and Marta Olejniczak. 2021. "Paving the way towards precise and safe CRISPR genome editing." *Biotechnology Advances* 49: 107737.

Stinson, Benjamin M, and Joseph J Loparo. 2021. "Repair of DNA double-strand breaks by the nonhomologous end joining pathway." *Annual Review of Biochemistry* 90 (1): 137–164.

Stomper, Julia, John Charles Rotondo, Gabriele Greve, and Michael Lübbert. 2021. "Hypomethylating agents (HMA) for the treatment of acute myeloid leukemia and myelodysplastic syndromes: Mechanisms of resistance and novel HMA-based therapies." *Leukemia* 35 (7): 1873–1889.

Tatwavedi, Dharamveer, Andrea Pellagatti, and Jacqueline Boultwood. 2023. "Recent advances in the application of induced pluripotent stem cell technology to the study of myeloid malignancies." *Advances in Biological Regulation* 100993.

van den Bos, Christian, Robert Keefe, Mary Keville, and Michael McCaman. 2024. "Manufacturing and quality of advanced therapeutics: Cellular therapies, CAR-T and gene therapies." In *Biopharmaceutical Manufacturing: Progress, Trends and Challenges*, 453–488. Springer.

Wang, Joy Y, and Jennifer A Doudna. 2023. "CRISPR technology: A decade of genome editing is only the beginning." *Science* 379 (6629): eadd8643.

Wells, Timothy J, Tyron Esposito, Ian R Henderson, and Larisa I Labzin. 2024. "Mechanisms of antibody-dependent enhancement of infectious disease." *Nature Reviews Immunology* 1–16.

Willbanks, Amber, Shaun Wood, and Jason X Cheng. 2021. "RNA epigenetics: Fine-tuning chromatin plasticity and transcriptional regulation, and the implications in human diseases." *Genes* 12 (5): 627.

Xu, Yuanyuan, and Zhanjun Li. 2020. "CRISPR-Cas systems: Overview, innovations and applications in human disease research and gene therapy." *Computational and Structural Biotechnology Journal* 18: 2401–2415.

Yannaki, Evangelia, Nikoletta Psatha, Anastasia Papadopoulou, Takis Athanasopoulos, Achilleas Gravanis, Maria G Roubelakis, Panagiotis Tsirigotis, Achilles Anagnostopoulos, Nicholas P Anagnou, and George Vassilopoulos. 2021. "Success stories and challenges ahead in hematopoietic stem cell gene therapy: Hemoglobinopathies as disease models." *Human Gene Therapy* 32 (19–20): 1120–1137.

Yesilipek, M. Akif. 2020. "Hematopoietic stem cell transplantation in patients with hemoglobinopathies." *Hemoglobin* 44 (6): 377–384.

Yetisgin, Abuzer Alp, Sibel Cetinel, Merve Zuvin, Ali Kosar, and Ozlem Kutlu. 2020. "Therapeutic nanoparticles and their targeted delivery applications." *Molecules* 25 (9): 2193.

Zaib, Sumera, Mushtaq A Saleem, and Imtiaz Khan. 2022. "CRISPR-Cas9 genome engineering: Trends in medicine and health." *Mini Reviews in Medicinal Chemistry* 22 (3): 410–421.

Zhang, Huimin, Chunhong Qin, Changming An, Xiwang Zheng, Shuxin Wen, Wenjie Chen, Xianfang Liu, Zhenghua Lv, Pingchang Yang, and Wei Xu. 2021. "Application of the CRISPR/Cas9-based gene editing technique in basic research, diagnosis, and therapy of cancer." *Molecular Cancer* 20: 1–22.

Zhao, Zhihan, Peng Shang, Prarthana Mohanraju, and Niels Geijsen. 2023. "Prime editing: Advances and therapeutic applications." *Trends in Biotechnology* 41 (8): 1000–1012.

9 CRISPR-Cas Systems in Neurological Diseases

Aarif Ali, Bashir Ahmad Malla, Mohammad Saleem Bhat, and Javeed Iqbal Ahmad Bhat

9.1 OVERVIEW OF NEUROLOGICAL DISEASES

Neurological disorders are medical conditions that disrupt the normal functions of the nervous system. This elaborate system includes the brain, spinal cord, and the numerous nerves that radiate from these structures. It is essentially the governing center for all bodily functions, including breathing, mobility, cognitive activities, and emotional sensations. Across worldwide, neurological diseases are the leading cause of disability (second one in case of mortality) (Feigin et al., 2017). The most prevalent neurological disorders significantly impair cognitive abilities, contributing to elevated levels of morbidity, disability, and mortality (Feigin et al., 2019). Through global population ages, neurodegenerative disorders (NDDs) are emerging as significant health challenges. Within 50 years, they are expected to be the second leading cause of mortality globally, even more than cancer (Nichols et al., 2022). The NDDs result into a gradual and progressive loss of neurons, hence affecting motor, sensory, and various other systems, including their cognitive functions (Blackhurst and Funk, 2023). Common symptoms of these disorders can manifest in different ways, such as difficulties with movement, sensory perception, cognitive abilities, memory, and abstract thinking, all of which may get worse as the disease advances (Cerejeira et al., 2012). Among the most prominent NDDs, Alzheimer's disease (AD), Parkinson's disease (PD), and amyotrophic lateral sclerosis (ALS) each contribute to the tragic loss of thousands of lives. Sadly, the number of patients suffering from these disorders is anticipated to rise steadily in the near-term years (Blackhurst and Funk, 2023). Recent estimates indicate that neurological diseases account for 3% of the worldwide burden according to the study of Global Burden of Disease (Murray et al., 2012). Predominantly, the NDD mentioned by the study includes AD, dementias, PD, epilepsy, several headache disorders (migraine, medication-overuse headache, and tension-type headache) (Murray et al., 2012). While this percentage may appear modest, conditions like migraine, stroke, epilepsy, and dementia feature amongst the topmost 50 causes of disability-adjusted life years (DALYs), impairing both individuals as well as their families (Murray et al., 2012). Migraine and epilepsy are significant contributors to the global neurological burden, accounting for one-third and one-fourth of this challenge (Murray et al., 2012). Looking ahead, the burden of neurological diseases is projected to rise dramatically in low- and middle-income

DOI: 10.1201/9781003537151-9

countries (LMICs) over the next decade (Murray et al., 2012). Despite the profound effects that these neurological disorders have on individuals and communities, our understanding of their epidemiology remains limited. This includes insights into how disease frequency varies across different regions and times, as well as the risk factors and outcomes associated with these disorders, particularly in LMICs. Patients with neurological conditions often find themselves in need of substantial social and economic support due to various physical, cognitive, and psychosocial challenges (WHO, 2006). Alarmingly, even though disability rates are high, the services and resources to address these needs are often lacking, particularly LMICs (WHO, 2006).

Thus, NDDs are defined by the accumulation of protein aggregates, and the significance of these aggregates in causing neuronal toxicity has been widely discussed. In most NDDs, there is notable evidence that pathological protein, especially amyloid-beta (Aβ), tau (τ), α-synuclein (α-syn), and transactive response DNA-binding protein 43 (TDP-43) proteins, forms spread-like structures throughout the brain in distinct patterns (Goedert, 2015). While each NDD typically shows the buildup of specific protein aggregates, there are numerous instances where multiple pathological proteins aggregate together. AD is primarily defined by the accumulation of the Aβ and τ-proteins, while PD and ALS result due to the aggregation of α-synuclein and TDP-43 proteins, respectively (Goutman et al., 2022). Among all these NDDs, AD accounts for about 50–60% of the cases of dementia and therefore is the most common form. This condition is characterized by notable changes in the brain, particularly marked gray matter loss, along with the buildup of amyloid plaques (APs) and neurofibrillary tangles (NFTs) (Hyman et al., 2012). The APs are formed from clusters of Aβ, which arise from the breakdown of amyloid precursor protein (APP) through the actions of two proteolytic β- and γ-secretase enzymes (also called presenilin 1) (Hardy and Selkoe, 2002), while as NFTs are clusters of abnormal τ-protein that form twisted filaments (Grundke-Iqbal et al., 1986). The process begins with the irregular phosphorylation of τ-protein, which causes it to detach from axonal microtubules. This disruption allows phosphorylated τ-protein to accumulate improperly in the cell body and dendrites and develops these pathological conditions. Somatodendritic accumulations of phosphorylated τ-protein, commonly called "pre-tangles," appear prior to the formation of fibrils (Alafuzoff et al., 2008). These pre-tangles act as precursors to NFTs, which consist of paired helical filaments of τ. The formation of NFTs occurs after the appearance of pre-tangles and is closely linked to neuronal degeneration. Interestingly, even after neurons have died, NFTs remain as ghostly remnants, known as ghost tangles (Gomez-Isla et al., 1996). The τ-protein is an intriguing component of the human brain, existing in six distinct isoforms, each featuring a unique microtubule-binding repeat region (Goedert et al., 1989). Among these isoforms, half possess a repeat region consisting of three repeats, while the remaining half boast with four repeats. In the context of AD, researchers have identified both three and four repeats τ-isoforms within the τ-aggregates, hence highlighting the complexity of this protein's role in neurodegenerative processes (Goedert et al., 1989). The *synucleinopathies* refer to a group of disorders, like multiple system atrophy (MSA) and Lewy body diseases (LBD), that occur due to the α-synuclein aggregates, in which MSA develops due to the accumulation of α-synuclein aggregates in glial cells, which later on forms cytosolic inclusions known as Papp–Lantos bodies. On the other hand, the key pathological features of LBD are the α-synuclein aggregates in neuronal cell bodies and later on results into Lewy bodies (LB) formation.

A provisional diagnosis of an NDD is normally done on the basis of their clinical signs, symptoms, and other pathological features. However, definitive confirmation necessitates postmortem pathological examination in order to characterize the abnormal protein deposits. The etiological and pathological bases of neurological diseases are still largely elusive on the top priorities facing medical and biological science, owing to the relatively high incidence, partial unidentified processes nature, as well as their profound burden for patients, families, and society.

9.2 ROLE OF GENETIC MUTATIONS IN NEUROLOGICAL DISORDERS

Molecular genetics is crucial for understanding the mechanisms behind NDDs, as it helps in identifying specific genes linked to these conditions and which can lead to more effective treatment (Brettschneider et al., 2015). Additionally, molecular genetics has been instrumental in pinpointing the proteins that form aggregates within neurological cells and contributes to the development of NDDs. The most prevalent NDD disorders include AD, PD, ALS, and Huntington's disease (HD), each associated with different genes but having the common characteristic of protein aggregation, which alters the physical and chemical properties of nerve cells (Kovacs, 2018).

9.2.1 Alzheimer's Disease

AD is one of the most common NDDs and affects about 35 million patients globally. AD is caused by the buildup of amyloid proteins and neurofibrillary degeneration, which result into APs and NFT formation, respectively (Schellenberg and Montine, 2012). Evidence suggests that 58–79% of dementia cases have AD genetic component, and therefore, AD is a leading cause of dementia (Gatz et al., 2006). Reports of inherited family traits dating back to the 1930s explained it as an autosomal dominant inheritance trait that indicates 50% chances of occurrence of this disease in the next generation if having a mutation in APP or presenilin 1 (PSEN1) and/or presenilin 2 (PSEN2) genes (Schellenberg and Montine, 2012).

9.2.1.1 Amyloid Precursor Protein

APP is an ancient and highly conserved protein that serves as a precursor to Aβ, a polypeptide consisting of 37 to 49 amino acid residues. APP can follow either the amyloidogenic or non-amyloidogenic pathway, undergoing cleavage by distinct secretase enzymes: alpha (α-secretase), beta (β-secretase), or gamma (γ-secretase) (Roberts et al., 1994). At normal physiological setting, and in a non-amyloidogenic pathway, the APP is typically cleaved by the α- and γ-secretases (Wu et al., 2017). Conversely, in an amyloidogenic pathway, the APP is cleaved by β- and γ-secretases and results in the formation of different sized Aβ polypeptides, like Aβ38, Aβ40, and Aβ42 (Okochi et al., 2013). Therefore, this pathway contributes to the formation of Aβ plaques, as among all Aβ polypeptides, Aβ42 exhibits the lowest solubility

(Hampel et al., 2023). Plaque formation begins with Aβ monomers, their aggregation, and progresses to the establishment of APs. In the brains of Alzheimer's patients, the amyloid fibrillar form of Aβ42 type, known as APs, has been observed (Wallon et al., 2012). Studies show that APP gene mutation contributes to familial susceptibility to AD. Other reports indicate that duplications of APP or β cleavage site mutation results in an overproduction of Aβ species and develops early-onset AD (Wallon et al., 2012). For example, a missense mutation in coding sequence 17 (exon 17) of APP gene was identified in 1991, resulting in the substitution of valine residue with isoleucine at the 717th position (Val717Ile) (Goate et al., 1991). As this mutation was first detected in English, therefore it is commonly termed as the "London mutation." From this discovery, Chartier-Harlin and Murrell and their co-workers independently reported other mutations (Val717Gly and Val717Phe, respectively) that affected the valine residue at the same position in other families with AD (Chartier-Harlin et al., 1991; Murrell et al., 1991). Interestingly, Murrell and his colleagues discovered this mutation in Romanian relatives and termed this the "Indiana mutation." Mullan and his co-workers, in 1993, presented their findings from two large Swedish families with early-onset AD that carried a double mutation (Lys670Asn and Met671Leu) in the N-terminus of APP and referred to them as the "Swedish mutations" (Mullan et al., 1993). After these discoveries, a variety of additional APP mutations have been found, leading to a heightened focus of scientists on APP research (Hardy, 2017). A study that included 1,795 subjects from Iceland established that the Ala673Thr substitution mutation, often referred to as the "Icelandic mutation," lowers the risk of AD. This Ala673Thr mutation results in the reduction of Aβ aggregates (about 40%) due to the suppression of Aβ40 and Aβ42 formation (Jonsson et al., 2012; Benilova et al., 2014).

9.2.1.2 Presenilin 1

The PSEN1 gene is situated on chromosome 14 and produces a PSEN1 protein of 467 amino acid residues long. PSEN1 exhibits a 67% similarity to the PSEN2. PSEN1 protein is described as a transmembrane protein, as it consists of nine transmembrane domains with four hydrophilic loops in the cytosolic, as well as the non-cytosolic, environment. The N-terminal is toward the intracellular (cytosolic) side, whereas the C-terminal is toward the extracellular side. The PSEN1 gene–encoded presenilin-1, also called the catalytic core of γ-secretase, acts as an intramembranous protease enzyme by cleaving various type I transmembrane proteins (Laudon et al., 2005). Thus PSEN1, besides APP cleavage, also serves other biological roles, like being involved in the Notch-Delta signaling pathway, β-cadherin processing, and the setup of proper calcium homeostasis.

Numerous motifs and residues within PSEN1, including wobbling nuclease finder (WNF), G″ for Glycine, "x" for any amino acid, and "D" for Aspartic acid (GxGD), and PALP, have been identified as potentially significant in the functioning of the γ-secretase enzyme. Thus, PSEN1 is characterized as having two catalytic aspartate (Asp) residues, at the 275th and the 385th position (Asp275 and Asp385) of the transmembrane domain number 6 and 7, respectively (Vetrivel et al., 2006). These residues are essential for the endoproteolytic function of PSEN1 and for its γ-secretase activity (Vetrivel et al., 2006). Furthermore, other significant residues have been identified within the transmembrane region, specifically between the 95 and the 98 positions. Deficient PSEN1 functionality might interfere with APP processing, causing variations in the production of amyloid peptides (De Strooper et al., 1998). About 6% of AD patients exhibit mutations in this gene (Bagyinszky et al., 2014). These mutations are recognized as the predominant cause of early-onset AD, often leading to a rapidly advancing form of the illness (Park et al., 2017). Mutations occurring in these areas can significantly impair or completely disrupt PSEN1 proteolysis and γ-secretase function (Brunkan et al., 2005). Over 300 mutations in PSEN1 gene have been documented, leading to a wide range of clinical phenotypes (De Strooper et al., 2012). Besides, early-onset AD involves individuals with PSEN1 mutations and manifests atypical symptoms, like spasticity, visual disturbances, and seizures. The relationship between PSEN1 mutations and the development of neurodegeneration and dementia in familial Alzheimer's disease (fAD) continues to be a subject of intense discussion. Researchers have proposed two distinct hypotheses, although they are not mutually exclusive but are used to elucidate how PSEN1 gene mutations lead to fAD (Hardy and Selkoe, 2002). The amyloid hypothesis suggests that certain mutations may kick-start AD by elevating Aβ42 levels (Hardy and Selkoe, 2002). In contrast, the presenilin hypothesis presents an alternative perspective on disease development, proposing that mutations in the PSEN1 gene lead to the decrease or loss of critical enzymatic functions of presenilin within the neuronal cells, ultimately resulting in neurodegeneration and dementia associated with fAD.

In the N-terminal fragment, 12 mutations have been found; however, their nature of pathogenicity remains uncertain (Perrone et al., 2020). Among those mutations, a strongest influence on the onset of disease is created by two mutations, like Asp40 deletion and Ala79Val substitution (Perrone et al., 2020). In sporadic frontotemporal dementia (FTD) and AD-affected individuals, Asp40 deletion mutation has been detected. Instances of the Ala79Val substitution mutation have been reported in various cases throughout Europe and the USA, with the phenotypes of the disease showing considerable variability. Furthermore, the Ala79Val mutation variant of PSEN1 is also linked with late-onset AD, irrespective of family history (Cruchaga et al., 2012). A variety of mutations have been discovered in the transmembrane domain number 1 region of the PSEN1, with the majority verified mutations having an impact on AD, as the initial transmembrane domain includes their multiple residues, playing a crucial role in γ-secretase activity. Several PSEN1 mutations have been associated with FTD or AD and presenting FTD-like symptoms, particularly in cases involving behavioral or language variant disorders. For instance, mutations like leu113Pro, Gly183Val, Leu226Phe, Met233Leu, or the insertion of Arg at the 352nd position has shown comparable FTD manifestation (Yang et al., 2023).

9.2.1.3 Presenilin 2

The year 1995 marked the initial identification of PSEN2 as a gene linked to AD, following the earlier discoveries of APP and PSEN1 (Levy-Lahad et al., 1995). The PSEN2 gene is present on the long arm of chromosome no. 1 (1q42.13), and it contains 12 coding sequences, with the first two exons representing untranslated regions (UTR). PSEN2 encodes a transmembrane protein known as PSEN2, which is composed up of 448 amino acids and has 55 kDa weight (De Strooper et al., 1997; Li et al., 1997). The PSEN2 protein cradles the plasma membrane about nine times and hence contains nine transmembrane domains, like PSEN1 protein (Li et al., 2013). It shares 67% sequence homology with PSEN1, such that mainly the N-terminus and differences in hydrophilic loop are apparent, whereas its transmembrane regions are rather conserved (Rademakers et al., 2003) The pro-peptide of PSEN2 tends to undergo autolytic cleavage in its large loop domain and becomes an unstable holoprotein, which then combines with β-catenin and forms a stable, biologically active heterodimer. The γ-secretase complex catalytic sites in PSEN2 protein include the cross of two transmembrane spans of the 6 and the 7 domains, with aspartate residues at positions 263rd and 366th. The role of this dimer is to cleave APP, and it produces different-sized Aβ polypeptides, including Aβ42. The Aβ42 variant is recognized for its greater propensity to form aggregates compared to Aβ40, and therefore, in the brain, Aβ accumulation is a key pathological indicator of AD (De Strooper et al., 2012). Different mutations in PSEN2 may increases the enzyme activity of γ-secretase. Therefore, presenilin gene mutations are considered a significant risk factor for AD development (Citron et al., 1997). Additionally, researchers have demonstrated that presenilin mutations related to AD can also affect intracellular calcium (Ca^{2+}) signaling, which further leads to an increase in Aβ aggregation, AP formation, and ultimately, neuronal cell death (Leissring et al., 2001). Currently, 38 mutations have been documented in PSEN2, but the frequency of mutations in PSEN1 gene exceeds those in PSEN2 by more than five times (Larner, 2011). Among these PSEN2 mutations, two are frameshift mutations, like Glu126 and Lys306, with the rest being nonsynonymous substitutions (Lee et al., 2014; Wallon et al., 2012). Out of 38 mutations identified, only 17 are predicted to contribute to disease causation, and about 16 mutations are reported in the transmembrane domains. Among the mutations identified, 10 are classified as non-pathogenic, and the pathogenicity of the rest remains ambiguous.

9.2.2 Parkinson's Disease

PD is recognized as the second most common NDD linked to aging, and it manifests as a progressive movement disorder. Approximately 7 to 10 million people globally are diagnosed with PD, and the condition is usually sporadic in nature. The principal causes of PD involve loss of dopaminergic neurons (neurotransmitter-forming neuron) in the substantia nigra pars compacta and the formation of LB inclusions within neurons. Both gene mutations and protein aggregation significantly influence the dopaminergic neurodegeneration process. Studies in the field of genetics associated with PD have discovered numerous monogenic types of the disorder and genetic factors that raise the probability of neurodegeneration. The contribution of genetic variation to the overall risk of developing PD is estimated to be roughly about 25% (Goldman et al., 2019). It has been identified that mutations in various loci of the parkin (PARK) gene, like PARK1 to PARK13 loci (13 chromosomal regions), are linked to PD (Schultheis et al., 2004). The autosomal dominant mutations in genes like SNCA (α-synuclein) and Leucine-rich repeat kinase 2 (LRRK2) and mutation in Parkinsonism-associated deglycase 7 (DJ-1), Phosphatase and tensin homolog (pten)–induced kinase 1 (PINK1), and F-box protein 7 (FBXO7) genes result in the development of autosomal recessive Parkinsonism (Papagiannakis et al., 2018).

9.2.2.1 SNCA

SNCA gene codes for protein called α-synuclein and is significantly present in neuronal cells. This highly conserved neuronal protein is vital for controlling the cellular vesicular neurotransmission and also regulating dopamine neurotransmission in humans (Ferreira et al., 2017). Furthermore, SNCA gene was reported as the first one to have mutations linked with autosomal dominant PD. In the entire SNCA, gene mutations are less overall, and till now, only three missense mutations, along with duplications and triplications, have been identified (Klein and Schlossmacher, 2006). The most frequently reported missense mutation found in SNCA gene is Ala53Thr, and it has been reported in Swedish, Italian, Korean, and Greek families (Puschmann et al., 2009; Ki et al., 2007). Similarly, in other families, Glu46Lys and Ala30Pro mutations were also found (Zarranz et al., 2004). In the entire exon of SNCA, 17 duplications were found, in which 4 are sporadic and 13 are familial PD cases (Ibanez et al., 2004; Chartier-Harlin et al., 2004). In three independent families, triplications in the SNCA gene were also reported, and remarkably, one of the families exhibited triplications (Ibanez et al., 2004; Farrer et al., 2004).

9.2.2.2 Parkin (PARK2)

Parkin (PARK) is recognized as the second largest gene within the human genome, encoding a PARK protein of 465 amino acid residues. The encoded protein acts like an E3 ubiquitin ligase, playing a crucial role in the ubiquitination-dependent proteasomal degradation process. PARK was the second gene associated with PD and the first gene definitively linked to an autosomal recessive PD. Numerous PARK gene mutations have been identified, encompassing alterations across all 12 exons and observed in diverse ethnic populations. These mutations represent the most prevalent known cause of early-onset PD and contribute toward about 77% of familial cases with an age of less than 30 years (Klein and Lohmann-Hedrich, 2007). To date, a total of 887 mutations in the coding sequence (exon) of this gene has been documented, which includes 147 distinct variations. Among these mutations, approximately 33% consist of single-nucleotide changes, 13% are characterized by small deletions, and 54% comprise both deletions and duplications (Grunewald and Klein, 2012).

9.2.2.3 Leucine-Rich Repeat Kinase 2 (LRRK2)

The LRRK2 gene directs the production of a cytoplasmic protein known as Dardarin, made up of 2,527 residues. This gene is active within the brain and is made up of 51 exons. More than 100 documented missense and nonsense mutations in this gene have been identified in families affected by late-onset PD, suggesting their role in the development of typical idiopathic PD (Nuytemans et al., 2010). Research studies also demonstrated that at least 16 mutations, which include the 6 recurrent variants, namely, Arg114Cys, Arg1441Gly, Arg1441His, Tyr1699Cys, Gly2019Ser, and Ile2020Thr, are likely pathogenic in nature (Nuytemans et al., 2010). These mutations are grouped within ten exons, mainly affecting the C-terminal region of Dardarin protein. Some specific mutations can cause a change in a single amino acid within the Dardarin protein, resulting in modifications to its structure and as well as functions. This mutation predominantly replaces arginine with glycine residue at a particular position, which has been linked to the occurrence of Parkinsonism in the population. So far, about 12 different mutational alterations have been recorded, among which Gly2019Ser is the most frequently identified, present approximately in 1–2% sporadic cases, while familial-autosomal dominant cases make up about 5–7% (Gegg et al., 2010).

9.2.2.4 Phosphatase and Tensin Homolog (PTEN)–Induced Kinase 1 (PINK1)

The PINK1 gene, linked with the phosphatase and tensin homolog, is implicated as the second most common genetic cause of autosomal recessive early-onset PD. Studies have shown that the mutation rate of PINK1 gene falls within the range of 1–9%, with variation noted among distinct ethnicities, underscoring the complexity of genetic factors influencing this disorder (Bonifati et al., 2005). The predominant types of mutations known in the PINK1 are missense and nonsense mutations. Until now, a small number of cases have been documented involving whole-exon deletion. For example, to date about three different families with deletions spanning in exons 4 to 8, exons 6 to 8, and exon 7 have been reported (Camargos et al., 2009). Additionally, Marongiu and his colleagues reported a case that involves whole-gene deletion with heterozygous nature (Marongiu et al., 2007).

9.2.2.5 DJ-1

The DJ-1 gene contains seven coding regions and encodes Parkinson protein-7, made up of 189 residues. The DJ-1 protein (Parkinson protein-7) is present in many tissues and serves as detector of oxidative stress. One of the main jobs of DJ-1 is to shield the neurons from harmful reactive oxygen species (ROS). Mutation or some change in DJ-1 gene are linked to 1–2% of early-onset PD with an autosomal recessive inheritance nature (Pankratz et al., 2006). In this gene, about ten different mutations, including deletions and point, have been identified (Klein and Westenberger, 2012). Normally, the DJ-1 protein is active in dimer form, and mutations such as Asp149Ala, Leu166Pro, Met26Ile, and Glu64Asp in the dimeric form result in the development of various diseases, including PD (Takahashi-Niki et al., 2004).

9.2.3 Huntington's Disease

HD is a significant NDD caused by an autosomal dominant mutation in the first exon of the huntingtin gene (HTT), which encodes huntingtin protein (HTT) (Malankhanova et al., 2017). The HD pathology is characterized by the expansion of the trinucleotide sequence: cytosine-guanine-adenine (CAG) repeats on short arm of chromosome no. 4 (4p16.3) (Dianov and Hubscher, 2013; Kumar et al., 2020). Normally, the healthy individuals' HTT gene typically possesses 16–20 CAG repeats, while individuals with HD often exhibit more than 40 CAG repeats (Depienne and Mandel, 2021). The mutant huntingtin protein (mHTT) features elongated polyglutamine residues, resulting in protein misfolding and aggregate formation. These aggregates are known to disrupt numerous molecular and cellular processes, which ultimately lead to the death of striatal and cortical neurons (Waldvogel et al., 2015). There is an inverse association between the number of CAG repeats in HTT gene and the timing of disease onset, hence indicating that there is a greater number of repeats linked to an earlier appearance of symptoms. HD is characterized by significant increases in motor or neurological symptoms, which usually manifest around the age of 40 years (Lee et al., 2012). Although the number of CAG repeats in HTT accounts for about 50–70% of the variability in the age of onset, additional factors remain to be explored, which involve environmental influencers or genetic modifiers (Langbehn et al., 2004). Recent advancements in genetic technologies have revealed several other genes that regulate HTT expression and hence recognized as genetic modifiers of HD.

9.2.4 Amyotrophic Lateral Sclerosis

ALS is a tragic and invariably fatal NDD affecting the motor neurons, and it often overlaps with FTD in its clinical presentation. Studies have demonstrated that, in ALS, about 90% of cases are sporadic amyotrophic lateral sclerosis (sALS), and only 10% are classified as familial amyotrophic lateral sclerosis (fALS). ALS typically acts as an autosomal dominant trait, but fALS may also be inherited in an X-linked recessive manner (Andersen and Al-Chalabi, 2011). Both the sALS and fALS display similar clinical manifestations, and either type can develop in conjunction with FTD. About 90% of cases found in fALS are also considered sporadic cases and are linked to pathogenic mutations. In ALS, 10% of patients have a history of the disease in their families, and approximately 30 genes have been identified in these families, which are found to be associated with ALS (Renton et al., 2014). The four common genes linked with ALS are the superoxide dismutase 1 (SOD1) gene, TAR DNA-binding protein gene, fused in sarcoma (FUS) gene, and chromosome 9 open reading frame 72 (C9ORF72) gene. The harbor pathogenic variants contribute to 60% fALS and 10% sALS cases, and this means that these four genes explain one out of every six instances of ALS (Renton et al., 2014).

9.2.4.1 Superoxide Dismutase 1

Thirty years prior, a groundbreaking application of linkage analysis in NDD uncovered a locus on the long arm of chromosome no. 21 (21q) associated with ALS in about 23 families. Later on, mutation and segregation analysis confirmed that pathogenic variants in SOD1 are the causative factors for the disease (Siddique et al., 1991). In the fALS SOD1 gene, over 170 mutations have been identified, and collectively these mutations contribute about 20% of diseases cases (Abel et al., 2012). Among these mutations, the majority are missense mutations and distributed throughout the exon without having specific mutation hotspots (Abel et al., 2012). A small number of mutations responsible for the truncation of SOD1 protein terminals have been found. However, none of the ALS-related SOD1 gene mutations are expected to completely halt SOD1 protein production. This indicates that the mutant SOD1 protein is likely necessary for the onset of motor neuron degeneration (Abel et al., 2012). Research study on North American individuals affected with ALS indicated that Ala5Val substitution mutation is the most common one in SOD1 gene (Cudkowicz et al., 1997), while other SOD1 mutations, like Ser106Leu and Gly73Ser, present themselves with a more severe phenotype (Berdynski et al., 2022). Though most of the ALS mutations are autosomal dominant, Asp91Ala mutation is an autosomal recessive form and has been frequently found in Scandinavian population (Feneberg et al., 2020).

9.2.4.2 TAR-DNA-Binding Protein (TARDBP)

The TARDBP gene codes TAR-DNA-binding protein, like TDP–43. The TDP-43 protein has 43 kDa molecular weight, is a highly conserved RNA-binding and widely expressed protein, and is primarily found in the nucleus under normal physiological conditions. In 2006, a pivotal study revealed that TDP-43 is the most prevalent ubiquitinated protein found in cytosolic inclusions and is associated with sALS and FTD (Neumann et al., 2006). After that, TDP-43 is recognized as a major contributor to the pathophysiology of ALS and FTLD, which has led to an increase in its prominence. Therefore, fALS and FTLD diseases are also caused by TARDBP gene mutations (Sreedharan et al., 2008). The degeneration and finally the death of motor neurons in the brain and spinal cord are often triggered by TARDBP mutations (Buratti, 2015), because this is linked with the buildup of aggregates and/or the loss of nuclear TDP-43 protein (Buratti, 2015). The accumulation process starts with a modified TDP-43, which is commonly ubiquitinated and acts as the main building block of these cytoplasmic aggregates (Buratti, 2015). About 30 mutations in the TARDBP gene are linked to the fALS, which constitutes about 4% of all fALS cases (Buratti, 2015).

9.2.4.3 Fused in Sarcoma

The year 2009 marked the identification of FUS as an additional gene associated with ALS (Vance et al., 2009). This finding is noteworthy because FUS protein exhibits both structural and functional characteristics akin to TDP-43. The FUS protein is encoded by the FUS gene and has a 75 kDa molecular weight. It is mainly found in the nucleus of neurons, but it is also mis-localized into the cell cytosol and results in cytoplasmic aggregate formation in ALS patients. FUS protein is involved in various processes related to RNA generation and trafficking. As it binds to both DNA and RNA, it therefore plays a crucial role in regulating transcription and mRNA processing within neurons (Wang et al., 2015). The occurrence of basophilic inclusions in spinal cord is associated with childhood ALS resulting from FUS gene mutations (Munoz et al., 2009). There are 19 mutations in the FUS gene linked with ALS development and are found in about 4% of ALS patients (Vance et al., 2009). The majority of these mutations occur within the nuclear localization signal (NLS) domain present at the C-terminal of the protein and which hinders nuclear translocation of FUS (Vance et al., 2009). Thus, the neurodegeneration associated with FUS mutations may result from both the loss of its nuclear functions and the acquisition of toxic properties in the cytoplasm (Shelkovnikova et al., 2014).

9.2.4.4 Chromosome 9 Open Reading Frame 72

C9ORF72 is the latest discovered gene associated with ALS, present on chromosome no. 9, and has an open reading frame (ORF). The gene features an expansion of hexanucleotide (GGGGCC) intronic repeats in about 30% of fALS, 8–10% of sALS, and 0.5% of control cases (Majounie et al., 2012). As this repeat is found in the intronic region of the gene, it is therefore anticipated that it produces a long RNA transcript rather than a protein. Individuals with fALS and having C9ORF72 mutation typically exhibit the pathology like the TDP-43 aggregate. Furthermore, some neurons in the hippocampus and the cerebellum are TDP-43-negative but show a unique form of inclusion bodies that are star-like or small dot-like. C9ORF72 gene mutations are the most common mutations linked with familial FTD, ALS, and FTD/ALS overlap (Majounie et al., 2012); hence, recognition of the C9ORF72 mutation association with ALS and FTD has inspired the scientific community to explore its potential involvement in additional neurodegenerative conditions (Beck et al., 2013).

9.3 POTENTIAL OF CRISPR-CAS9 SYSTEM IN TARGETING NEUROLOGICAL DISEASES

Neurodegeneration involves the gradual decline in structure and functions of neurons, which finally leads to the loss of these essential cells. This process worsens over with time and can ultimately result in dysfunction within the nervous system and finally develops into an NDD (Kovacs, 2019). The CRISPR-Cas9 system is a ribonucleic acid (RNA)–mediated selective deoxyribonucleic acid (DNA) editing or degradation system. The CRISPR-Cas9 system consists of the CRISPR and Cas9. The *CRISPR* is a short-sized guide RNA and is called sgRNA or crRNA, approximately 20–30 nts and complementary to the target DNA, while Cas9 is an endo-DNase protein. The main function of crRNA is guiding Cas9 to degrade or edit the target DNA. Thus, the CRISPR-Cas9 system has emerged as a promising technique for treating different diseases, including NDDs. Some of the applications of the CRISPR-Cas9 technique in treating various NDDs are mentioned in what follows.

9.3.1 Alzheimer's Disease

The most frequently occurring NDD is AD, featuring the accumulation of Aβ and τ substances. The established Aβ hypothesis has provided a basis for the development of potential therapies, aimed at modifying the disease by inhibiting Aβ production and promoting the clearance of these toxic proteins from neurons (brain) (Rohn et al., 2018). However, despite numerous attempts to create disease-modifying treatments through animal models, many have proven unsuccessful (Cummings et al., 2014). In light of this, CRISPR-Cas9 technology has recently gained significant attention in the field of AD, appreciated for its short experimental duration and low resource usage. The technology is now widely utilized for constructing AD models, screening for pathogenic genes, and developing novel therapeutic targets (Lu et al., 2021). Majority of AD cases are sporadic, and some associated with autosomal dominant mutations in APP or PSEN1 or/and PSEN2 genes (Zhang et al., 2020). Among these three genes, the majority of mutations is found in the APP gene and thus develops this disease, but interestingly, some variations actually offer protection against AD (Jonsson et al., 2012). For example, research scientists in 2018 found protective deletions in 3′-UTR of this gene (Nagata et al., 2018) after introducing these deletion mutations into the APP-KI of mice (a model for AD) causes a significant decrease in Aβ pathology, similar to human mutations in APP gene (Nagata et al., 2018). Therefore, different animal models of AD were generated through the CRISPR-Cas9 system by inserting three point mutation, namely, Gly676Arg, Phe681Tyr, and

Arg684His, in the APP gene independently (Serneels et al., 2020). Another animal model involving τ-protein knockout mice was generated by CRISPR-Cas9 method to study the role of τ-protein in AD (Tan et al., 2018). In this model, short deletion was introduced in exon 1 of the Mapt gene, which produces τ-protein, and the model was created on a C57Bl/6J background, which exhibits resistance to excitotoxicity and demonstrates no signs of memory impairment (Tan et al., 2018). Various alleles of the Plcγ2 gene exist, with some linked to an elevated risk of AD and others correlated with a diminished risk (Sims et al., 2017). Notably, the Plcγ2-P522R variant causes a reduction in the likelihood of developing AD through the CRISPR-Cas9 method (Takalo et al., 2020). During this process, the catalytically inactive Cas9 protein was combined with activators or inhibitors and used to decrease the production of Aβ, as the crRNA directs this Cas9 towards the target gene and, moreover, does not create double-strand breaks. This approach has been found to elevate or decrease the target gene expression (Konermann et al., 2015). In summary, the CRISPR-Cas9 system in the field of AD has been effectively evaluated across various animal models, employing diverse methodologies. But further research is required to thoroughly investigate this technology in AD and decrease the occurrence in non-specific target of the CRISPR-Cas9 system.

9.3.2 Parkinson's Disease

The advent of CRISPR-Cas9 technology offers a remarkable potential to accelerate basic NDD research, mostly investigating the pathogenicity of NDD, which may ultimately lead to novel therapeutic interventions, especially for PD. Kantor and his co-workers have focused on developing an epigenetic-based therapeutic strategy that regulates SNCA expression (Kantor et al., 2018). They have developed a CRISPR method for the specific editing of DNA methylation within non-coding sequence 1, employing a comprehensive lentiviral vector delivery system. The delivery system mainly contains a deactivated Cas9 (dCas9) protein with DNA-methyltransferase-3A catalytic domain; results show that they successfully downregulated SNCA at mRNA as well as at protein level (Kantor et al., 2018). Predominantly, numerous *in vitro* studies have been conducted to develop different cell models for PD. For instance, Vermilyea and his co-workers created a CRISPR-Cas9 system to develop the Gly2019Ser mutation in LRRK2 gene by using induced pluripotent stem cells (iPSCs) derived from marmosets (Vermilyea et al., 2020). After following the differentiation of these iPSCs into dopaminergic neurons, they observed high ROS, reduced neuronal viability, and diminished neurite complexity, all these characteristics are associated with PD (Vermilyea et al., 2020). After this discovery, the CRISPR-Cas9 technology was instrumental in generating different *in vivo* models for PD. Like in 2016, a specific case of using the CRISPR system in creating the Asp620Asn Vps35 knock-in mice model produced various key characteristics of PD (Ishizu et al., 2016). In 2022, another research study was done by eliminating the Ala53Thr SNCA gene through CRISPR-Cas9 system in *in vitro* as well as *in vivo* conditions (Yoon et al., 2022). The *in vitro* results indicated a significant reduction in the expression of Ala53Thr SNCA protein, whereas in the *in vivo* rat model study of PD, the deletion of Ala53Thr SNCA resulted in a notable decrease in α-synuclein expression, motor symptoms, and dopaminergic degeneration (Yoon et al., 2022). In 2021, two distinct teams of research scientists made significant strides by generating non-human primate models of PD (Chen et al., 2021; Li et al., 2021). To specifically focus on the PINK1 gene, they implemented a CRISPR-Cas9 variant (nCas9) with Asp10Ala mutation, which successfully eliminates or decreases the double-strand cleavage needs. The formed nCas9 variant is capable of cleaving only one strand of target DNA, resulting in a marked decrease in off-target editing events of the CRISPR technology (Chen et al., 2021; Li et al., 2021). A separate study effectively employed a conventional CRISPR-Cas9 system, administered through adeno-associated virus vector (AAV), like AAV9, to achieve simultaneous knockout of the two PD-associated genes, PINK1 and DJ-1, within the substantia nigra of monkeys (Dianov and Hubscher, 2013). This technique facilitated the manifestation of various critical features of PD, including alterations in behavior and neuropathology (Dianov and Hubscher, 2013).

9.3.3 Huntington's Disease

The exploration of CRISPR-Cas9 system in context of HD is currently at an early stage of preclinical investigation. There are several potential methods, including the CRISPR method, through which HTT gene editing can be achieved. To cause HTT gene editing, these methods involve different mechanisms, like excision of trinucleotide (CAG) repeats mutation and correcting it by yielding wild-type HTT, or the inactivation of mutant allele to create a hemizygous null state, or specific non-allele targeting, all of which ultimately results in the reduction of HTT protein. Additionally, it is conceivable that the epigenetic changes mediated by CRISPR-Cas9 technology could also provide an effective treatment avenue for HD, by influencing HTT expression without effecting or alternating the whole genome. Different HD animal models have been developed to examine the pathogenesis of the disease. The main focus is on reducing the mHTT protein level and modifying the affected neurons through experimental interventions in HD treatment (Al-Gharaibeh et al., 2017). Findings from different studies using these animal models have indicated that lowering mHTT levels can enhance disease phenotypes and potentially reverse neuropathological changes (Tabrizi et al., 2019). One notable model is the YAC128 HD mouse, which harbors the complete human mutant HTT gene within its genome, resulting in mHTT gene production, characterized with only 128 CAG repeats (Slow et al., 2003). This HD model demonstrates selective characteristics, like age-dependent atrophy, neurodegeneration in the striatum as well as in the cortex, and progressive decrease in motor and cognitive functions (Ehrnhoefer et al., 2009). In 2017, Yang and his colleagues reported a similar method by involving CRISPR-Cas9 system in developing the Q140 transgenic HD mouse model. They found a selective reduction of mHTT protein, which results in the decrease in HD pathological manifestations and an improvement in motor skills. The use of the RNAi method has proven effective in reducing mHTT, which results in enhancing behavioral outcomes in these selected animal models (Carroll et al., 2011). Due to the encouraging results observed in these HD animal models, different clinical trials involving stem cell transplants in humans have been initiated to evaluate their effectiveness

(Clelland et al., 2008). While some positive outcomes have been reported after the transplantation of various cell types, along with some challenges, like graft immune rejection and the development of adverse effects (Cicchetti et al., 2014), a research study found that iPSC transplantation into the striata of rats followed by the 3-nitropropionic acid treatment resulted in increasing motor function and transplanted cell differentiation into a specific neuron region within the striatum (Fink et al., 2014). Following this experiment, an improvement in motor functions has been found when neural stem cells (NSCs) derived from iPSCs of a juvenile HD patient was transplanted into the striata of YAC128 mice (Jeon et al., 2014). Thus, the implementation of CRISPR-Cas9-mediated approaches for targeted HTT gene modification, followed by the reduction of HTT protein, demonstrates substantial potential for treating HD by using CRISPR technology.

9.3.4 Amyotrophic Lateral Sclerosis

For ALS treatment, the CRISPR-Cas9 technology has also emerged as a hot study topic for researchers, owing to its possible application in treatment and management of this disease (Yamashita et al., 2013). For example, a study utilizing the Gly93Ala SOD1 mouse model showed silencing of mutant SOD1 gene through the RNAi method, resulting into a delay in the onset of ALS and an increase in mice lifespan (Borel et al., 2016). Another study found that using the CRISPR-Cas9 system can alter the mutant SOD1 (Gly93Ala) expression in mouse model in *in vivo* experiments, achieved via an AAV delivery system and resulting in the delay of the onset of ALS (Gaj et al., 2017). Another research study achieved correction of Ala272Cys mutation of the SOD1 gene in the iPSCs obtained from ALS patients (Wang et al., 2017). In a comparable study, the CRISPR-Cas9 gene-editing method was applied to Glu100Gly SOD1 mutant in an isogenic cell line, which resulted in an enhanced motor neuron proportion, increased soma dimensions, longer neurite lengths, and a reduction in cell death (Bhinge et al., 2017). In 2021, Perkins and his co-workers successfully utilized the CRISPR-Cas9 method to correct C9ORF72 mutations, restoring synaptic function and providing new mechanism insights into early ALS, along with the identification of new therapeutic targets (Perkins et al., 2021), whereas in 2017, Wang and his colleagues were the first to illustrate the successful correction of FUS Gly1566Ala mutation in iPSCs derived from ALS patients, through the CRISPR-Cas9 technology. In a separate investigation in iPSCs, the CRISPR-Cas9 system was also employed to revert the recessive FUS His517Gln mutation into its normal form. From the results it was found that the corrected form was subsequently differentiated into Isl1/Tuj1-positive motor neurons as assessed through the MAPK pathway (Bhinge et al., 2017). Liu and his co-workers recently employed the CRISPR-Cas9 method to develop an astrocyte elevated gene-1-deficient HT22 cell model, facilitating the investigation of the role of this gene in ALS (Liu et al., 2022). Notably, the CRISPR-iPSC-based disease model has the potential to accelerate ALS research by investigating the four main ALS-associated genes, namely, SOD1, FUS, TDP-43, and C9ORF72 (Sen and Thummer, 2022; Wang et al., 2017). Therefore, within ALS research, the CRISPR-Cas9 system has primarily been applied in the creation of small animal models and the improvement of *in vitro* models. Nonetheless, recent evidence showcasing the importance of CRISPR-Cas9 system in treating ALS in mice has developed great hope among ALS patients. But in the future, they may be important for new therapeutic developments for both familial and sporadic types of the ALS disease.

9.4 DELIVERING THE CRISPR-CAS SYSTEM INTO THE BRAIN

Due to its proficiency in altering genomes in diverse cell types, including human cells, the CRISPR-Cas9 system is utilized to edit different types of genes, including brain cell gene. However, the delivery of CRISPR-Cas components into the brain remains a complex task. A significant challenge in developing therapies aimed at the brain is the necessity to traverse the blood–brain barrier (BBB). The BBB serves to protect the neural tissue by regulating the entry of various molecules, which complicates the delivery of therapeutic agents. Research work from the Comprehensive Medicinal Chemistry database reveals that out of 7,000 drugs assessed, only 5% can successfully cross the BBB and enter the central nervous system (CNS) (Pardridge, 2005). In 2014, a research study has detailed different delivery strategies aimed at improving the delivery of therapeutic agents to neural tissue and classified them as less or more invasive methods (Agustín-Pavón and Isalan 2014).

The use of intranasal access through nanoparticle-assisted drug delivery methods presents a less-invasive approach for crossing the BBB. This method utilizes solid colloidal particles, known as nanoparticles (NPs), with 1 to 1,000 nm in size (Zhou et al., 2018). It has been noted that successful passage through the BBB, along with effective cellular labeling, can be achieved when NPs are conjugated with non-ionic surfactants or cationic stabilizers, but the inclusion of anionic molecules may hinder this process. Furthermore, studies indicates that NPs' size and charge do not significantly affect their ability to traverse the BBB (Voigt et al., 2014). Additionally, other studies highlight that exosomes, cell-penetrating peptides (also called "Trojan horse" peptides), and the protein transduction domain with a size range of 5–30 amino acids are capable of transferring across the cell membrane (Shi et al., 2014).

The more invasive methods involve direct injections into the brain's parenchyma or ventricles during stereotactic surgery, like ultrasound-activated microbubbles and laser irradiation (Agustín-Pavón and Isalan, 2014). Such techniques can function as transfer vehicles to safeguard the integrity of CRISPR-Cas system components, thereby enhancing the opportunities for precise editing at the specific target site, while benefiting from the capability to cross the BBB. Research conducted by Yanagida and his co-workers demonstrated that the pharmacological agent FTY720, along with the 1-phosphosphingosine receptor 1 gene, promotes the selective passage of small compounds across the BBB without significant inflammation or damage to the CNS (Yanagida et al., 2017). This phenomenon was linked to alterations in the cytoskeletal structure of tight junction proteins.

Another promising method is cell-based therapy, which encompasses the administration of cells into the bloodstream or tissues to address a variety of medical conditions. In the context of neurological health, such therapies have been formulated to combat ailments such as PD, AD, traumatic brain injuries, and various other NDDs. The utilization of immune cell–based delivery is increasingly recognized as a viable approach in combating brain cancers/tumor (Turk et al., 2021). These immune cells can be re-engineered to specifically recognize and target antigens associated with the tumor cells (Turk et al., 2021). Due to the CNS's strict control over immune cell access to the brain, effective delivery methods are essential. Despite these challenges, clinical applications of T, dendritic, and natural killer cells have been realized (Weber et al., 2020). The application of stem cells offers a promising strategy for NDD, particularly in the context of replacing damaged neurons in the specific regions of the brain. This therapeutic approach typically involves the use of iPSCs and NSCs, which can be administered directly to the affected areas of the brain.

Similarly, viral vectors represent a viable option for the delivery of various transgenes, including the CRISPR-Cas9 components. Lentiviruses possess the capability to infect non-dividing cells, with a packaging capacity of 8.5 kb, which adequately accommodates most Cas9 genes and guide RNAs (Kumar et al., 2001). The successful administration of lentiviral Cas9 into murine models has been demonstrated to effectively regulate lung cancer progression (Sanchez-Rivera et al., 2014). In contrast, adenoviruses are able to infect both dividing and non-dividing cells; however, they tend to provoke a robust host immune response (Wang et al., 2004). The AAVs represent a highly promising option for gene delivery due to their lack of genomic integration capability and their ability to elicit a safe, non-immune response. The Food and Drug Administration has granted permission for AAV utilization in various human clinical trials. In brain, for an *in vivo* gene-editing method, smaller-sized Cas9 gene variants like spCas9 (4.2 kb) and saCas9 (3.2 kb) are used, as they can be effectively packaged within an AAV delivery vector (Ran et al., 2015). Research teams have successfully altered specific genes within neurons by administering viral vectors that express CRISPR-Cas9 components into the mouse brain (Platt et al., 2014).

Non-viral vectors, when compared to their viral counterparts, demonstrate lower immunogenicity and do not promote the integration of genomic DNA. The methodologies for non-viral delivery encompass bio-reducible lipids, liposome-mediated transfection, and electroporation. In a study, a bio-reducible lipid delivery system packed with Cas9 was directly delivered into the brain of target mouse for the treatment of neurological conditions (Wang et al., 2016). Scientists also successfully delivered liposome-containing Cas9-gRNAs complexes into the genome of inner ear hair cells and neuronal cell of the target mouse (Wang et al., 2016). However, the potential of these non-viral delivery systems for direct CRISPR-Cas9 delivery into the brain is yet to be explored. Furthermore, concerns exist regarding the stability of expression produced by these non-viral delivery methods, as well as the safety implications of their injection into the brain.

9.5 POTENTIAL RISKS AND FUTURE DIRECTIONS

CRISPR-Cas-based gene editing has gained significant traction in the field of neuroscience due to its remarkable efficiency, straightforward application, affordability, and versatility. This technology has enabled researchers to explore the complexities of various neurons and neurological disorders. Nevertheless, certain potential risks warrant attention. Firstly, the action of the Cas endonuclease can lead to unintended DNA cleavage, resulting in small insertions or deletions at off-target sites (Zhang et al., 2015). Additionally, cytosine/adenine deaminases may inadvertently affect non-specific RNA (Zhou et al., 2020). Such off-target effects could skew experimental outcomes in CRISPR-based animal models and may cause irreversible disruptions to normal gene function in patients undergoing CRISPR treatments. Secondly, the relatively large size of the Cas gene poses challenges for direct cellular entry. Currently, methods such as liposomes, viral vectors, and ribonucleoproteins are employed for delivering Cas-sgRNA into cells, each with limitations regarding cargo size, delivery efficiency, and safety (Lino et al., 2018). Future research will likely focus on addressing complex neurobiological challenges, correcting genetic disorders, and advancing cell therapies in the nervous system through CRISPR-Cas-based approaches while mitigating associated risks.

9.6 CONCLUSION

In conclusion, the CRISPR-Cas9 method offers a significant potential for addressing NDD. CRISPR-Cas technologies have rapidly emerged as a leading method for precise gene modification and are widely acknowledged as some of the most effective tools for therapeutic gene editing across various conditions. The capacity to selectively modify specific genes linked to these disorders opens new pathways for the development of targeted therapies. Although still in the preliminary phases, preclinical investigations have yielded promising outcomes in animal models, highlighting the potential of this technology to combat various neurological conditions. To sum, the merits of this technique far exceed its possible drawbacks, establishing the CRISPR-Cas9 technology as the leading gene-editing tool for research into NDD. A variety of challenges still exist that must be overcome before CRISPR-Cas9 can be effectively translated into human therapies. Considerations regarding safety, off-target effects, and delivery methods require careful scrutiny and enhancement. Despite these challenges, the potential of CRISPR-Cas9 to benefit individuals suffering from neurodegenerative diseases is considerable. Continued research and development are essential to unlock the full capabilities of this technology and provide hope to the millions impacted by these serious disorders.

REFERENCES

Abel, Olubunmi, John F. Powell, Peter M. Andersen, and Ammar Al-Chalabi. "ALSoD: A user-friendly online bioinformatics tool for amyotrophic lateral sclerosis genetics." *Human Mutation* 33, no. 9 (2012): 1345–1351.

Agustin-Pavon, Carmen, and Mark Isalan. "Synthetic biology and therapeutic strategies for the degenerating brain: Synthetic biology approaches can transform classical cell and gene therapies, to provide new cures for neurodegenerative diseases." *Bioessays* 36, no. 10 (2014): 979–990.

Alafuzoff, Irina, Thomas Arzberger, Safa Al-Sarraj, Istvan Bodi, Nenad Bogdanovic, Heiko Braak, Orso Bugiani et al. "Staging of neurofibrillary pathology in Alzheimer's disease: A study of the BrainNet Europe Consortium." *Brain Pathology* 18, no. 4 (2008): 484–496.

Al-Gharaibeh, Abeer, Rebecca Culver, Andrew N. Stewart, Bhairavi Srinageshwar, Kristin Spelde, Laura Frollo, Nivya Kolli et al. "Induced pluripotent stem cell-derived neural stem cell transplantations reduced behavioral deficits and ameliorated neuropathological changes in YAC128 mouse model of Huntington's disease." *Frontiers in Neuroscience* 11 (2017): 628.

Andersen, Peter M., and Ammar Al-Chalabi. "Clinical genetics of amyotrophic lateral sclerosis: What do we really know?." *Nature Reviews Neurology* 7, no. 11 (2011): 603–615.
Bagyinszky, Eva, Young Chul Youn, Seong Soo A. An, and SangYun Kim. "The genetics of Alzheimer's disease." *Clinical Interventions in Aging* (2014): 535–551.
Beck, Jon, Mark Poulter, Davina Hensman, Jonathan D. Rohrer, Colin J. Mahoney, Gary Adamson, Tracy Campbell et al. "Large C9orf72 hexanucleotide repeat expansions are seen in multiple neurodegenerative syndromes and are more frequent than expected in the UK population." *The American Journal of Human Genetics* 92, no. 3 (2013): 345–353.
Benilova, Iryna, Rodrigo Gallardo, Andreea-Alexandra Ungureanu, Virginia Castillo Cano, An Snellinx, Meine Ramakers, Carmen Bartic, Frederic Rousseau, Joost Schymkowitz, and Bart De Strooper. "The Alzheimer disease protective mutation A2T modulates kinetic and thermodynamic properties of amyloid-β (Aβ) aggregation." *Journal of Biological Chemistry* 289, no. 45 (2014): 30977–30989.
Berdynski, Mariusz, Przemysław Miszta, Krzysztof Safranow, Peter M. Andersen, Mitsuya Morita, Sławomir Filipek, Cezary Żekanowski, and Magdalena Kuźma-Kozakiewicz. "SOD1 mutations associated with amyotrophic lateral sclerosis analysis of variant severity." *Scientific Reports* 12, no. 1 (2022): 103.
Bhinge, Akshay, Seema C. Namboori, Xiaoyu Zhang, Antonius M. J. VanDongen, and Lawrence W. Stanton. "Genetic correction of SOD1 mutant iPSCs reveals ERK and JNK activated AP1 as a driver of neurodegeneration in amyotrophic lateral sclerosis." *Stem Cell Reports* 8, no. 4 (2017): 856–869.
Blackhurst, Britanie M., and Kristen E. Funk. "Viral pathogens increase risk of neurodegenerative disease." *Nature Reviews Neurology* 19, no. 5 (2023): 259–260.
Bonifati, Vincenzo, C. F. Rohe, G. J. Breedveld, Edito Fabrizio, M. De Mari, Cristina Tassorelli, A. Tavella et al. "Early-onset parkinsonism associated with PINK1 mutations: Frequency, genotypes, and phenotypes." *Neurology* 65, no. 1 (2005): 87–95.
Borel, Florie, Gwladys Gernoux, Brynn Cardozo, Jake P. Metterville, Gabriela T. Cabrera, Lina Song, Qin Su et al. "Therapeutic rAAVrh10 mediated SOD1 silencing in adult SOD1 G93A mice and nonhuman primates." *Human Gene Therapy* 27, no. 1 (2016): 19–31.
Brettschneider, Johannes, Kelly Del Tredici, Virginia M.-Y. Lee, and John Q. Trojanowski. "Spreading of pathology in neurodegenerative diseases: A focus on human studies." *Nature Reviews Neuroscience* 16, no. 2 (2015): 109–120.
Brunkan, A. L., M. Martinez, J. Wang, E. S. Walker, D. Beher, M. S. Shearman, and A. M. Goate. "Two domains within the first putative transmembrane domain of presenilin 1 differentially influence presenilinase and γ-secretase activity." *Journal of Neurochemistry* 94, no. 5 (2005): 1315–1328.
Buratti, Emanuele. "Functional significance of TDP-43 mutations in disease." *Advances in Genetics* 91 (2015): 1–53.
Camargos, Sarah Teixeira, Leonardo Oliveira Dornas, Parastoo Momeni, Andrew Lees, John Hardy, Andrew Singleton, and Francisco Cardoso. "Familial Parkinsonism and early onset Parkinson's disease in a Brazilian movement disorders clinic: Phenotypic characterization and frequency of SNCA, PRKN, PINK1, and LRRK2 mutations." *Movement Disorders: Official Journal of the Movement Disorder Society* 24, no. 5 (2009): 662–666.
Carroll, Jeffrey B., Simon C. Warby, Amber L. Southwell, Crystal N. Doty, Sarah Greenlee, Niels Skotte, Gene Hung, C. Frank Bennett, Susan M. Freier, and Michael R. Hayden. "Potent and selective antisense oligonucleotides targeting single-nucleotide polymorphisms in the Huntington disease gene/allele-specific silencing of mutant huntingtin." *Molecular Therapy* 19, no. 12 (2011): 2178–2185.
Cerejeira, J., L. Lagarto, and E. B. Mukaetova-Ladinska. "Behavioral and psychological symptoms of dementia." *Frontiers in Neurology* 3 (2012): 73.

Chartier-Harlin, Marie-Christine, Fiona Crawford, Henry Houlden, Andrew Warren, David Hughes, Liana Fidani, Alison Goate et al. "Early-onset Alzheimer's disease caused by mutations at codon 717 of the β-amyloid precursor protein gene." *Nature* 353, no. 6347 (1991): 844–846.

Chartier-Harlin, Marie-Christine, Jennifer Kachergus, Christophe Roumier, Vincent Mouroux, Xavier Douay, Sarah Lincoln, Clotilde Levecque et al. "α-synuclein locus duplication as a cause of familial Parkinson's disease." *The Lancet* 364, no. 9440 (2004): 1167–1169.

Chen, Zhen-Zhen, Jian-Ying Wang, Yu Kang, Qiao-Yan Yang, Xue-Ying Gu, Da-Long Zhi, Li Yan, Cheng-Zu Long, Bin Shen, and Yu-Yu Niu. "PINK1 gene mutation by pair truncated sgRNA/Cas9-D10A in cynomolgus monkeys." *Zoological Research* 42, no. 4 (2021): 469.

Cicchetti, Francesca, Steve Lacroix, Giulia Cisbani, Nicolas Vallières, Martine Saint-Pierre, Isabelle St-Amour, Ranna Tolouei et al. "Mutant huntingtin is present in neuronal grafts in Huntington disease patients." *Annals of Neurology* 76, no. 1 (2014): 31–42.

Citron, Martin, David Westaway, Weiming Xia, George Carlson, Thekla Diehl, Georges Levesque, Kelly Johnson-Wood et al. "Mutant presenilins of Alzheimer's disease increase production of 42-residue amyloid β-protein in both transfected cells and transgenic mice." *Nature Medicine* 3, no. 1 (1997): 67–72.

Clelland, Claire D., Roger A. Barker, and Colin Watts. "Cell therapy in Huntington disease." *Neurosurgical Focus* 24, no. 3–4 (2008): E9.

Cruchaga, Carlos, Sumitra Chakraverty, Kevin Mayo, Francesco L. M. Vallania, Robi D. Mitra, Kelley Faber, Jennifer Williamson et al. "Rare variants in APP, PSEN1 and PSEN2 increase risk for AD in late-onset Alzheimer's disease families." *PLoS One* 7, no. 2 (2012): e31039.

Cudkowicz, M. E., D. McKenna-Yasek, P. E. Sapp, W. Chin, B. Geller, D. L. Hayden, D. A. Schoenfeld, B. A. Hosler, H. R. Horvitz, and R. H. Brown. "Epidemiology of mutations in superoxide dismutase in amyotrophic lateal sclerosis." *Annals of Neurology: Official Journal of the American Neurological Association and the Child Neurology Society* 41, no. 2 (1997): 210–221.

Cummings, Jeffrey L., Travis Morstorf, and Kate Zhong. "Alzheimer's disease drug-development pipeline: Few candidates, frequent failures." *Alzheimer's Research & Therapy* 6 (2014): 1–7.

De Strooper, Bart, Monique Beullens, Bart Contreras, Lyne Levesque, Katleen Craessaerts, Barbara Cordell, Dieder Moechars et al. "Phosphorylation, subcellular localization, and membrane orientation of the Alzheimer's disease-associated presenilins." *Journal of Biological Chemistry* 272, no. 6 (1997): 3590–3598.

De Strooper, Bart, Takeshi Iwatsubo, and Michael S. Wolfe. "Presenilins and γ-secretase: Structure, function, and role in Alzheimer disease." *Cold Spring Harbor Perspectives in Medicine* 2, no. 1 (2012): a006304.

De Strooper, Bart, Paul Saftig, Katleen Craessaerts, Hugo Vanderstichele, Gundula Guhde, Wim Annaert, Kurt Von Figura, and Fred Van Leuven. "Deficiency of presenilin-1 inhibits the normal cleavage of amyloid precursor protein." *Nature* 391, no. 6665 (1998): 387–390.

Depienne, Christel, and Jean-Louis Mandel. "30 years of repeat expansion disorders: What have we learned and what are the remaining challenges?." *The American Journal of Human Genetics* 108, no. 5 (2021): 764–785.

Dianov, Grigory L., and Ulrich Hubscher. "Mammalian base excision repair: The forgotten archangel." *Nucleic Acids Research* 41, no. 6 (2013): 3483–3490.

Ehrnhoefer, Dagmar E., Stefanie L. Butland, Mahmoud A. Pouladi, and Michael R. Hayden. "Mouse models of Huntington disease: Variations on a theme." *Disease Models & Mechanisms* 2, no. 3–4 (2009): 123–129.

Farrer, Matt, Jennifer Kachergus, Lysia Forno, Sarah Lincoln, Deng-Shun Wang, Mary Hulihan, Demetrius Maraganore et al. "Comparison of kindreds with parkinsonism and α-synuclein genomic multiplications." *Annals of Neurology: Official Journal of the American Neurological Association and the Child Neurology Society* 55, no. 2 (2004): 174–179.

Feigin, Valery L., Amanuel Alemu Abajobir, Kalkidan Hassen Abate, Foad Abd-Allah, Abdishakur M. Abdulle, Semaw Ferede Abera, Gebre Yitayih Abyu et al. "Global, regional, and national burden of neurological disorders during 1990–2015: A systematic analysis for the Global Burden of Disease Study 2015." *The Lancet Neurology* 16, no. 11 (2017): 877–897.

Feigin, Valery L., Emma Nichols, Tahiya Alam, Marlena S. Bannick, Ettore Beghi, Natacha Blake, William J. Culpepper et al. "Global, regional, and national burden of neurological disorders, 1990–2016: A systematic analysis for the Global Burden of Disease Study 2016." *The Lancet Neurology* 18, no. 5 (2019): 459–480.

Feneberg, Emily, Martin R. Turner, Olaf Ansorge, and Kevin Talbot. "Amyotrophic lateral sclerosis with a heterozygous D91A SOD1 variant and classical ALS-TDP neuropathology." *Neurology* 95, no. 13 (2020): 595–596.

Ferreira, M., and J. Massano. "An updated review of Parkinson's disease genetics and clinicopathological correlations." *Acta Neurologica Scandinavica* 135, no. 3 (2017): 273–284.

Fink, Kyle D., Andrew T. Crane, Xavier Lévêque, Dylan J. Dues, Lucas D. Huffman, Allison C. Moore, Darren T. Story et al. "Intrastriatal transplantation of adenovirus-generated induced pluripotent stem cells for treating neuropathological and functional deficits in a rodent model of Huntington's disease." *Stem Cells Translational Medicine* 3, no. 5 (2014): 620–631.

Gaj, Thomas, David S. Ojala, Freja K. Ekman, Leah C. Byrne, Prajit Limsirichai, and David V. Schaffer. "In vivo genome editing improves motor function and extends survival in a mouse model of ALS." *Science Advances* 3, no. 12 (2017): eaar3952.

Gatz, Margaret, Chandra A. Reynolds, Laura Fratiglioni, Boo Johansson, James A. Mortimer, Stig Berg, Amy Fiske, and Nancy L. Pedersen. "Role of genes and environments for explaining Alzheimer disease." *Archives of General Psychiatry* 63, no. 2 (2006): 168–174.

Gegg, Matthew E., J. Mark Cooper, Kai-Yin Chau, Manuel Rojo, Anthony H. V. Schapira, and Jan-Willem Taanman. "Mitofusin 1 and mitofusin 2 are ubiquitinated in a PINK1/parkin-dependent manner upon induction of mitophagy." *Human Molecular Genetics* 19, no. 24 (2010): 4861–4870.

Goate, Alison, Marie-Christine Chartier-Harlin, Mike Mullan, Jeremy Brown, Fiona Crawford, Liana Fidani, Luis Giuffra et al. "Segregation of a missense mutation in the amyloid precursor protein gene with familial Alzheimer's disease." *Nature* 349, no. 6311 (1991): 704–706.

Goedert, M. G. S. M., Maria Grazia Spillantini, R. Jakes, D. Rutherford, and R. A. Crowther. "Multiple isoforms of human microtubule-associated protein tau: Sequences and localization in neurofibrillary tangles of Alzheimer's disease." *Neuron* 3, no. 4 (1989): 519–526.

Goedert, Michel. "Alzheimer's and Parkinson's diseases: The prion concept in relation to assembled Aβ, tau, and α-synuclein." *Science* 349, no. 6248 (2015): 1255555.

Goldman, Samuel M., Kenneth Marek, Ruth Ottman, Cheryl Meng, Kathleen Comyns, Piu Chan, Jinghong Ma et al. "Concordance for Parkinson's disease in twins: A 20-year update." *Annals of Neurology* 85, no. 4 (2019): 600–605.

Gomez-Isla, Teresa, Joseph L. Price, Daniel W. McKeel Jr, John C. Morris, John H. Growdon, and Bradley T. Hyman. "Profound loss of layer II entorhinal cortex neurons occurs in very mild Alzheimer's disease." *Journal of Neuroscience* 16, no. 14 (1996): 4491–4500.

Goutman, Stephen A., Orla Hardiman, Ammar Al-Chalabi, Adriano Chió, Masha G. Savelieff, Matthew C. Kiernan, and Eva L. Feldman. "Recent advances in the diagnosis and prognosis of amyotrophic lateral sclerosis." *The Lancet Neurology* 21, no. 5 (2022): 480–493.

Grundke-Iqbal, Inge, Khalid Iqbal, Maureen Quinlan, Yunn-Chyn Tung, Masooma S. Zaidi, and Henryk M. Wisniewski. "Microtubule-associated protein tau. A component of Alzheimer paired helical filaments." *Journal of Biological Chemistry* 261, no. 13 (1986): 6084–6089.

Grunewald, Anne, and Christine Klein. "13 Parkin-associated Parkinson's disease." *Parkinson's Disease* (2012): 161.

Hampel, Harald, Yan Hu, John Hardy, Kaj Blennow, Christopher Chen, George Perry, Seung Hyun Kim et al. "The amyloid-β pathway in Alzheimer's disease: A plain language summary." *Neurodegenerative Disease Management* 13, no. 3 (2023): 141–149.

Hardy, John, and Dennis J. Selkoe. "The amyloid hypothesis of Alzheimer's disease: Progress and problems on the road to therapeutics." *Science* 297, no. 5580 (2002): 353–356.

Hardy, John. "The discovery of Alzheimer-causing mutations in the APP gene and the formulation of the "amyloid cascade hypothesis"." *The FEBS Journal* 284, no. 7 (2017): 1040–1044.

Hyman, Bradley T., Creighton H. Phelps, Thomas G. Beach, Eileen H. Bigio, Nigel J. Cairns, Maria C. Carrillo, Dennis W. Dickson et al. "National Institute on Aging–Alzheimer's Association guidelines for the neuropathologic assessment of Alzheimer's disease." *Alzheimer's & Dementia* 8, no. 1 (2012): 1–13.

Ibanez, P., A. M. Bonnet, B. Debarges, E. Lohmann, F. Tison, Y. Agid, A. Dürr, A. Brice, and P. Pollak. "Causal relation between α-synuclein locus duplication as a cause of familial Parkinson's disease." *The Lancet* 364, no. 9440 (2004): 1169–1171.

Ishizu, Nobutaka, Daishi Yui, Akira Hebisawa, Hidenori Aizawa, Wanpeng Cui, Yuko Fujita, Kenji Hashimoto et al. "Impaired striatal dopamine release in homozygous Vps35 D620N knock-in mice." *Human Molecular Genetics* 25, no. 20 (2016): 4507–4517.

Jeon, Iksoo, Chunggab Choi, Nayeon Lee, Wooseok Im, Manho Kim, Seung-Hun Oh, In-Hyun Park, Hyun Sook Kim, and Jihwan Song. "In vivo roles of a patient-derived induced pluripotent stem cell line (HD72-iPSC) in the YAC128 model of Huntington's disease." *Int J Stem Cells* 7, no. 1 (2014): 43–47.

Jonsson, Thorlakur, Jasvinder K. Atwal, Stacy Steinberg, Jon Snaedal, Palmi V. Jonsson, Sigurbjorn Bjornsson, Hreinn Stefansson et al. "A mutation in APP protects against Alzheimer's disease and age-related cognitive decline." *Nature* 488, no. 7409 (2012): 96–99.

Kantor, Boris, Lidia Tagliafierro, Jeffrey Gu, Madison E. Zamora, Ekaterina Ilich, Carole Grenier, Zhiqing Y. Huang, Susan Murphy, and Ornit Chiba-Falek. "Downregulation of SNCA expression by targeted editing of DNA methylation: A potential strategy for precision therapy in PD." *Molecular Therapy* 26, no. 11 (2018): 2638–2649.

Ki, C-S., E. F. Stavrou, N. Davanos, W. Y. Lee, E. J. Chung, J-Y. Kim, and A. Athanassiadou. "The Ala53Thr mutation in the α-synuclein gene in a Korean family with Parkinson disease." *Clinical Genetics* 71, no. 5 (2007): 471–473.

Klein, Christine, and Katja Lohmann-Hedrich. "Impact of recent genetic findings in Parkinson's disease." *Current Opinion in Neurology* 20, no. 4 (2007): 453–464.

Klein, Christine, and Michael G. Schlossmacher. "The genetics of Parkinson disease: Implications for neurological care." *Nature Clinical Practice Neurology* 2, no. 3 (2006): 136–146.

Klein, Christine, and Ana Westenberger. "Genetics of Parkinson's disease." *Cold Spring Harbor Perspectives in Medicine* 2, no. 1 (2012): a008888.

Konermann, Silvana, Mark D. Brigham, Alexandro E. Trevino, Julia Joung, Omar O. Abudayyeh, Clea Barcena, Patrick D. Hsu et al. "Genome-scale transcriptional activation by an engineered CRISPR-Cas9 complex." *Nature* 517, no. 7536 (2015): 583–588.

Kovacs, Gabor G. "Concepts and classification of neurodegenerative diseases." In *Handbook of Clinical Neurology*, vol. 145, pp. 301–307. Elsevier, 2018.

Kovacs, Gabor G. "Molecular pathology of neurodegenerative diseases: Principles and practice." *Journal of Clinical Pathology* 72, no. 11 (2019): 725–735.

Kumar, Ashok, Vijay Kumar, Kritanjali Singh, Sukesh Kumar, You-Sam Kim, Yun-Mi Lee, and Jong-Joo Kim. "Therapeutic advances for Huntington's disease." *Brain Sciences* 10, no. 1 (2020): 43.

Kumar, Mukesh, Brian Keller, Ndeye Makalou, and Richard E. Sutton. "Systematic determination of the packaging limit of lentiviral vectors." *Human Gene Therapy* 12, no. 15 (2001): 1893–1905.

Langbehn, Douglas R., Ryan R. Brinkman, Daniel Falush, Jane S. Paulsen, M. R. Hayden, and an International Huntington's Disease Collaborative Group. "A new model for prediction of the age of onset and penetrance for Huntington's disease based on CAG length." *Clinical Genetics* 65, no. 4 (2004): 267–277.

Larner, A. J. "Presenilin-1 mutation Alzheimer's disease: A genetic epilepsy syndrome?." *Epilepsy & Behavior* 21, no. 1 (2011): 20–22.

Laudon, Hanna, Emil M. Hansson, Karin Melén, Anna Bergman, Mark R. Farmery, Bengt Winblad, Urban Lendahl, Gunnar Von Heijne, and Jan Näslund. "A nine-transmembrane domain topology for presenilin 1." *Journal of Biological Chemistry* 280, no. 42 (2005): 35352–35360.

Lee, J.-M., E. M. Ramos, J.-H. Lee, T. Gillis, J. S. Mysore, M. R. Hayden, S. C. Warby et al. "CAG repeat expansion in Huntington disease determines age at onset in a fully dominant fashion." *Neurology* 78, no. 10 (2012): 690–695.

Lee, Joseph H., Amanda Kahn, Rong Cheng, Christiane Reitz, Badri Vardarajan, Rafael Lantigua, Martin Medrano et al. "Disease-related mutations among Caribbean Hispanics with familial dementia." *Molecular Genetics & Genomic Medicine* 2, no. 5 (2014): 430–437.

Leissring, Malcolm A., Frank M. LaFerla, Nick Callamaras, and Ian Parker. "Subcellular mechanisms of presenilin-mediated enhancement of calcium signaling." *Neurobiology of Disease* 8, no. 3 (2001): 469–478.

Levy-Lahad, Ephrat, Wilma Wasco, Parvoneh Poorkaj, Donna M. Romano, Junko Oshima, Warren H. Pettingell, Chang-en Yu et al. "Candidate gene for the chromosome 1 familial Alzheimer's disease locus." *Science* 269, no. 5226 (1995): 973–977.

Li, Hao, Shihao Wu, Xia Ma, Xiao Li, Tianlin Cheng, Zhifang Chen, Jing Wu et al. "Co-editing PINK1 and DJ-1 genes via adeno-associated virus-delivered CRISPR/Cas9 system in adult monkey brain elicits classical parkinsonian phenotype." *Neuroscience Bulletin* 37, no. 9 (2021): 1271–1288.

Li, Jinhe, Min Xu, Hui Zhou, Jianyi Ma, and Huntington Potter. "Alzheimer presenilins in the nuclear membrane, interphase kinetochores, and centrosomes suggest a role in chromosome segregation." *Cell* 90, no. 5 (1997): 917–927.

Li, Xiaochun, Shangyu Dang, Chuangye Yan, Xinqi Gong, Jiawei Wang, and Yigong Shi. "Structure of a presenilin family intramembrane aspartate protease." *Nature* 493, no. 7430 (2013): 56–61.

Lino, Christopher A., Jason C. Harper, James P. Carney, and Jerilyn A. Timlin. "Delivering CRISPR: A review of the challenges and approaches." *Drug Delivery* 25, no. 1 (2018): 1234–1257.

Liu, Kunmei, Panpan Wan, Yue Huang, Bin Wang, Xuequan Wang, Rui Zhang, and Le Guo. "Transcriptome analysis in an AEG-1-deficient neuronal HT22 cell line.-" *Experimental and Therapeutic Medicine* 24, no. 5 (2022): 1–14.

Lu, Likui, Xi Yu, Yongle Cai, Miao Sun, and Hao Yang. "Application of CRISPR/Cas9 in Alzheimer's disease." *Frontiers in Neuroscience* 15 (2021): 803894.

Majounie, Elisa, Alan E. Renton, Kin Mok, Elise G. P. Dopper, Adrian Waite, Sara Rollinson, Adriano Chiò et al. "Frequency of the C9orf72 hexanucleotide repeat expansion in patients with amyotrophic lateral sclerosis and frontotemporal dementia: A cross-sectional study." *The Lancet Neurology* 11, no. 4 (2012): 323–330.

Malankhanova, Tuyana B., Anastasia A. Malakhova, Sergey P. Medvedev, and Suren M. Zakian. "Modern genome editing technologies in Huntington's disease research." *Journal of Huntington's Disease* 6, no. 1 (2017): 19–31.

Marongiu, Roberta, Francesco Brancati, Angelo Antonini, Tamara Ialongo, Caterina Ceccarini, Oronzo Scarciolla, Anna Capalbo et al. "Whole gene deletion and splicing mutations expand the PINK1 genotypic spectrum." *Human Mutation* 28, no. 1 (2007): 98–98.

Mullan, M., S. Tsuji, T. Miki, T. Katsuya, S. Naruse, K. Kaneko, T. Shimizu et al. "Clinical comparison of Alzheimer's disease in pedigrees with the codon 717 Val→ Ile mutation in the amyloid precursor protein gene." *Neurobiology of Aging* 14, no. 5 (1993): 407–419.

Munoz, David G., Manuela Neumann, Hirofumi Kusaka, Osamu Yokota, Kenji Ishihara, Seishi Terada, Shigetoshi Kuroda, and Ian R. Mackenzie. "FUS pathology in basophilic inclusion body disease." *Acta Neuropathologica* 118 (2009): 617–627.

Murray, Christopher J. L., Theo Vos, Rafael Lozano, Mohsen Naghavi, Abraham D. Flaxman, Catherine Michaud, Majid Ezzati et al. "Disability-adjusted life years (DALYs) for 291 diseases and injuries in 21 regions, 1990–2010: A systematic analysis for the Global Burden of Disease Study 2010." *The Lancet* 380, no. 9859 (2012): 2197–2223.

Murrell, Jill, Martin Farlow, Bernardino Ghetti, and Merrill D. Benson. "A mutation in the amyloid precursor protein associated with hereditary Alzheimer's disease." *Science* 254, no. 5028 (1991): 97–99.

Nagata, Kenichi, Mika Takahashi, Yukio Matsuba, Fumi Okuyama-Uchimura, Kaori Sato, Shoko Hashimoto, Takashi Saito, and Takaomi C. Saido. "Generation of App knock-in mice reveals deletion mutations protective against Alzheimer's disease-like pathology." *Nature Communications* 9, no. 1 (2018): 1800.

Neumann, Manuela, Deepak M. Sampathu, Linda K. Kwong, Adam C. Truax, Matthew C. Micsenyi, Thomas T. Chou, Jennifer Bruce et al. "Ubiquitinated TDP-43 in frontotemporal lobar degeneration and amyotrophic lateral sclerosis." *Science* 314, no. 5796 (2006): 130–133.

Nichols, Emma, Jaimie D. Steinmetz, Stein Emil Vollset, Kai Fukutaki, Julian Chalek, Foad Abd-Allah, Amir Abdoli et al. "Estimation of the global prevalence of dementia in 2019 and forecasted prevalence in 2050: An analysis for the Global Burden of Disease Study 2019." *The Lancet Public Health* 7, no. 2 (2022): e105–e125.

Nuytemans, Karen, Jessie Theuns, Marc Cruts, and Christine Van Broeckhoven. "Genetic etiology of Parkinson disease associated with mutations in the SNCA, PARK2, PINK1, PARK7, and LRRK2 genes: A mutation update." *Human Mutation* 31, no. 7 (2010): 763–780.

Okochi, Masayasu, Shinji Tagami, Kanta Yanagida, Mako Takami, Takashi S. Kodama, Kohji Mori, Taisuke Nakayama, Yasuo Ihara, and Masatoshi Takeda. "γ-Secretase modulators and presenilin 1 mutants act differently on presenilin/γ-secretase function to cleave Aβ42 and Aβ43." *Cell Reports* 3, no. 1 (2013): 42–51.

Pankratz, Nathan, Michael W. Pauciulo, Veronika E. Elsaesser, Diane K. Marek, Cheryl A. Halter, Joanne Wojcieszek, Alice Rudolph et al. "Mutations in DJ-1 are rare in familial Parkinson disease." *Neuroscience Letters* 408, no. 3 (2006): 209–213.

Papagiannakis, Nikolaos, Christos Koros, Maria Stamelou, Athina-Maria Simitsi, Matina Maniati, Roubina Antonelou, Dimitra Papadimitriou et al. "Alpha-synuclein dimerization in erythrocytes of patients with genetic and non-genetic forms of Parkinson's disease." *Neuroscience Letters* 672 (2018): 145–149.

Pardridge, William M. "The blood-brain barrier: Bottleneck in brain drug development." *NeuroRx* 2 (2005): 3–14.

Park, Jiyun, Seong Soo A. An, Vo Van Giau, Kyuhwan Shim, Young Chul Youn, Eva Bagyinszky, and SangYun Kim. "Identification of a novel PSEN1 mutation (Leu232Pro) in a Korean patient with early-onset Alzheimer's disease and a family history of dementia." *Neurobiology of Aging* 56 (2017): 212-e11.

Perkins, Emma M., Karen Burr, Poulomi Banerjee, Arpan R. Mehta, Owen Dando, Bhuvaneish T. Selvaraj, Daumante Suminaite et al. "Altered network properties in C9ORF72 repeat expansion cortical neurons are due to synaptic dysfunction." *Molecular Neurodegeneration* 16 (2021): 1–16.

Perrone, Federica, Maria Bjerke, Elisabeth Hens, Anne Sieben, Maarten Timmers, Arne De Roeck, Rik Vandenberghe et al. "Amyloid-β 1–43 cerebrospinal fluid levels and the interpretation of APP, PSEN1 and PSEN2 mutations." *Alzheimer's Research & Therapy* 12 (2020): 1–14.

Platt, Randall J., Sidi Chen, Yang Zhou, Michael J. Yim, Lukasz Swiech, Hannah R. Kempton, James E. Dahlman et al. "CRISPR-Cas9 knockin mice for genome editing and cancer modeling." *Cell* 159, no. 2 (2014): 440–455.

Puschmann, Andreas, Owen A. Ross, Carles Vilariño-Güell, Sarah J. Lincoln, Jennifer M. Kachergus, Stephanie A. Cobb, Suzanne G. Lindquist et al. "A Swedish family with de novo α-synuclein A53T mutation: Evidence for early cortical dysfunction." *Parkinsonism & Related Disorders* 15, no. 9 (2009): 627–632.

Rademakers, Rosa, Marc Cruts, and Christine Van Broeckhoven. "Genetics of early-onset Alzheimer dementia." *The Scientific World Journal* 3, no. 1 (2003): 497–519.

Ran, F. Ann, Le Cong, Winston X. Yan, David A. Scott, Jonathan S. Gootenberg, Andrea J. Kriz, Bernd Zetsche et al. "In vivo genome editing using Staphylococcus aureus Cas9." *Nature* 520, no. 7546 (2015): 186–191.

Renton, Alan E., Adriano Chiò, and Bryan J. Traynor. "State of play in amyotrophic lateral sclerosis genetics." *Nature Neuroscience* 17, no. 1 (2014): 17–23.

Roberts, Susan Boseman, J. A. Ripellino, K. M. Ingalls, N. K. Robakis, and K. M. Felsenstein. "Non-amyloidogenic cleavage of the beta-amyloid precursor protein by an integral membrane metalloendopeptidase." *Journal of Biological Chemistry* 269, no. 4 (1994): 3111–3116.

Rohn, Troy T., Nayoung Kim, Noail F. Isho, and Jacob M. Mack. "The potential of CRISPR/Cas9 gene editing as a treatment strategy for Alzheimer's disease." *Journal of Alzheimer's Disease & Parkinsonism* 8, no. 3 (2018).

Sanchez-Rivera, Francisco J., Thales Papagiannakopoulos, Rodrigo Romero, Tuomas Tammela, Matthew R. Bauer, Arjun Bhutkar, Nikhil S. Joshi et al. "Rapid modelling of cooperating genetic events in cancer through somatic genome editing." *Nature* 516, no. 7531 (2014): 428–431.

Schellenberg, Gerard D., and Thomas J. Montine. "The genetics and neuropathology of Alzheimer's disease." *Acta Neuropathologica* 124 (2012): 305–323.

Schultheis, Patrick J., Tamara T. Hagen, Kate K. O'Toole, Akiko Tachibana, Charles R. Burke, Diana L. McGill, Gbolahan W. Okunade, and Gary E. Shull. "Characterization of the P5 subfamily of P-type transport ATPases in mice." *Biochemical and Biophysical Research Communications* 323, no. 3 (2004): 731–738.

Sen, Tirthankar, and Rajkumar P. Thummer. "CRISPR and iPSCs: Recent developments and future perspectives in neurodegenerative disease modelling, research, and therapeutics." *Neurotoxicity Research* 40, no. 5 (2022): 1597–1623.

Serneels, Lutgarde, Dries T'Syen, Laura Perez-Benito, Tom Theys, Matthew G. Holt, and Bart De Strooper. "Modeling the β-secretase cleavage site and humanizing amyloid-beta precursor protein in rat and mouse to study Alzheimer's disease." *Molecular Neurodegeneration* 15 (2020): 1–11.

Shelkovnikova, Tatyana A., Hannah K. Robinson, Joshua A. Southcombe, Natalia Ninkina, and Vladimir L. Buchman. "Multistep process of FUS aggregation in the cell cytoplasm involves RNA-dependent and RNA-independent mechanisms." *Human Molecular Genetics* 23, no. 19 (2014): 5211–5226.

Shi, Nian-Qiu, Xian-Rong Qi, Bai Xiang, and Yong Zhang. "A survey on "Trojan Horse" peptides: Opportunities, issues and controlled entry to "Troy"." *Journal of Controlled Release* 194 (2014): 53–70.

Siddique, Teepu, Denise A. Figlewigz, Margaret A. Pericak-Vance, Jonathan L. Haines, Guy Rouleau, Anita J. Jeffers, Peter Sapp et al. "Linkage of a gene causing familial amyotrophic lateral sclerosis to chromosome 21 and evidence of genetic-locus heterogeneity." *New England Journal of Medicine* 324, no. 20 (1991): 1381–1384.

Sims, Rebecca, Sven J. Van Der Lee, Adam C. Naj, Céline Bellenguez, Nandini Badarinarayan, Johanna Jakobsdottir, Brian W. Kunkle et al. "Rare coding variants in PLCG2, ABI3, and TREM2 implicate microglial-mediated innate immunity in Alzheimer's disease." *Nature Genetics* 49, no. 9 (2017): 1373–1384.

Slow, Elizabeth J., Jeremy van Raamsdonk, Daniel Rogers, Sarah H. Coleman, Rona K. Graham, Yu Deng, Rosemary Oh et al. "Selective striatal neuronal loss in a YAC128 mouse model of Huntington disease." *Human Molecular Genetics* 12, no. 13 (2003): 1555–1567.

Sreedharan, Jemeen, Ian P. Blair, Vineeta B. Tripathi, Xun Hu, Caroline Vance, Boris Rogelj, Steven Ackerley et al. "TDP-43 mutations in familial and sporadic amyotrophic lateral sclerosis." *Science* 319, no. 5870 (2008): 1668–1672.

Tabrizi, Sarah J., Rhia Ghosh, and Blair R. Leavitt. "Huntingtin lowering strategies for disease modification in Huntington's disease." *Neuron* 101, no. 5 (2019): 801–819.

Takahashi-Niki, Kazuko, Takeshi Niki, Takahiro Taira, Sanae M. M. Iguchi-Ariga, and Hiroyoshi Ariga. "Reduced anti-oxidative stress activities of DJ-1 mutants found in Parkinson's disease patients." *Biochemical and Biophysical Research Communications* 320, no. 2 (2004): 389–397.

Takalo, Mari, Rebekka Wittrahm, Benedikt Wefers, Samira Parhizkar, Kimmo Jokivarsi, Teemu Kuulasmaa, Petra Mäkinen et al. "The Alzheimer's disease-associated protective Plcγ2-P522R variant promotes immune functions." *Molecular Neurodegeneration* 15 (2020): 1–14.

Tan, Daniel, Sherilyn Yao, Arne Ittner, Josefine Bertz, Yazi D. Ke, Lars M. Ittner, and Fabien Delerue. "Generation of a new tau knockout (tau Δex1) line using CRISPR/Cas9 genome editing in mice." *Journal of Alzheimer's Disease* 62, no. 2 (2018): 571–578.

Turk, Olivia M., Ryan C. Woodall, Margarita Gutova, Christine E. Brown, Russell C. Rockne, and Jennifer M. Munson. "Delivery strategies for cell-based therapies in the brain: Overcoming multiple barriers." *Drug Delivery and Translational Research* 11 (2021): 2448–2467.

Vance, Caroline, Boris Rogelj, Tibor Hortobágyi, Kurt J. De Vos, Agnes Lumi Nishimura, Jemeen Sreedharan, Xun Hu et al. "Mutations in FUS, an RNA processing protein, cause familial amyotrophic lateral sclerosis type 6." *Science* 323, no. 5918 (2009): 1208–1211.

Vermilyea, Scott C., Alexander Babinski, Nina Tran, Samantha To, Scott Guthrie, Jillian H. Kluss, Jenna Kropp Schmidt et al. "In vitro CRISPR/Cas9-directed gene editing to model LRRK2 G2019S Parkinson's disease in common marmosets." *Scientific Reports* 10, no. 1 (2020): 3447.

Vetrivel, Kulandaivelu S., Yun-wu Zhang, Huaxi Xu, and Gopal Thinakaran. "Pathological and physiological functions of presenilins." *Molecular Neurodegeneration* 1 (2006): 1–12.

Voigt, Nadine, Petra Henrich-Noack, Sarah Kockentiedt, Werner Hintz, Jürgen Tomas, and Bernhard A. Sabel. "Surfactants, not size or zeta-potential influence blood–brain barrier passage of polymeric nanoparticles." *European Journal of Pharmaceutics and Biopharmaceutics* 87, no. 1 (2014): 19–29.

Waldvogel, Henry J., Eric H. Kim, Lynette J. Tippett, Jean-Paul G. Vonsattel, and Richard L. M. Faull. "The neuropathology of Huntington's disease." *Behavioral Neurobiology of Huntington's Disease and Parkinson's Disease* (2015): 33–80.

Wallon, David, Stéphane Rousseau, Anne Rovelet-Lecrux, Muriel Quillard-Muraine, Lucie Guyant-Maréchal, Olivier Martinaud, Jérémie Pariente et al. "The French series of autosomal dominant early onset Alzheimer's disease cases: Mutation spectrum and cerebrospinal fluid biomarkers." *Journal of Alzheimer's Disease* 30, no. 4 (2012): 847–856.

Wang, Alfred Y., Peter D. Peng, Anja Ehrhardt, Theresa A. Storm, and Mark A. Kay. "172. Comparison of adenoviral and adeno-associated viral vectors for pancreatic gene delivery in vivo." *Molecular Therapy* 9, no. S1 (2004): S66.

Wang, Lixia, Fei Yi, Lina Fu, Jiping Yang, Si Wang, Zhaoxia Wang, Keiichiro Suzuki et al. "CRISPR/Cas9-mediated targeted gene correction in amyotrophic lateral sclerosis patient iPSCs." *Protein & Cell* 8, no. 5 (2017): 365–378.

Wang, Ming, John A. Zuris, Fantao Meng, Holly Rees, Shuo Sun, Pu Deng, Yong Han et al. "Efficient delivery of genome-editing proteins using bioreducible lipid nanoparticles." *Proceedings of the National Academy of Sciences* 113, no. 11 (2016): 2868–2873.

Wang, Xueyin, Jacob C. Schwartz, and Thomas R. Cech. "Nucleic acid-binding specificity of human FUS protein." *Nucleic Acids Research* 43, no. 15 (2015): 7535–7543.

Weber, Evan W., Marcela V. Maus, and Crystal L. Mackall. "The emerging landscape of immune cell therapies." *Cell* 181, no. 1 (2020): 46–62.

World Health Organization (WHO). (2006). *Neurological Disorders: Public Health Challenges*. World Health Organization.

Wu, Qin, Jin-xia Sun, Xiang-he Song, Jing Wang, Cun-quan Xiong, Fei-xiang Teng, and Cui-xiang Gao. "Blocking beta 2-adrenergic receptor inhibits dendrite ramification in a mouse model of Alzheimer's disease." *Neural Regeneration Research* 12, no. 9 (2017): 1499–1506.

Yamashita, Takenari, Hui Lin Chai, Sayaka Teramoto, Shoji Tsuji, Kuniko Shimazaki, Shinichi Muramatsu, and Shin Kwak. "Rescue of amyotrophic lateral sclerosis phenotype in a mouse model by intravenous AAV 9-ADAR 2 delivery to motor neurons." *EMBO Molecular Medicine* 5, no. 11 (2013): 1710–1719.

Yanagida, Keisuke, Catherine H. Liu, Giuseppe Faraco, Sylvain Galvani, Helen K. Smith, Nathalie Burg, Josef Anrather, Teresa Sanchez, Costantino Iadecola, and Timothy Hla. "Size-selective opening of the blood–brain barrier by targeting endothelial sphingosine 1–phosphate receptor 1." *Proceedings of the National Academy of Sciences* 114, no. 17 (2017): 4531–4536.

Yang, Su, Renbao Chang, Huiming Yang, Ting Zhao, Yan Hong, Ha Eun Kong, Xiaobo Sun et al. "CRISPR/Cas9-mediated gene editing ameliorates neurotoxicity in mouse model of Huntington's disease." *The Journal of Clinical Investigation* 127, no. 7 (2017): 2719–2724.

Yang, Youngsoon, Eva Bagyinszky, and Seong Soo A. An. "Presenilin-1 (PSEN1) mutations: Clinical phenotypes beyond Alzheimer's disease." *International Journal of Molecular Sciences* 24, no. 9 (2023): 8417.

Yoon, Hyung Ho, Sunghyeok Ye, Sunhwa Lim, Ara Jo, Hawon Lee, Felix Hong, Seung Eun Lee et al. "CRISPR-Cas9 gene editing protects from the A53T-SNCA overexpression-induced pathology of Parkinson's disease in vivo." *The CRISPR Journal* 5, no. 1 (2022): 95–108.

Zarranz, Juan J., Javier Alegre, Juan C. Gómez-Esteban, Elena Lezcano, Raquel Ros, Israel Ampuero, Lídice Vidal et al. "The new mutation, E46K, of α-synuclein causes parkinson and Lewy body dementia." *Annals of Neurology: Official Journal of the American Neurological Association and the Child Neurology Society* 55, no. 2 (2004): 164–173.

Zhang, Lili, Chen Chen, Marvin S. H. Mak, Junfeng Lu, Zeqing Wu, Qiuhe Chen, Yifan Han, Yuefeng Li, and Rongbiao Pi. "Advance of sporadic Alzheimer's disease animal models." *Medicinal Research Reviews* 40, no. 1 (2020): 431–458.

Zhang, Xiao-Hui, Louis Y. Tee, Xiao-Gang Wang, Qun-Shan Huang, and Shi-Hua Yang. "Off-target effects in CRISPR/Cas9-mediated genome engineering." *Molecular Therapy Nucleic acids* 4 (2015): 1–8.

Zhou, Yiqun, Zhili Peng, Elif S. Seven, and Roger M. Leblanc. "Crossing the blood-brain barrier with nanoparticles." *Journal of Controlled Release* 270 (2018): 290–303.

Zhou, Haibo, Jinlin Su, Xinde Hu, Changyang Zhou, He Li, Zhaorong Chen, Qingquan Xiao et al. "Glia-to-neuron conversion by CRISPR-CasRx alleviates symptoms of neurological disease in mice." *Cell* 181, no. 3 (2020): 590–603.

10 The CRISPR-Cas System in Microbes

Tooba Mehboob, Naila Tabassam, Huma Hameed, Muhammad Zaman, Mahtab Ahmad Khan, Imtiaz Majeed, and Tehseen Riaz

10.1 OVERVIEW OF THE CRISPR-CAS SYSTEM IN MICROBES

One of the breakthroughs in advanced biology is the innovation of the clustered regularly interspersed short palindromic repeat (CRISPR) and the CRISPR-associated proteins (Cas proteins). CRISPRs basically are DNA regions which are found in the genomes of prokaryotes and are made up of distinct spacer sequences of the same length that separate similar short repeats (30 to 40 bp) (Hille & Charpentier, 2016). Small palindromic repeats are rather prevalent; approximately, 50% of all known bacteria and 90% of archaea include CRISPR regions in their genomes (Grissa et al., 2007; Hille et al., 2018). This could suggest that prokaryotes place a high value on them. In recognition of their innovative work on the CRISPR-Cas genetic editing method, the Nobel prize was awarded to Jenifer Doudna and Emmanuelle Charpentier in the field of chemistry in the year 2020 respectively (Shmakova et al., 2022). The unearthing of a sequence of 29 nucleotide (nt) repeats in *Escherichia coli* (*E. coli*) by Nakata and his colleagues in 1987 during their investigation of the isozyme of alkaline phosphatase (iap) gene signified the inception in the historical perspective of CRISPR-Cas systems (Ishino et al., 1987).

Subsequently, it was shown that these repetitive sequences were found in 40% of sequenced bacterial genetic codes and 90% of archaea (Haurwitz et al., 2010). However, the exact nature of their function had not been determined. In the next decade, the sequencing of many microbial genomes revealed more repetitive elements derived from the genetic material of diverse bacterial and archaeal species. Later, this diverse set of interspaced repeat sequences was referred to as collected repeat elements (Mojica et al., 2000). When spacer sequences were identified as extrachromosomal origins or phage association after being separated from direct repetitions in 2005, it was a significant breakthrough (Mojica et al., 2005). The fundamental workings and mechanism of the CRISPR-Cas system were cleared in 2010. This system consists of a cluster of 6 to 20 adjacent genes that code for CRISPR-associated (Cas) proteins, along with a genomic region containing unique spacer sequences (Shabbir et al., 2016). A diverse group of scientists leveraging the CRISPR-Cas system to craft dairy cultures that are immune to phages, as well as for multiple biotechnological applications (Hsu et al., 2014). In archaea and bacteria, the CRISPR-Cas system is an adaptive defense system that is challenged by invaders like bacteriophages and

DOI: 10.1201/9781003537151-10

mobile genetic elements (MGEs) (Barrangou et al., 2007). For nearly 90 years, the advent of antibiotic substances has played a crucial role in transforming modern medical practices. Antibiotics can treat infections that were previously deadly, but their excessive use in veterinary and agricultural settings presents a serious risk to public health since they release a large amount of antibiotics into the environment. Due to the high concentration of antibiotics, both commensal and pathogenic bacteria are subject to strong selection pressures that facilitate the development and dissemination of genes responsible for antimicrobial resistance.

Through a variety of strategies, including the employment of the efflux pump, the destruction of antibiotic molecules via enzymes, and biochemical alteration in ribosomes and cell wall to shield the cell-level targets of antibiotics, these antibiotic-resistant genes let bacteria resist antibiotics. When combined, these resistance pathways endanger the therapeutic efficacy of antibiotics (Shabbir et al., 2019). A study carried out by the Centers for Disease Control and Prevention in 2013 indicated that antimicrobial-resistant (AMR) microorganisms affect over 2 million individuals annually, leading to approximately 23,000 deaths in the United States. It is expected that by the year 2050, drug-resistant illnesses will result in nearly ten million deaths annually. This implies that deaths from drug-resistant infections will surpass those from diabetes, cancer, and auto accidents. The capacity of bacteria to show extraordinary phenotypic and genotypic variability to survive in unfavorable conditions is a feature that contributes to the development of antibiotic resistance (Erickson et al., 2017). The primary cause of antimicrobial reduction is the lower production rate of novel antibiotics; in the last 45 years, there have been no innovative classes of approved antibiotics for treating gram-negative bacterial infections, and only 37 antibiotics are currently undergoing phase 2 or 3 clinical trials. These factors motivate the exploration of alternative strategies for addressing AMR infections, leading to the development of a specialized antibacterial toolkit designed for specific targeting. In pursuit of this goal, researchers have formulated a range of antibacterial therapies, such as phage treatments, bacterial toxins, antibodies, and antivirulence agents, utilizing distinct peptide and nucleic acid foundations (de la Fuente-Nunez & Lu, 2017).

Based on the initial findings of the up-to-date breakthrough in CRISPR-Cas technology and science, which has prospered, developed, and kept on bearing fruit in the past 30 years, with numerous examples of success, the scientific spotlight has gradually turned to getting the advantages from this editing genes on the internet, while the research that underpins it continues to be ignored. As a result, in addition to ongoing developments in the use of the technology known as CRISPR, there is an urgent desire to keep examining its fundamental features; however, current information on these aspects is dispersed across numerous papers in the literature. This chapter encourages aspiring CRISPR experts and then explains on the benefits of more precise investigation, rather than the previous academic apathy of this innovative and unexplored area (Nidhi et al., 2021)

10.1.1 Classification

Bacteria and archaea naturally have a CRISPR-Cas system that serves as an adaptive immune mechanism (Makarova et al., 2020). They are best recognized for being essential parts of a newfangled generation in the process of genome editing. The most prevalent type of intrinsic defense system is the restriction-modification enzymes which evolve a vast array of specificities. CRISPR-Cas is involved in a ceaseless confrontation with viruses as like any other defense mechanism and causes the Cas genes to evolve rapidly along with the gene pool and architectures of the CRISPR-Cas loci to represent significant diversity, which in turn causes the defense mechanisms themselves to diversify. In particular, the viable coevolution of CRISPR-Cas systems along with virus-encoded anti-CRISPR proteins is probably a contributing factor in their diversification (Koonin et al., 2017). The CRISPR-Cas system showcases significant variation in its gene composition, the structural arrangement of genomic loci, and the actual sequences involved, particularly within the core genes that are shared among various CRISPR-Cas types.

A logical and relatively straightforward taxonomy of CRISPR-Cas systems is integral to clarify the origins and development of this variety, as well as, more practically, to monitor novel variants and accomplish a comprehensive description of CRISPR-Cas loci within microorganisms' genetic material. The repeated sequences show some clustering in the sequence space, but it is clear that there is not enough data to establish a reliable classification foundation. A phylogenetic classification based on a single tree is not feasible due to the absence of universal c gene (Makarova et al., 2018). All CRISPR-Cas systems can be classified primarily into one of two groups based on how the effector modules are designed. In class 2 systems, the effector is a single large multidomain protein, while in class 1 systems, the effector complexes comprise numerous Cas proteins (Koonin & Makarova, 2019). Additional research and analysis found two new categories and numerous subcategories of the class 2 CRISPR-Cas system. The type VI systems were the only CRISPR type to target RNA out of the two recently identified and defined types. In certain instances, class 2 systems exhibit a distinctive characteristic wherein the effector protein is also identified as the cause of the pre-crRNA processing (Koonin et al., 2017). Further divergence has led to the division of the class 1 CRISPR-Cas system into subtypes I, III, and IV, while the second class has also been split into kinds II, V, and VI, each of which has further subtypes. The CRISPR-Cas system's mode of action is further categorized into three stages: adaptation, expression and maturation, and interference.

The combination of these stages constitutes the effector module, forming the foundation for this classification (Chaudhuri et al., 2022). Types I and III are most widespread among archaea, but not as common in bacteria, while the rare type IV, which contains primitive CRISPR-Cas loci devoid of the adaptability module, are all members of the class 1 system. Currently, there are seven subtypes of type I systems: I-A through I-F and I-U. Type I contains Cas1 through Cas8 proteins in addition to the distinctive Cas3 protein. There is a known Cas gene combination for each subtype (Makarova et al., 2015; Javed et al., 2018). The type I and type III effector complexes of CRISPR-Cas have complicated architectures, with a paralogous repeat associated mysterious protein (RAMP) backbone made up of proteins, for example, Cas7 and Cas5, which have extra "large" and "small" and the RNA recognition motif (RRM) fold. One Cas5 subunit, in addition to many Cas7 subunits, is present in these effector complexes (Koonin et al., 2017; Zhao et al., 2014). The complex holds the guide RNA, which is made up of a repeat segment and a spacer. The large subunit

Cas8 in type I and Cas10 in type III or the Cas5 subunit interact by binding the crRNA's 50 handles. The tiny subunit interacts with the crRNA backbone, coupled to Cas7, and is usually found in many copies. Interestingly, the amount of Cas7 subunits in the complex's backbone coincides with the bound spacer's length (Koonin et al., 2017). The operational modules of class 2 are characterized by an individual big multidomain protein each so that the corresponding CRISPR-Cas loci have a simpler and more consistent organization compared to class 1 during the development of the present official CRISPR-Cas classification (Makarova et al., 2015). The classification of class 2 encompasses three subtypes of the extensively studied type II, which includes the Cas9 endonuclease enzyme, a prominent tool in genome editing. Furthermore, type V has recently been incorporated into the classification system, highlighting the expected effector protein Cpf1, yet it is still largely undefined (Koonin et al., 2017; Makarova et al., 2015).

Similarly, Table 10.2 shows the summary of CRISPR-Cas system class 2.

TABLE 10.1
Summary of CRISPR-Cas System Class 1

Class	Type	Subtype	Effector	Target	Tracer RNA Requirement
1	I	A, B, C, D, E, F, U	Cascade	Ds-DNA	Not at all
1	III	A, B, C, D	Cascade	ss-RNA	Not at all
1	IV	A, B,	Cascade	Ds-DNA	Not at all

Source: Xu and Li (2020), Hille et al. (2018), and Jiang and Doudna (2017).

TABLE 10.2
Summary of CRISPR-Cas System Class 2

Class	Type	Subtype	Effector	Target	Tracer RNA
2	II	A	Sp-Cas9	Ds-DNA	Yes
2	II	A	Sa-Cas9	Ds-DNA	Yes
2	II	B	Fn-Cas9	Ds-DNA	Yes
2	II	C	Nm-Cas9	Ds-DNA/ss-RNA	Yes
2	V	A	Cas-12a (Cpf1)	Ds-DNA	No
2	V	B	Cas-12b (C2c1)	Ds-DNA	Yes
2	V	C	Cas-12c (C2c3)	Ds-DNA	Yes
2	VI	A	Cas-13a (C2c2)	ss-RNA	No
2	VI	B	Cas-13b (C2c4)	ss-RNA	No
2	VI	C	Cas-13c (C2c7)	ss-RNA	No
2	VI	D	Cas-13d	ss-RNA	No

Source: Jiang and Doudna (2017), Koonin et al. (2017), Makarova et al. (2011), and Xu and Li (2020).

Cells have evolved many strategies to combat viruses; thus, viruses have also devised counteractions to these strategies. To neutralize CRISPR-Cas systems, various measures have been outlined. By randomly mutating the basic elements involved in the interaction with cr-RNA or the PAM recognition, viruses can escape CRISPR-Cas activity. In *Pseudomonas aeruginosa* phages, sophisticated countermeasures have been found that encode many proteins impacting the effectiveness of type I-E and I-F systems. The expression of Cas proteins or the cr-RNA is not affected by the proteins, because the activities of proteins are unclear. Probably, it seems they interfere with the activity of CRISPR-Cas complexes. Evidence suggests that viruses employing the CRISPR-Cas system contribute to the promotion of infestation. *Vibrio cholerae* ICP1 phages have a type I-F CRISPR-Cas system which focuses on the specific site of the host, known as phage-inducible chromosomal island-like element (PLE), or having an anti-phage system. Once the phage genetic material is inserted inside the cell, to facilitate the infection of the *Vibrio cholerae* in the host, the cr-RNAs and Cas genes of the virus are expressed. If there are modifications to either the host or the CRISPR, the viral CRISPR-Cas system becomes misaligned with the PLE, greatly reducing the ability of ICP1 to infect. The few phages that manage to infect do so by obtaining a new spacer that targets the host site, demonstrating that these viruses can take advantage of the adaptable nature of the CRISPR-Cas system (Rath et al., 2015).

10.2 ROLE OF CRISPR-CAS9 SYSTEM IN BACTERIAL VIRULENCE AND ANTIMICROBIAL RESISTANCE

The CRISPR-Cas technique governs endogenous transcription in bacteria, protecting them against outside intruders and regulating their pathogenicity. Because the *Francisella novicida* (*F. novicida*) bacteria carry the Cas9 protein, which inhibits the activation of toll-like receptor 2 (TLR2), the bacteria can evade the host immune system (Shabbir et al., 2016). The recognition of bacterial lipoproteins by TLR2 prompts a pro-inflammatory response in the host, thereby engaging the immune system to eradicate bacteria (Mukherjee et al., 2016). It has been theorized that the type II CRISPR-Cas system can suppress the bacterial lipoprotein (BLP) genes Cas9, trans-activating CRISPR RNA (tracrRNA), and short CRISPR-associated RNA (sacRNA), which in turn control the suppression of *F. novicida* BLP (Barrangou, 2015). Cas9 is abundant in both commensal and pathogenic bacteria. It has been documented that Cas9 is involved in the adhesion of pathogenic microbes (*Neisseria meningitides* and *Campylobacter jejuni*) to the surface of host cells as well as their multiplication within those cells. The Cas2 component of *Legionella pneuomphila*'s CRISPR-Cas system is thought to be responsible for the bacteria's pathogenicity, which makes it the causal agent of Legionnaire's illness (Saha et al., 2020a).

10.2.1 Viral Escape from the System: CRISPR-Cas

To evade the CRISPR-Cas-driven immunity of bacteria, bacteriophages have altered their genomes through deletion or insertion. The acquisition and application of the bacterial CRISPR system for targeting the viral defense mechanism produced by the bacteriophage is considered to be an evolutionary significance of this system. It has been found that the CRISPR-Cas defense system has enabled mutations in viral and bacterial genomes (Qaisar et al., 2017).

10.2.2 Using CRISPR-CAS Technology in Genome Editing

The concept of genetic modification comprises a variety of techniques that scientists use to modify the DNA of organisms. These methods enable the insertion, removal, or modification of genetic sequences at specific locations within the genome (Hartung & Schiemann, 2014). Using CRISPR-Cas9 represents the most effective method for modifying genomes (Rizwan et al., 2016). It is a bacterial genome editing mechanism that evolved spontaneously (Liu et al., 2020a). Small DNA fragments known as CRISPR arrays are created by bacteria using phage genome fragments as templates (Nussenzweig & Marraffini, 2020). These CRISPR arrays let bacteria create memory against viral genomes to fend off future invasions (Westra & Levin 2020). *In vitro* genome modification may be accomplished using CRISPR-Cas technology (Schulze & Lammers, 2020). A tiny RNA fragment containing a brief "guide" sequence that binds to a specific target region in the genome has been developed. This directed RNA can bind to the Cas9 protein (Rizwana et al., 2016). These guided RNAs aid the Cas9 enzyme's recognition of targeted DNA *in vivo* and *in vitro*. The Cpf1 or Cas9 enzyme functions as a nuclease, causing double breaks in the targeted genome. Scientists employ biological repair of DNA mechanism to insert or delete genetic material fragments after cutting DNA (Huynh et al., 2020). Bacteria such as *E. coli*, *Lactobacillus reuteri* (used in the synthesis of probiotics), *Streptococcus pneumoniae*, *Clostridium beijerinckii* (used in the industrial manufacture of alcohol), and *Streptomyces* species (used in the synthesis of antimicrobials) have adapted to the CRISPR-Cas system's genome editing capacity (Barrangou & van Pijkeren, 2016). In *Campylobacter jejuni*, the CRISPR-Cas mechanism increases antibiotic insensitivity. The integrity of the bacterial envelope, which protects against harm from membrane targeting, is essential for the continued existence of bacterial pathogens inside the host body antibiotics (Hurdle et al., 2011). It is commonly known that the CRISPR-Cas controls the permeability or improves the integrity of the bacterial membrane. Prokaryotes are protected against foreign nucleic acids and invading viruses by their CRISPR-Cas system. Its CRISPR-Cas system controls the permeability of the bacterial membrane (Chen et al., 2016). It causes bacteria to become immune to antibiotics that target membranes (Ahmad et al., 2019). Controlling membrane permeability makes germs more harmful, by preventing them from being recognized by many host receptors (Chen et al., 2016).

10.2.3 Intracellular Delivery of Antimicrobial Agents Using CRISPR-Cas9

When a colony with a functioning CRISPR-Cas system was exposed to genome-targeting guide RNAs, the cell cycle was stopped, which ultimately resulted in the death of the bacterium or the loss of significant genomic portions containing the target sequence (Selle et al., 2015; Vercoe et al., 2013). These insights opened the door for the creation of CRISPR enzymes, which are programmable antibiotics that can target certain bacteria and permanently break their DNA. This method was created mainly to deal with the growing problem of resistance to antibiotics and the dwindling availability of novel small-molecule medications (Koch et al., 2014). A major obstacle to the creation of CRISPR antibiotics is likely to be effective delivery to a variety of bacteria. Thus far, the majority of CRISPR antibiotics have been supplied via bacteriophage particles and encoded as DNA (Park et al., 2017).

Phage, also known as *bacteriophage*, are able to insert their heritable material into the cytoplasm of their host by piercing the bacterial cell wall. Phagemids and the phage genome are the two forms in which CRISPR antimicrobial agents have been encoded to guarantee phage particle packing. Since phagemids only encode the origin of replication, packing signals, and selection marker, they are carriers that lack any other phage components that may inadvertently change cell physiology. Using common *in vitro* assembly methods, phagemids may be modified to carry CRISPR antimicrobials as plasmids. As a result, several teams have delivered CRISPR antibiotics to *E. coli* and *Staphylococcus aureus* (*S. aureus*) using phagemids (Bikard et al., 2014; Cui & Bikard 2016).

The second method involves inserting CRISPR antimicrobial DNA into the phage genome. Phage genomes may include undesirable phage components and require more work to manipulate DNA than phagemids. However, phage genomes often surpass phagemids in the packaging race and contain components that can defy the immunological responses of bacteria, like restriction-modification systems that damage the invader's DNA. Yosef et al. effectively developed a type I CRISPR-Cas system by using the λ phages as genetic code, which incorporates into the bacterial genome, to allow the inheritable target of multidrug-resistant plasmids and stop the propagation of resistance to multiple drugs within a bacterial population (Yosef et al., 2015). One feature of certain novel targeting tactics is the injection of the phage CRISPR-Cas9 systems into nanoparticles. In order to eradicate the issue of resistance to antibiotics in the near distant future, innovative focused approaches are being looked for (Allemailem et al., 2022). **Table 10.3** shows examples of some novel CRISPR-Cas9-based antimicrobial drug delivery strategies in various bacterial species.

10.3 CRISPR-CAS SYSTEMS INVOLVING MICROBIAL PROCESSES OTHER THAN IMMUNITY

10.3.1 CRISPR-Cas in Microbial Biotechnology

In industrial microbiology, significant advancements have been observed recently. These advancements are because of the progression in genomic engineering techniques. Significant developments have been seen in the manipulation of genomic sequences through the mutation, deletion, and insertion of nucleotides, as well as in the utilization of the transcriptome and epigenome. By expanding the list of chemical products, metabolites, and biomolecules produced by microbial fermentation and being dependent on these techniques, the industry of biotechnology fulfils its growing requirements (David & Siewers, 2015). Due to the typically clean and renewable

TABLE 10.3
Novel CRISPR-Cas9-Based Antimicrobial Drug Delivery Strategies

Delivery System	CRISPR-Cas 9 Locus	Bacteria	Summary of Results	References
Phage	CRISPR-Cas9	*E. coli* *S. aureus*	RNA-guided enzymes in blaNDM-1 or blaSHV-18 cause double-strand breaks, which nearly 1,000 times lower the transduction effectiveness of plasmids carrying these genes. Designed mecA to resensitize MRSA to methicillin by targeting the phagemid pDB91 encased in phage ΦNM1. φSaBov temperate phage genome has integrated the CRISPR-Cas9 system, the phage's specificity for the host has been increased by enhancement of the tail fiber protein, and the nuc genes have been extracted from the host chromosome.	Bikard et al., 2014; Park et al., 2017
Conjugative plasmid	CRISPR-Cas9	*E. coli* *Salmonella*	Polymyxin's antibiotic resistance is restored by the delivery of the CRISPR-Cas9 system via the conjugative plasmid that addresses the mcr 1-gene. A novel technique based on TAPs that uses bacterial conjugation to give strain-specific antibacterial activity using CRISPR-Cas9 genome editing technologies. TAP-directed plasmid-borne carbapenem-resistant gene may resensitize the medication to this strain. Nik-A, sop-A, vag-C, mcr1, and hic-B are examples of antitoxin genes that were employed as target genes to remove drug-resistant plasmids. The mcr1 gene is the target of the CRISPR-Cas9 genome-editing system, which is conjugated to *E. coli* and integrated on the p-Mob Cas9 plasmid. Plasmids based on the Inc-P RK-2 conjugative mechanism can be used to deliver TevSpCas9 dual nuclease. High *S. enterica* death rates were observed when single as well as multiplexed sgRNAs were directed towards non-essential genes.	Wan et al., 2020; Reuter et al., 2021; Wang et al., 2019
Mobile genomic island	CRISPR dCas9, CRISPR-Cas9	*S. aureus*	*S. aureus* infections are treated with highly mobile SAPIs, and ABD-2003 eliminates *S. aureus* by causing a double-strand break to occur in the agr-A region of the chromosome.	Ram et al., 2018
Hydrogel	CRISPR-Cas9	*S. aureus*	Quantitative antibiofilm supplied using alginate hydrogel improved the therapeutic effects of fosfomycin-phage (dual) over time. This module effectively decreased soft tissue infection.	Kim et al., 2021
Nanoparticle	CRISPR-Cas9	*S. aureus*	Creating a nano-sized CRISPR combination by combining sgRNA with SpCas9-bPEI greatly increased the transfection capacity of MRSA. The purpose of this complex was to specifically target mecA, a key gene linked to methicillin resistance.	Kang et al., 2017

nature of traditional petrol-dependent methods, microbial cell factories provide a promising alternative by producing fuels, chemicals, and pharmaceutics. Genome editing is a time-consuming process which proposes many challenges and barriers in microbial production. The CRISPR-Cas system has quickly become a valuable tool for fundamental research and biotechnological applications, including metabolic engineering, genome editing, molecular diagnostics, and gene function analysis in microorganisms, plants, and mammals. A variety of model and non-model organisms, like *Saccharomyces cerevisiae*, *Lactobacillus*, *E. coli*, *Mycobacterium*, *Pseudomonas*, and *Bacillus*, utilized CRISPR-Cas systems for the field of microbial biotechnology. Twenty-six unique strains of *Lactobacillus buchneri*, isolated from the fermentation of spoiled pickle, were genotyped on the basis of spacer compositions, along with CRISPR repeat sequences (Briner & Barrangou, 2014). In *S. thermophilus*, the processes of utilizing, expressing, and interfering with CRISPR-Cas systems can be applied to vaccinate bacteria against possible phage threats. Bacteria equipped with endogenous CRISPR-Cas systems can experience repeated phage challenges to obtain spacer sequences that grant immunity (Barrangou et al., 2013). On the other hand, an external CRISPR-Cas system can be integrated into industrial strains to create a synthetic immune system (Donohoue et al., 2018).

10.3.2 CRISPR-Cas System in Genome Engineering

The fabrication of biogas, biological products, organic molecules, and their by-products is heavily influenced by bacterial or fungal microorganism. Therefore, the aptness to meticulously modify their genomes to augment results of desired characteristics is of considerable importance. Pertaining to adequacy, accuracy, and ingenuity, CRISPR, TALENs, ZFNs, and CRISPR-Cas have contributed in part in biogenetic innovations (Gaj et al., 2013). In genetic engineering of commercially significant microbes, the laborious availability of molecular tools promotes the conducive use of CRISPR-Cas instead of engineered nuclease platforms due to its hasty performance, economical benefits, and basic composition. The emergence of infelicitous genetic scars, or commonly referred to as markers, in *Clostridium* and *Streptomyces* spp. and other commercially salient microorganisms is owed to the conventional gene engineering of relying on incompetent homologous recombination. The precise and "scarless" genetic alterations for *Clostridium* spp. that govern the synthesis of ethanol and other associated channels with substantial upside for enhanced biofuel production are possible with recently developed CRISPR-Cas-based methods (Huang et al., 2016; Wang et al., 2016). The integrated genetic modification of biologically active substances comprising of antibiotics and chemotherapy medications formulated from *Streptomyces* bacteria also concerns the CRISPR-Cas9 technique. In recent developments, to elevate the functionality of gene editing and to execute the systematic genetic knock-ins that trigger silent gene clusters involved in biosynthesis and boost metabolic efficiency in *Streptomyces* species, CRISPR-Cas has been utilized profoundly (Zhang et al., 2017). Additionally, CRISPR-based methods are employed to enhance the genetic engineering of technologically essential filamentous fungi, such as *Penicillium chrysogenum*, *Aspergillus* sp., and *Trichoderma reesei* (Liu et al., 2015; Pohl et al., 2016), along with other fungal yeast species, including *Pichia*

TABLE 10.4
Application of CRISPR-Cas in Biotechnology

Microorganism	Product	Type of Toolbox	Advantages	References
Clostridium ljungdahlii	3-hydroxybutyrate (3HB)	CRISPRi	The reduction of pta expression enhances 3HB production by 2.3 times, yielding a titer of 21 mM.	Woolston et al., 2018
Corynebacterium glutamicum	Glutamate	Cas9 nickase (D10A) with activation-induced cytidine deaminase	Increased production by threefold.	Yu et al., 2021
E. coli	Succinate	Orthogonal CRISPR and CRISPRi systems	The titer experienced a 178% growth, starting at 2.5 g/L and advancing to 15 g/L during fermentation.	Sung et al., 2019
Bacillus subtilis	Hyaluronic acid (HA)	CRISPRi	An increase in HA titer of up to 108% was observed at a level of 2.26 g/L, along with a rise in molecular mass.	Westbrook et al., 2018
***Synechocystis* sp.**	n-Butanol	Inducible CRISPRi	Fivefold increase of carbon partitioning to n-butanol relative to a non-repression strain.	Shabestary et al., 2018
E. coli	Uridine	CRISPR-Cas9	A concentration of 70.3 g/L was recorded during the fed-batch fermentation.	Wu et al., 2018
Saccharomyces cerevisiae	2,3-butanediol (BDO)	The development of tRNA-sgRNA operons utilizing modular ligation extension of guide RNA operons (LEGO) technology in CRISPRi	Increased BDO titer twofold.	Deaner et al., 2018

pastoris and *Yarrowia lipolytica*, that are vital for biosynthesis (Weninger et al., 2016). CRISPR-Cas9 system for genetic manipulation has been utilized to engineer four loci in *Myceliophthora thermophila*, a thermophilic biomass-degrading fungus that generates commercially significant thermostable enzymes. This allowed to increase cellulase production and get over the challenges posed by numerous gene editing in this species at the same time (Liu et al., 2017). This work depicted that the CRISPR genome editing tool was promptly versatile for the hyperthermophile fungi; in addition to focusing on industrial biotechnology along the production of chemicals and enzymes, it will concurrently amplify the genetic manipulation of various fungal organisms (Shapiro et al., 2018).

10.3.3 CRISPR-Cas Enhancing Plant–Microbe Interactions for Sustainable Agriculture

Natural interactions exist between organisms (microbes) and their host (plant) which have evolved in tandem. The *microbiota*, also known as *botanical-associated micro-organism*, is an indispensable element of the plant life. Plant–microbe interactions (PMs) may result in beneficial, detrimental, or harmful effects on the health of the host. The critical role of microbiota in supporting plant growth and enhancing stress tolerance is thoroughly documented. The use of multidimensional PM interactions necessitates the employment of modern methods to identify essential genetic and molecular components. A novel technique known as CRISPR-Cas possesses tremendous outcomes to investigate PM interactions (Shelake et al., 2019a). To enhance the potential of microbes or plants for pragmatic trait modification, genetic engineering tools can be employed that will require a thorough comprehension of PM interactions (Shelake et al., 2019a). The process of carbon fixation in soil is governed by PM interactions via modifying the terrestrial carbon cycle (Shelake et al., 2019b). Emphasizing PM interactions, many studies have illustrated the valuable effects of CRISPR-based technologies that promote disease resistance (Zaidi et al., 2018; Chen et al., 2019).

10.3.4 CRISPR-Cas Mechanism for Modified Bacterial Functions and Adaptive Immunity

In the domains of bacteria and archaea, CRISPR-Cas systems operate as an adaptive immune defense, facilitating the recognition and neutralization of invasive MGEs, such as viruses and plasmids. The system is designed to capture fragments of foreign DNA and embed them as spacers within the host's CRISPR array. These spacers provide a form of molecular memory, enabling the detection of previously encountered invaders. The regulation of this process is achieved through the specific binding of CRISPR RNA (crRNA) to particular DNA or RNA sequences. In certain CRISPR systems, there is also a requirement for a surrounding DNA motif, called the protospacer adjacent motif (PAM). The bacteria use a similar basic method of RNA-DNA and/or RNA-RNA complementarity to control two different processes: defending against incoming genetic material and regulating various physiological processes. The creation of biofilms, adhesion, programmed cell death, quorum sensing, and bacterial pathogenicity are the best-documented instances of alternative functions. Although partial complementarity seems to be crucial for certain alternative activities reported, a strong complementarity among the crRNA and the target DNA/RNA appears to serve as an effective mechanism for the protection of phage. Furthermore, the regulation of RNA cleavage can be both sequence-specific and non-specific;

along with the modulation of transcriptional regulator expression, they are mediated by Cas proteins, the exact mechanisms of which are still unknown. In the last ten years, various Cas proteins' auxiliary roles and RNA-guided targeting methods have been developed into potent GE and biotechnological tools. The exploration of various CRISPR forms reveals substantial potential for future technological developments, leading to further progress in biotechnology and a deeper understanding of underlying mechanisms.

10.3.5 Emerging Therapeutic Applications of the CRISPR System

In the modern era, antibiotic resistance, which is made worse by the scarcity of medicines for ailments that are resilient to drugs, is one of the extreme concerns of public well-being. In the absence of novel strategies, antimicrobial resistance is predicted to cost $100 trillion and entail 10 million deaths by 2050. Genetic modifications frequently facilitate the normal aging process that leads to antibiotic resistance (de la Fuente-Nunez & Lu, 2017). Misuse or prescribing of the wrong antibiotic is a major contributory factor to the rise of antimicrobial resistance. Other things, such as inappropriate disease management and poor hygiene, also cause antibiotic resistance. Antibacterial efflux, alteration of a drug interaction site, modification, deactivation of a drug, and restricting the absorption of a drug are all methods through which antimicrobial resistance can prevail (Shukla et al., 2021).

The internal environment of the bacteria is regulated by lowering the antibiotic concentration and removing the toxic compounds from it in the antibiotic efflux (Livermore, 2003). By altering the mechanism by which the drug acts on them, bacteria can alter their desired target site, causing the weakened, altered attachment or zero amount of drug at all (Peterson & Kaur, 2018). Nano-antibiotics (synthesized virus-like nanoparticles), bioengineered synthetic peptides, and engineered bacteriophages are several ways through which antibiotic-resistant bacteria can be dealt with, by utilizing the eubiotic as growth promoter, but the effectiveness and precision of such methods can be implemented and established (Gomaa et al., 2014). The discovery of CRISPR-Cas9 has completely changed the field of gene editing and modulation. Utilizing its power to develop precise division in desired inherited material in bacteria resistant to drugs may apprise this procedure. The advantage of that particular method over others is that it is much more precise and accurate than other techniques. Therefore, a similar single guide RNA (sg-RNA) for a desired gene can be developed, incorporating into the system the ability to cause the death of bacteria (Serajian et al., 2021).

Type I CRISPR-Cas9 technique engineered for *E. coli* was one of the initial models of this system. Due to this, the genome deteriorated with the 3′ to 5′ exonuclease Cas-3. Investigations parallel to the aforementioned study have been executed for the type II CRISPR-Cas system. Initial evaluation of the CRISPR-Cas9 system against antibiotic-resistant strains was carried out in 2014 (Bikard et al., 2014; Citorik et al., 2014), exhibiting results indicating that the engineered CRISPR-Cas9 system can cause caspase-mediated cellular degradation or damage to intracellular organelles due to the exposure of the genetic sequence related to the antibiotic resistance. A study established that maneuvering of the multiple types of microorganisms via the CRISPR-Cas type II system of *Streptococcus pyogenes* is an effective means (Serajian et al., 2021). Detailed studies deduced that type I CRISPR-Cas systems have better ability to cause apoptosis or DNA damage, owing to the system-intrinsic property of exonuclease activity or because of the large-scale DNA. The majority of the bacteria possess this system, but it induces partial death of the bacterial colony; the sole reason for this may be the inappropriate conveyance of the CRISPR-Cas system in the microbe (Trevisan et al., 2017). The CRISPR-Cas system is also utilized to regulate gene transcription. The use of the inactive Cas9 enzyme enables scientists to perform upregulation or downregulation of gene transcription, as it retains its capacity to recognize and bind to the intended DNA sequence. Application of dCas9 for gene repression is denoted by the term CRISPR interference (*CRISPR-i*), whereas it is called CRISPR activation (*CRISPR-a*) when applied for the activation of genes. For this purpose, the enzyme of Cas9 is nurtured at the desired DNA sequence, which damages the transcription factor binding or the RNA polymerase, which differentiate from its typical method, which is DNA cleavage (Strich & Chertow, 2018).

10.3.6 CRISPR-Cas and Its Application in Parasitology

As per the WHO report, infestation caused by parasites continues to be one of the biggest causes of mortality globally (Torgerson et al., 2015). The lack of a vaccine, drug inefficiency, and antimicrobic resistance highlight the requirement for new parasite-based treatment selections, including the use of new tools like CRISPR (Hewitson & Maizels, 2014; McVeigh & Maule, 2019). Over the last several decades, extraordinary efforts have been made to sequence the genomes of over 150 parasites (Howe et al., 2017). Genome sequencing aids in the dissection and identification of the genome's function. CRISPR technology enables scientists to modify genes and introduce new patterns into the genome, as demonstrated in the schistosome genome (Beckmann & Grevelding 2012).

The use of a genetic modification system has evolved as a potential therapy for parasitic infections, with the ability to modify host genetic material required by the parasite for replication (Kirti et al., 2021; Bryant et al., 2019). In the genome, to induce a DSB at a particular position, this system uses Cas9 endonuclease. When Cas9 is attached to a sg-RNA, it becomes specific. The DSB will then be fixed by one of three pathways, including homology-directed repair (HDR), non-homologous end joining (NHEJ), and error-prone microhomology-mediated end joining (MMEJ). The CRISPR-Cas9 system facilitates the targeted insertion, deletion, or modification of DNA with minimal residual genetic impact (Bryant et al., 2019).

This method was implemented to modify the genomes of parasites that cause an increase in death rates, like *Plasmodium* spp., *Toxoplasma gondii*, and *Cryptosporidium* spp. ZFN has been used extensively in *Plasmodium falciparum* genome modification; some limitations to its use involve focusing capacity, expense, hard design, and application (Singer et al., 2015). Ghorbal and Wagner published an adaptation of the CRISPR-Cas9 method in 2014, allowing for the alteration of *Plasmodium falciparum* using diverse methods to generate sgRNA (Ghorbal et al., 2014; Wagner et al., 2014). Janssen et al. (2018) used CRISPR-Cas9 techniques to change the genetic material of *Trichomonas vaginalis*, one of the most prevalent venereal diseases, which affects approximately 250 million people worldwide. In this study, two endogenous genes were deleted by substituting the drug obstruction genes cassette (Janssen et al., 2018).

10.3.7 Advancements of CRISPR-Cas in Micro-Biotechnology

Currently, bacteria-specific CRISPR-Cas tools face difficulties, like Cas9's high off-target rate, Cas-12a's poor cleavage activity, and insufficient creation of endogenous systems. As a result, considering the challenges and obstacles, certain viewpoints in the subsequent areas can be provided. To promote precision, investigators altered the Cas9 protein by overcoming the positively charged residues with the neutral amino acids and obtained eSp-Cas9 (K810A/K1003A/R1060A or K848A/K1003A/R1060A), in addition to an entirely new hyperaccurate Cas9 variation (N692A/M694A/Q695A/H698A, called HypaCas9) that has no impact on the main targeted activity. Moreover, to increase cutting action, the investigators focused on CRISPR-Cas-12a platforms, specifically AsCas-12a and Lb-Cas-12a, and started an HDV-ribozyme to the 3' end of the cr-RNA. This raised editing effectiveness from 1.1 to 5.2 times. To broaden the modification repertoire, scientists optimized existing Cas proteins like AsCas-12a (enAsCas-12a) and FnCas-12a, recognizing a wide range of PAM sequences. Furthermore, scientists discovered novel Cas-X molecules of protein, including Casl2b and Casl2e. Casl2b showed less or non-specific effects in multicellular species, along with a broader PAM specificity. To cleave the targeted genes, Cas-X is an innovative or RNA-directed DNA endonuclease within *E. coli.* Bacteria have the most common single–base editor CBE, which introduces a stop codon to terminate gene expression and inactivate protein function. This method greatly streamlines useful gene determination and metabolic base engineering studies. BE-PACE, or phage-assisted ongoing development of the base editor, was developed to improve modifying effectiveness and gene target integration without GC target restriction. It should be emphasized that the electrical current CBE structures employed in bacteria tend to have a significant off-target rate. Future advancements in the design of sgRNAs or the implementation of protein engineering could help reduce this rate (Liu et al., 2020b).

10.4 PHYSIOLOGICAL ROLES OF CRISPR-CAS SYSTEMS

10.4.1 Physiological Role of the CRISPR-Cas System in Bacterial Virulence, Adherence, Cytotoxicity, and Group Behavior

Various pathogenic bacteria that affect humans, plants, and animals have a CRISPR-Cas system which controls pathogenicity via endogenous gene regulation. As the Cas3 gene from the type I-E system of *Salmonella enterica* subsp. *Enterica* serovar Enteritidis is deleted, its ability to infect murine, human cell line, and chicken after oral incubation is also minimized (Mohanraju et al., 2022). *Francisella novicida* was the first direct and well-documented evidence of virulence and pathogenicity regulation. It possesses the mechanism that alters the human immune response by moderating gene expression via the CRISPR-Cas system. During phagocytosis, *Francisella novicida* enters the macrophage and forms phagosomes that have numerous receptors for the recognition of pathogens. One of those receptors is TLR2, which identifies the bacterial lipoprotein and initiates the host immune response. *Francisella novicida* uses Cas9, tracer RNA, and small CRISPR-associated RNA (scaRNA) to lower the TLR2 activation and lower the immune response. *Neisseria meningitidis* also uses Cas9 to infect the lung epithelial cells of the host (Wu et al., 2022).

The CRISPR-Cas system affects bacterial adherence and cytotoxicity by modulating bacterial–host interactions, which are still under investigation. Pathogenic strains of *Neisseria*, *Streptococcus*, and *Campylobacter* Cas9 poorly adhere to mammalian host cells (Heidrich et al., 2019). *Campylobacter jejuni* and *Streptococcus agalactiae* Cas9 can enhance bacterial cytotoxic levels along with adherence and existence in host cells (Saha et al., 2020b). Research proves that the CRISPR system can also boost bacterial pathogenicity by regulating group behaviors. A study on the transcriptome of Cas3 *S. mutans* UA159 (type I-C) revealed that a reduction of biofilm-forming genes can be controlled by the VicR/K TCS that further influences the tissue specificity of *streptococcal* bacteria (Tang et al., 2019). In *Pseudomonas aeruginosa*, crRNA hybridization also controlled its biofilm formation and motility (Heussler et al., 2015). Other studies showed Cas9 decreased biofilm formation among pathogenic group behaviors, including *Campylobacter jejuni* (type II-C) (Shabbir et al., 2018), *Streptococcus pyogenes* (type II-A) (Gao et al., 2019), *Neisseria meningitidis* (type II-C), and *Streptococcus agalactiae* (type II-A) (Ma et al., 2018).

10.4.2 Relationship of CRISPR-Cas System in Terms of Genome Repair, Non-Virulent Gene Regulation, and Programmed Cell Death

Recent studies show that the CRISPR-Cas system also monitors the genes that are not a direct part of bacterial virulence. Experimental data has shown Casl in *E. Coli* can cleave branched DNA substrates and is more sensitive to DNA damage. In a recent study, it was found that type II-A systems and NHEJ systems cannot coexist due to competition for DNA substrates between NHEJ and type II-A proteins, specifically Csn2. It was also discovered that RecBCD, a different repair system of DNA, is essential for *E. coli* type I-E spacer acquisition (Levy et al., 2015). Such findings indicate a complicated interconnection of CRISPR with DNA repair systems, but the mechanisms are still difficult to find (Newsom et al., 2021).

10.5 CONCLUSION

CRISPR-Cas systems have increased in appeal in the past decade due to their adaptability and effectiveness compared to conventional defense systems. Bacterial applications have been used to include simplified genetic technology, programmed transcriptional control, strain writing, and the field of epidemiology via bacteria vaccination towards genetic components that are mobile, and smart antibiotics, though these methods need broad gene editing of bacteria, which is contingent on the accessibility of reliable genetic modification parameters. CRISPR-Cas systems have developed a new set of standards for genetically modifying bacteria. Cas9 and Casl2a amino acids have been established into effective tools for studying bacterial inherited processes, improving commercial bacteria metabolic processes, and doing other genetic modifications. The majority of these programs are still in their early stages, so there is plenty of room for improvement. Nonetheless, the advances made thus far have begun to reshape how we conduct both applied and basic studies in bacteria, with significant implications.

REFERENCES

Ahmad, Shafiq, Shujait Ali, Asghar Abbas, Rao Zahid Abbas, Muhammad Zubair, Muhammad Asif Raza, Naheed Bano, Muhammd Usman Shafi, Syed Nadeem Badar, and Kashif J Hussain. 2019. "Effects of dietary supplementation of linseed oil (Omega-3) on quality parameters of Nili Ravi bull spermatozoa." *Livestock Science* 224: 57–59.

Allemailem, Khaled S, Mohammed A Alsahli, Ahmad Almatroudi, Faris Alrumaihi, Fahd Khaleefah Alkhaleefah, Arshad Husain Rahmani, and Amjad Ali J Khan. 2022. "Current updates of CRISPR/Cas9-mediated genome editing and targeting within tumor cells: An innovative strategy of cancer management." *Cancer Communications* 42 (12): 1257–1287.

Barrangou, Rodolphe, Anne-Claire Coûté-Monvoisin, Buffy Stahl, Isabelle Chavichvily, Florian Damange, Dennis A Romero, Patrick Boyaval, Christophe Fremaux, and Philippe Horvath. 2013. "Genomic impact of CRISPR immunization against bacteriophages." *Biochemical Society Transactions* 41 (6): 1383–1391.

Barrangou, Rodolphe, Christophe Fremaux, Hélène Deveau, Melissa Richards, Patrick Boyaval, Sylvain Moineau, Dennis A Romero, and Philippe Horvath. 2007. "CRISPR provides acquired resistance against viruses in prokaryotes." *Science* 315 (5819): 1709–1712.

Barrangou, Rodolphe, and Jan-Peter van Pijkeren. 2016. "Exploiting CRISPR–Cas immune systems for genome editing in bacteria." *Current Opinion in Biotechnology* 37: 61–68.

Barrangou, Rodolphe J. 2015. "The roles of CRISPR–Cas systems in adaptive immunity and beyond." *Current Opinion in Immunology* 32: 36–41.

Beckmann, S, and CG Grevelding. 2012. "Paving the way for transgenic schistosomes." *Parasitology* 139 (5): 651–668.

Bikard, David, Chad W Euler, Wenyan Jiang, Philip M Nussenzweig, Gregory W Goldberg, Xavier Duportet, Vincent A Fischetti, and Luciano A Marraffini. 2014. "Exploiting CRISPR-Cas nucleases to produce sequence-specific antimicrobials." *Nature Biotechnology* 32 (11): 1146–1150.

Briner, Alexandra E, and Rodolphe Barrangou. 2014. "Lactobacillus buchneri genotyping on the basis of clustered regularly interspaced short palindromic repeat (CRISPR) locus diversity." *Applied and Environmental Microbiology* 80 (3): 994–1001.

Bryant, Jessica M, Sebastian Baumgarten, Lucy Glover, Sebastian Hutchinson, and Najma Rachidi. 2019. "CRISPR in parasitology: Not exactly cut and dried!" *Trends in Parasitology* 35 (6): 409–422.

Chaudhuri, Abira, Koushik Halder, and Asis Datta. 2022. "Classification of CRISPR/Cas system and its application in tomato breeding." *Theoretical and Applied Genetics* 135 (2): 367–387.

Chen, Baohui, Juan Guan, and Bo Huang. 2016. "Imaging specific genomic DNA in living cells." *Annual Review of Biophysics* 45 (1): 1–23.

Chen, Kunling, Yanpeng Wang, Rui Zhang, Huawei Zhang, and Caixia Gao. 2019. "CRISPR/Cas genome editing and precision plant breeding in agriculture." *Annual Review of Plant Biology* 70 (1): 667–697.

Citorik, Robert J, Mark Mimee, and Timothy K Lu. 2014. "Sequence-specific antimicrobials using efficiently delivered RNA-guided nucleases." *Nature Biotechnology* 32 (11): 1141–1145.

Cui, Lun, and David Bikard. 2016. "Consequences of Cas9 cleavage in the chromosome of Escherichia coli." *Nucleic Acids Research* 44 (9): 4243–4251.

David, Florian, and Verena Siewers. 2015. "Advances in yeast genome engineering." *FEMS Yeast Research* 15 (1): 1–14.

de la Fuente-Nunez, César, and Timothy K Lu. 2017. "CRISPR-Cas9 technology: Applications in genome engineering, development of sequence-specific antimicrobials, and future prospects." *Integrative Biology* 9 (2): 109–122.

Deaner, Matthew, Allison Holzman, and Hal S Alper. 2018. "Modular ligation extension of guide RNA operons (LEGO) for multiplexed dCas9 regulation of metabolic pathways in Saccharomyces cerevisiae." *Biotechnology Journal* 13 (9): 1700582.

Donohoue, Paul D, Rodolphe Barrangou, and Andrew P May. 2018. "Advances in industrial biotechnology using CRISPR-Cas systems." *Trends in Biotechnology* 36 (2): 134–146.

Erickson, Keesha E, Peter B Otoupal, and Anushree Chatterjee. 2017. "Transcriptome-level signatures in gene expression and gene expression variability during bacterial adaptive evolution." *Msphere* 2 (1).

Gaj, Thomas, Charles A Gersbach, and Carlos F Barbas. 2013. "ZFN, TALEN, and CRISPR/Cas-based methods for genome engineering." *Trends in Biotechnology* 31 (7): 397–405.

Gao, Nina J, Mahmoud M Al-Bassam, Saugat Poudel, Jacob M Wozniak, David J Gonzalez, Joshua Olson, Karsten Zengler, Victor Nizet, and J Andrés Valderrama. 2019. "Functional and proteomic analysis of Streptococcus pyogenes virulence upon loss of its native Cas9 nuclease." *Frontiers in Microbiology* 10: 1967.

Ghorbal, Mehdi, Molly Gorman, Cameron Ross Macpherson, Rafael Miyazawa Martins, Artur Scherf, and Jose-Juan Lopez-Rubio. 2014. "Genome editing in the human malaria parasite Plasmodium falciparum using the CRISPR-Cas9 system." *Nature Biotechnology* 32 (8): 819–821.

Gomaa, Ahmed A, Heidi E. Klumpe, Michelle L. Luo, Kurt Selle, Rodolphe Barrangou, and Chase L. Beisel. 2014. "Programmable removal of bacterial strains by use of genome-targeting CRISPR-Cas systems." *mBio* 5 (1). https://doi.org/10.1128/mbio.00928-13. https://journals.asm.org/doi/abs/10.1128/mbio.00928-13.

Grissa, Ibtissem, Gilles Vergnaud, and Christine J Pourcel. 2007. "The CRISPRdb database and tools to display CRISPRs and to generate dictionaries of spacers and repeats." *BMC Bioinformatics* 8: 1–10.

Hartung, Frank, and Joachim Schiemann. 2014. "Precise plant breeding using new genome editing techniques: Opportunities, safety and regulation in the EU." *The Plant Journal* 78 (5): 742–752.

Haurwitz, Rachel E, Martin Jinek, Blake Wiedenheft, Kaihong Zhou, and Jennifer A Doudna. 2010. "Sequence-and structure-specific RNA processing by a CRISPR endonuclease." *Science* 329 (5997): 1355–1358.

Heidrich, Nadja, Antony Hagmann, Saskia Bauriedl, Jörg Vogel, and Christoph Schoen. 2019. "The CRISPR/Cas system in Neisseria meningitidis affects bacterial adhesion to human nasopharyngeal epithelial cells." *RNA Biology* 16 (4): 390–396.

Heussler, Gary E, Kyle C Cady, Katja Koeppen, Sabin Bhuju, Bruce A Stanton, and George A O'Toole. 2015. "Clustered regularly interspaced short palindromic repeat-dependent, biofilm-specific death of Pseudomonas aeruginosa mediated by increased expression of phage-related genes." *MBio* 6 (3).

Hewitson, James P, and Rick M Maizels. 2014. "Vaccination against helminth parasite infections." *Expert Review of Vaccines* 13 (4): 473–487.

Hille, Frank, and Emmanuelle Charpentier. 2016. "CRISPR-Cas: Biology, mechanisms and relevance." *Philosophical Transactions of the Royal Society B: Biological Sciences* 371 (1707): 20150496.

Hille, Frank, Hagen Richter, Shi Pey Wong, Majda Bratovič, Sarah Ressel, and Emmanuelle Charpentier. 2018. "The biology of CRISPR-Cas: Backward and forward." *Cell* 172 (6): 1239–1259.

Howe, Kevin L, Bruce J Bolt, Myriam Shafie, Paul Kersey, and Matthew Berriman. 2017. "WormBase ParaSite–a comprehensive resource for helminth genomics." *Molecular and Biochemical Parasitology* 215: 2–10.

Hsu, Patrick D, Eric S Lander, and Feng Zhang. 2014. "Development and applications of CRISPR-Cas9 for genome engineering." *Cell* 157 (6): 1262–1278.

Huang, He, Changsheng Chai, Ning Li, Pete Rowe, Nigel P Minton, Sheng Yang, Weihong Jiang, and Yang Gu. 2016. "CRISPR/Cas9-based efficient genome editing in Clostridium ljungdahlii, an autotrophic gas-fermenting bacterium." *ACS Synthetic Biology* 5 (12): 1355–1361.

Hurdle, Julian G, Alex J O'neill, Ian Chopra, and Richard E Lee. 2011. "Targeting bacterial membrane function: An underexploited mechanism for treating persistent infections." *Nature Reviews Microbiology* 9 (1): 62–75.

Huynh, Nhan, Noah Depner, Raegan Larson, and Kirst King-Jones. 2020. "A versatile toolkit for CRISPR-Cas13-based RNA manipulation in Drosophila." *Genome Biology* 21 (1): 279.

Ishino, Yoshizumi, Hideo Shinagawa, Kozo Makino, Mitsuko Amemura, and Atsuo Nakata. 1987. "Nucleotide sequence of the iap gene, responsible for alkaline phosphatase isozyme conversion in Escherichia coli, and identification of the gene product." *Journal of Bacteriology* 169 (12): 5429–5433.

Janssen, Brian D, Yi-Pei Chen, Brenda M Molgora, Shuqi E Wang, Augusto Simoes-Barbosa, and Patricia J Johnson. 2018. "CRISPR/Cas9-mediated gene modification and gene knock out in the human-infective parasite Trichomonas vaginalis." *Scientific Reports* 8 (1): 270.

Javed, Muhammad R, Maria Sadaf, Temoor Ahmed, Amna Jamil, Marium Nawaz, Hira Abbas, and Anam Ijaz. 2018. "CRISPR-Cas system: History and prospects as a genome editing tool in microorganisms." *Current Microbiology* 75: 1675–1683.

Jiang, Fuguo, and Jennifer A Doudna. 2017. "CRISPR–Cas9 structures and mechanisms." *Annual Review of Biophysics* 46 (1): 505–529.

Kang, Sangrim, Jinsoo Kim, Junho K Hur, and Sang-Seob Lee. 2017. "CRISPR-based genome editing of clinically important Escherichia coli SE15 isolated from indwelling urinary catheters of patients." *Journal of Medical Microbiology* 66 (1): 18–25.

Kim, Hyun Young, Rachel Yoon Kyung Chang, Sandra Morales, and Hak-Kim Chan. 2021. "Bacteriophage-delivering hydrogels: Current progress in combating antibiotic resistant bacterial infection." *Antibiotics* 10 (2): 130.

Kirti, Apurva, Manish Sharma, Komal Rani, and Abhisheka Bansal. 2021. "CRISPRing protozoan parasites to better understand the biology of diseases." *Progress in Molecular Biology and Translational Science* 180: 21–68.

Koch, Gudrun, Ana Yepes, Konrad U Förstner, Charlotte Wermser, Stephanie T Stengel, Jennifer Modamio, Knut Ohlsen, Kevin R Foster, and Daniel Lopez. 2014. "Evolution of resistance to a last-resort antibiotic in Staphylococcus aureus via bacterial competition." *Cell* 158 (5): 1060–1071.

Koonin, Eugene V, and Kira S Makarova. 2019. "Origins and evolution of CRISPR-Cas systems." *Philosophical Transactions of the Royal Society B* 374 (1772): 20180087.

Koonin, Eugene V, Kira S Makarova, and Feng Zhang. 2017. "Diversity, classification and evolution of CRISPR-Cas systems." *Current Opinion in Microbiology* 37: 67–78.

Levy, Asaf, Moran G Goren, Ido Yosef, Oren Auster, Miriam Manor, Gil Amitai, Rotem Edgar, Udi Qimron, and Rotem Sorek. 2015. "CRISPR adaptation biases explain preference for acquisition of foreign DNA." *Nature* 520 (7548): 505–510.

Liu, Hualan, David S Robinson, Zong-Yen Wu, Rita Kuo, Yasuo Yoshikuni, Ian K Blaby, and Jan-Fang Cheng. 2020a. "Bacterial genome editing by coupling Cre-lox and CRISPR-Cas9 systems." *PLoS One* 15 (11): E0241867.

Liu, Qian, Ranran Gao, Jingen Li, Liangcai Lin, Junqi Zhao, Wenliang Sun, and Chaoguang Tian. 2017. "Development of a genome-editing CRISPR/Cas9 system in thermophilic fungal Myceliophthora species and its application to hyper-cellulase production strain engineering." *Biotechnology for Biofuels* 10: 1–14.

Liu, Rui, Ling Chen, Yanping Jiang, Zhihua Zhou, and Gen Zou. 2015. "Efficient genome editing in filamentous fungus Trichoderma reesei using the CRISPR/Cas9 system." *Cell Discovery* 1 (1): 1–11.

Liu, Zhenquan, Huina Dong, Yali Cui, Lina Cong, and Dawei Zhang. 2020b. "Application of different types of CRISPR/Cas-based systems in bacteria." *Microbial Cell Factories* 19: 1–14.

Livermore, David M. 2003. "Linezolid in vitro: Mechanism and antibacterial spectrum." *Journal of Antimicrobial Chemotherapy* 51 (suppl_2): ii9–ii16.

Ma, Ke, Qing Cao, Su Luo, Zhaofei Wang, Guangjin Liu, Chengping Lu, and Yongjie Liu. 2018. "Cas9 enhances bacterial virulence by repressing the regR transcriptional regulator in Streptococcus agalactiae." *Infection and Immunity* 86 (3).

Makarova, Kira S, Daniel H Haft, Rodolphe Barrangou, Stan JJ Brouns, Emmanuelle Charpentier, Philippe Horvath, Sylvain Moineau, Francisco JM Mojica, Yuri I Wolf, and Alexander F Yakunin. 2011. "Evolution and classification of the CRISPR–Cas systems." *Nature Reviews Microbiology* 9 (6): 467–477.

Makarova, Kira S, Yuri I Wolf, Omer S Alkhnbashi, Fabrizio Costa, Shiraz A Shah, Sita J Saunders, Rodolphe Barrangou, Stan JJ Brouns, Emmanuelle Charpentier, and Daniel H Haft. 2015. "An updated evolutionary classification of CRISPR–Cas systems." *Nature Reviews Microbiology* 13 (11): 722–736.

Makarova, Kira S, Yuri I Wolf, Jaime Iranzo, Sergey A Shmakov, Omer S Alkhnbashi, Stan JJ Brouns, Emmanuelle Charpentier, David Cheng, Daniel H Haft, and Philippe Horvath. 2020. "Evolutionary classification of CRISPR–Cas systems: A burst of class 2 and derived variants." *Nature Reviews Microbiology* 18 (2): 67–83.

Makarova, Kira S, Yuri I Wolf, and Eugene V Koonin. 2018. "Classification and nomenclature of CRISPR-Cas systems: Where from here?" *The CRISPR Journal* 1 (5): 325–336.

McVeigh, Paul, and Aaron G Maule. 2019. "Can CRISPR help in the fight against parasitic worms?" *Elife* 8: e44382.

Mohanraju, Prarthana, Chinmoy Saha, Peter van Baarlen, Rogier Louwen, Raymond HJ Staals, and John van der Oost. 2022. "Alternative functions of CRISPR–Cas systems in the evolutionary arms race." *Nature Reviews Microbiology* 20 (6): 351–364.

Mojica, Francisco JM, Cesar Díez-Villaseñor, Jesús García-Martínez, and Elena Soria. 2005. "Intervening sequences of regularly spaced prokaryotic repeats derive from foreign genetic elements." *Journal of Molecular Evolution* 60: 174–182.

Mojica, Francisco JM, Cesar Díez-Villaseñor, Elena Soria, and Guadalupe Juez. 2000. "Biological significance of a family of regularly spaced repeats in the genomes of Archaea, Bacteria and mitochondria." *Molecular Microbiology* 36 (1): 244–246.

Mukherjee, Suprabhat, Subhajit Karmakar, and Santi Prasad Sinha Babu. 2016. "TLR2 and TLR4 mediated host immune responses in major infectious diseases: A review." *Brazilian Journal of Infectious Diseases* 20 (2): 193–204.

Newsom, Sydney, Hari Priya Parameshwaran, Lindsie Martin, and Rakhi Rajan. 2021. "The CRISPR-Cas mechanism for adaptive immunity and alternate bacterial functions fuels diverse biotechnologies." *Frontiers in Cellular and Infection Microbiology*10: 619763.

Nidhi, Sweta, Uttpal Anand, Patrik Oleksak, Pooja Tripathi, Jonathan A Lal, George Thomas, Kamil Kuca, and Vijay Tripathi. 2021. "Novel CRISPR–Cas systems: An updated review of the current achievements, applications, and future research perspectives." *International Journal of Molecular Sciences* 22 (7): 3327.

Nussenzweig, Philip M, and Luciano A Marraffini. 2020. "Molecular mechanisms of CRISPR-Cas immunity in bacteria." *Annual Review of Genetics* 54 (1): 93–120.

Park, Joo Youn, Bo Youn Moon, Juw Won Park, Justin A Thornton, Yong Ho Park, and Keun Seok Seo. 2017. "Genetic engineering of a temperate phage-based delivery system for CRISPR/Cas9 antimicrobials against Staphylococcus aureus." *Scientific Reports* 7 (1): 44929.

Peterson, Elizabeth, and Parjit Kaur. 2018. "Antibiotic resistance mechanisms in bacteria: Relationships between resistance determinants of antibiotic producers, environmental bacteria, and clinical pathogens." *Frontiers in Microbiology* 9: 2928.

Pohl, Carsten, Jan AKW Kiel, Arnold JM Driessen, Roel AL Bovenberg, and Y Nygard. 2016. "CRISPR/Cas9 based genome editing of Penicillium chrysogenum." *ACS Synthetic Biology* 5 (7): 754–764.

Qaisar, Hafiz Muhammad Umer, Tanveer Ahmad, Muhammad Rizwan, and Muhammad Saqib. 2017. "Antimicrobial efficacy of combination of lincomycin and spiramycin (Lispiracin™) as systemic dry cow therapy for controlling bovine mastitis." *Punjab University Journal of Zoology* 32 (2): 197–201.

Ram, Geeta, Hope F Ross, Richard P Novick, Ivelisse Rodriguez-Pagan, and Dunrong Jiang. 2018. "Conversion of staphylococcal pathogenicity islands to CRISPR-carrying antibacterial agents that cure infections in mice." *Nature Biotechnology* 36 (10): 971–976.

Rath, Devashish, Lina Amlinger, Archana Rath, and Magnus Lundgren. 2015. "The CRISPR-Cas immune system: Biology, mechanisms and applications." *Biochimie* 117: 119–128.

Reuter, Audrey, Cécile Hilpert, Annick Dedieu-Berne, Sophie Lematre, Erwan Gueguen, Guillaume Launay, Sarah Bigot, and Christian Lesterlin. 2021. "Targeted-antibacterial-plasmids (TAPs) combining conjugation and CRISPR/Cas systems achieve strain-specific antibacterial activity." *Nucleic Acids Research* 49 (6): 3584–3598.

Rizwan, M, AZ Durrani, M. Ijaz, M. Kashif, and S Firyal. 2016. "Clinio-bacterialogical investigation of sub-clinical and clinical mastitis in dairy goats." *Veterinaria* 4 (1): 4–6.

Rizwana, Muhammad, Muhammad Alia, Osman Ahmad Khanb, Tanveer Ahmedc, and Aneela Zameer Durranid. 2016. "Swayback disease in ruminants." *Veterinari* 4 (2): 20–24.

Saha, Chinmoy, Deborah Horst-Kreft, Inez Kross, Peter J van der Spek, Rogier Louwen, and Peter van Baarlen. 2020a. "Campylobacter jejuni Cas9 modulates the transcriptome in Caco-2 intestinal epithelial cells." *Genes* 11 (10): 1193.

Saha, Chinmoy, Prarthana Mohanraju, Andrew Stubbs, Gaurav Dugar, Youri Hoogstrate, Gert-Jan Kremers, Wiggert A Van Cappellen, Deborah Horst-Kreft, Charlie Laffeber, and Joyce HG Lebbink. 2020b. "Guide-free Cas9 from pathogenic Campylobacter jejuni bacteria causes severe damage to DNA." *Science Advances* 6 (25): eaaz4849.

Schulze, Sabrina, and Michael Lammers. 2020. "The development of genome editing tools as powerful techniques with versatile applications in biotechnology and medicine: CRISPR/Cas9, ZnF and TALE nucleases, RNA interference, and Cre/loxP." *ChemTexts* 7 (1): 3.

Selle, Kurt, Todd R Klaenhammer, and Rodolphe Barrangou. 2015. "CRISPR-based screening of genomic island excision events in bacteria." *Proceedings of the National Academy of Sciences* 112 (26): 8076–8081.

Serajian, Sahar, Ehsan Ahmadpour, Sonia M Rodrigues Oliveira, Maria de Lourdes Pereira, and Siamak Heidarzadeh. 2021. "CRISPR-Cas technology: Emerging applications in clinical microbiology and infectious diseases." *Pharmaceuticals* 14 (11): 1171.

Shabbir, Muhammad AB, Yanping Tang, Zihui Xu, Mingyue Lin, Guyue Cheng, Menghong Dai, Xu Wang, Zhengli Liu, Zonghui Yuan, and Haihong Hao. 2018. "The involvement of the Cas9 gene in virulence of Campylobacter jejuni." *Frontiers in Cellular and Infection Microbiology* 8: 285.

Shabbir, Muhammad Abu Bakr, Haihong Hao, Muhammad Zubair Shabbir, Hafiz Iftikhar Hussain, Zahid Iqbal, Saeed Ahmed, Adeel Sattar, Mujahid Iqbal, Jun Li, and Zonghui Yuan. 2016. "Survival and evolution of CRiSPR–Cas system in prokaryotes and its applications." *Frontiers in Immunology* 7: 219697.

Shabbir, Muhammad Abu Bakr, Muhammad Zubair Shabbir, Qin Wu, Sammina Mahmood, Abdul Sajid, Muhammad Kashif Maan, Saeed Ahmed, Umer Naveed, Haihong Hao, and Zonghui Yuan. 2019. "CRISPR-cas system: Biological function in microbes and its use to treat antimicrobial resistant pathogens." *Annals of Clinical Microbiology and Antimicrobials* 18: 1–9.

Shabestary, Kiyan, Josefine Anfelt, Emil Ljungqvist, Michael Jahn, Lun Yao, and Elton P Hudson. 2018. "Targeted repression of essential genes to arrest growth and increase carbon partitioning and biofuel titers in cyanobacteria." *ACS Synthetic Biology* 7 (7): 1669–1675.

Shapiro, Rebecca S, Alejandro Chavez, and James J Collins. 2018. "CRISPR-based genomic tools for the manipulation of genetically intractable microorganisms." *Nature Reviews Microbiology* 16 (6): 333–339.

Shelake, Rahul Mahadev, Dibyajyoti Pramanik, and Jae-Yean Kim. 2019a. "Exploration of plant-microbe interactions for sustainable agriculture in CRISPR era." *Microorganisms* 7 (8): 269.

Shelake, Rahul Mahadev, Rajesh Ramdas Waghunde, Pankaj Prakash Verma, Chandrakant Singh, and Jae-Yean Kim. 2019b. "Carbon sequestration for soil fertility management: Microbiological perspective." *Soil Fertility Management for Sustainable Development* 25–42.

Shmakova, AA, OP Shmakova, AA Karpukhina, and YS Vassetzky. 2022. "CRISPR/Cas: History and perspectives." *Russian Journal of Developmental Biology* 53 (4): 272–282.

Shukla, Arpit, Nistha Jani, Monika Polra, Anushree Kamath, and Dhara Patel. 2021. "CRISPR: The multidrug resistance endgame?" *Molecular Biotechnology* 63 (8): 676–685.

Singer, Mirko, Jennifer Marshall, Kirsten Heiss, Gunnar R Mair, Dirk Grimm, Ann-Kristin Mueller, and Friedrich Frischknecht. 2015. "Zinc finger nuclease-based double-strand breaks attenuate malaria parasites and reveal rare microhomology-mediated end joining." *Genome Biology* 16: 1–18.

Strich, Jeffrey R, and Daniel S Chertow. 2018. "CRISPR-Cas Biology and Infectious Diseases Applications." *Journal of Clinical Microbiology* 01307–01318.

Sung, Li-Yu, Meng-Ying Wu, Mei-Wei Lin, Mu-Nung Hsu, Vu Anh Truong, Chih-Che Shen, Yi Tu, Kuen-Yuan Hwang, An-Pang Tu, and Yu-Han Chang. 2019. "Combining orthogonal CRISPR and CRISPRi systems for genome engineering and metabolic pathway modulation in Escherichia coli." *Biotechnology and Bioengineering* 116 (5): 1066–1079.

Tang, Boyu, Tao Gong, Xuedong Zhou, Miao Lu, Jumei Zeng, Xian Peng, Shida Wang, and Yuqing Li. 2019. "Deletion of cas3 gene in Streptococcus mutans affects biofilm formation and increases fluoride sensitivity." *Archives of Oral Biology* 99: 190–197.

Torgerson, Paul R, Brecht Devleesschauwer, Nicolas Praet, Niko Speybroeck, Arve Lee Willingham, Fumiko Kasuga, Mohammad B Rokni, Xiao-Nong Zhou, Eric M Fèvre, and Banchob Sripa. 2015. "World Health Organization estimates of the global and regional disease burden of 11 foodborne parasitic diseases, 2010: A data synthesis." *PLoS Medicine* 12 (12): e1001920.

Trevisan, Marta, Giorgio Palù, and Luisa Barzon. 2017. "Genome editing technologies to fight infectious diseases." *Expert Review of Anti-infective Therapy* 15 (11): 1001–1013.

Vercoe, Reuben B, James T Chang, Ron L Dy, Corinda Taylor, Tamzin Gristwood, James S Clulow, Corinna Richter, Rita Przybilski, Andrew R Pitman, and Peter C Fineran. 2013. "Cytotoxic chromosomal targeting by CRISPR/Cas systems can reshape bacterial genomes and expel or remodel pathogenicity islands." *PLoS Genetics* 9 (4): e1003454.

Wagner, Jeffrey C, Randall J Platt, Stephen J Goldfless, Feng Zhang, and Jacquin C Niles. 2014. "Efficient CRISPR-Cas9–mediated genome editing in Plasmodium falciparum." *Nature Methods* 11 (9): 915–918.

Wan, Peng, Shiyun Cui, Zhenbao Ma, Lin Chen, Xiaoshen Li, Ruonan Zhao, Wenguang Xiong, and Zhenling Zeng. 2020. "Reversal of mcr-1-mediated colistin resistance in Escherichia coli by CRISPR-Cas9 system." *Infection and Drug Resistance* 1171–1178.

Wang, Pengxia, Dongmei He, Baiyuan Li, Yunxue Guo, Weiquan Wang, Xiongjian Luo, Xuanyu Zhao, and Xiaoxue Wang. 2019. "Eliminating mcr-1-harbouring plasmids in clinical isolates using the CRISPR/Cas9 system." *Journal of Antimicrobial Chemotherapy* 74 (9): 2559–2565.

Wang, Yi, Zhong-Tian Zhang, Seung-Oh Seo, Patrick Lynn, Ting Lu, Yong-Su Jin, and Hans P Blaschek. 2016. "Bacterial genome editing with CRISPR-Cas9: Deletion, integration, single nucleotide modification, and desirable "clean" mutant selection in Clostridium beijerinckii as an example." *ACS Synthetic Biology* 5 (7): 721–732.

Weninger, Astrid, Anna-Maria Hatzl, Christian Schmid, Thomas Vogl, and Anton Glieder. 2016. "Combinatorial optimization of CRISPR/Cas9 expression enables precision genome engineering in the methylotrophic yeast Pichia pastoris." *Journal of Biotechnology* 235: 139–149.

Westbrook, Adam W, Xiang Ren, Jaewon Oh, Murray Moo-Young, and C Perry Chou. 2018. "Metabolic engineering to enhance heterologous production of hyaluronic acid in Bacillus subtilis." *Metabolic Engineering* 47: 401–413.

Westra, Edze R, and Bruce R Levin. 2020. "It is unclear how important CRISPR-Cas systems are for protecting natural populations of bacteria against infections by mobile genetic elements." *Proceedings of the National Academy of Sciences* 117 (45): 27777–27785.

Woolston, Benjamin M, David F Emerson, Devin H Currie, and Gregory Stephanopoulos. 2018. "Rediverting carbon flux in Clostridium ljungdahlii using CRISPR interference (CRISPRi)." *Metabolic Engineering* 48: 243–253.

Wu, Heyun, Yanjun Li, Qian Ma, Qiang Li, Zifan Jia, Bo Yang, Qingyang Xu, Xiaoguang Fan, Chenglin Zhang, and Ning Chen. 2018. "Metabolic engineering of Escherichia coli for high-yield uridine production." *Metabolic Engineering* 49: 248–256.

Wu, Qun, Luqing Cui, Yingying Liu, Rongpeng Li, Menghong Dai, Zhenwei Xia, and Min Wu. 2022. "CRISPR-Cas systems target endogenous genes to impact bacterial physiology and alter mammalian immune responses." *Molecular Biomedicine* 3 (1): 22.

Xu, Yuanyuan, and Zhanjun Li. 2020. "CRISPR-Cas systems: Overview, innovations and applications in human disease research and gene therapy." *Computational and Structural Biotechnology Journal* 18: 2401–2415.

Yosef, Ido, Miriam Manor, Ruth Kiro, and Udi Qimron. 2015. "Temperate and lytic bacteriophages programmed to sensitize and kill antibiotic-resistant bacteria." *Proceedings of the National Academy of Sciences* 112 (23): 7267–7272.

Yu, Shengzhu, Bo Zheng, Zhenya Chen, and Yi-Xin Huo. 2021. "Metabolic engineering of Corynebacterium glutamicum for producing branched chain amino acids." *Microbial Cell Factories* 20 (1): 230.

Zaidi, Syed Shan-e-Ali, M Shahid Mukhtar, and Shahid Mansoor. 2018. "Genome editing: Targeting susceptibility genes for plant disease resistance." *Trends in Biotechnology* 36 (9): 898–906.

Zhang, Mingzi M, Fong Tian Wong, Yajie Wang, Shangwen Luo, Yee Hwee Lim, Elena Heng, Wan Lin Yeo, Ryan E Cobb, Behnam Enghiad, and Ee Lui Ang. 2017. "CRISPR–Cas9 strategy for activation of silent Streptomyces biosynthetic gene clusters." *Nature Chemical Biology* 13 (6): 607–609.

Zhao, Hongtu, Gang Sheng, Jiuyu Wang, Min Wang, Gabor Bunkoczi, Weimin Gong, Zhiyi Wei, and Yanli Wang. 2014. "Crystal structure of the RNA-guided immune surveillance Cascade complex in Escherichia coli." *Nature* 515 (7525): 147–150.

11 Current Vectors for *In Vivo* Delivery of CRISPR-Cas9 System

Nausheen Shafi, Tasaduq Manzoor, Irfan Hussain, Junaid Nazir, Aarif Ali, Ubaid Gani, Lateef Ahmad Dar, Yaawar Bashir Mir, Syed Mudasir Ahmad, Meena Godha, and Priyanka Jha

11.1 INTRODUCTION

Clustered regularly interspaced short palindromic repeats and CRISPR-associated protein (CRISPR-Cas) are powerful tools for genome editing, offering precision and adaptability. Advances like base and prime editing use modified CRISPR-associated protein 9 (Cas9) enzymes paired with deaminases or reverse transcriptase to make specific changes in deoxyribonucleic acid (DNA), though these methods have limitations. Originally part of bacterial immune systems, CRISPR-Cas9 cuts DNA at targeted sites, guided by a single guide RNA (sgRNA). DNA repair mechanisms like non-homologous end joining (NHEJ) and homology-directed repair (HDR) fix these breaks, with HDR being more precise. Modified Cas9 versions further expand genome and epigenome editing possibilities beyond traditional repair pathways (Porto et al., 2020). The CRISPR-Cas9 components can be delivered into mammalian cell nuclei as DNA, ribonucleic acid (RNA), or preassembled ribonucleoprotein (RNP) complexes (Amirkhanov and Stepanov, 2019). Although RNA-based delivery methods address many of the challenges associated with DNA-based approaches, their effectiveness is constrained by the inherent instability of messenger RNA (mRNA) and the dependence on the host cell's translation machinery (Lino et al., 2018). Different kinds of Cas9 are used in CRISPR-based genome editing to accomplish diverse outcomes. The Cas9 variant decreases off-target consequences by producing single-strand nicks instead of double-strand breaks (DSBs), while dead Cas9 (dCas9), lacking catalytic activity, adjusts gene expression without causing DSBs. Conventional methods for editing genomes involve processes such as NHEJ and HDR pathways (Shen et al., 2013). While wild-type Cas9 and certain variations can be purchased, custom Cas9 fusion proteins necessary for complex editing methods, such as base editing, prime editing, and epigenome editing, are typically made using *Escherichia coli* (*E. coli*) expression systems. Chimeric RNA molecules, called sgRNAs, combine a CRISPR RNA (crRNA) with a trans-activating CRISPR RNA (tracrRNA) in CRISPR systems and can be obtained or synthesized *in vitro* (Zuris et al., 2015).

DOI: 10.1201/9781003537151-11

The molecular mechanism of gene editing involves the NHEJ and HDR pathways (Liu et al., 2019). NHEJ becomes active after RNPs are delivered, causing targeted DSBs, and is known for its tendency to introduce errors during DNA repair. It is highly efficient for introducing transgenes through the homology-independent targeted integration (HITI) technique (Kowalski et al., 2019). RNPs offer a hopeful method for effective NHEJ-driven genome editing because they have better ratios of on-target to off-target editing in comparison to plasmids, mRNA, and viruses. Additionally, RNP delivery is approximately half as toxic as plasmid delivery (Wang et al., 2019). In canonical NHEJ, the DSB triggers a rapid cellular response that joins the broken DNA ends with minimal end resection (Davis and Maizels, 2014). When CRISPR-Cas9 induces a DSB in DNA, the NHEJ pathway typically repairs the break, though it often introduces insertions or deletions (indels) due to Cas9 repeatedly cleaving the DNA until the protospacer adjacent motif (PAM) is altered. NHEJ is active throughout the cell cycle but lacks precision, limiting its usefulness for accurate genome editing (Betermier et al., 2014).

HDR is an accurate yet less-effective method of fixing DSBs by incorporating targeted deletions, insertions, or base substitutions at particular genomic locations with the help of an external DNA donor template. This particular mechanism for fixing DNA primarily works in the later stages of the cell cycle, specifically during the S and G2 phases, with sister chromatids being used as the main template for repair (Liu et al., 2019). RNP-mediated HDR editing was first observed in mammalian cells in 2014, and has since been utilized in various cell types to increase its impact in mammalian genome engineering (Lin et al., 2014). RNPs are particularly successful in HDR editing due to their ability to cause Cas9-mediated DSBs with decreased cellular toxicity (Zaboikin et al., 2017). Several molecular strategies have been created to enhance the efficiency of HDR in CRISPR-Cas9 genome editing (Chu et al., 2015). Some of these methods involve using drugs to manipulate DNA repair processes, control cell division, and bring important biomolecules closer to the Cas9 complex (Chu et al., 2015).

11.2 CARGOES FOR CRISPR-CAS9-BASED GENE EDITING

The use of CRISPR-Cas9, sourced from the immune system of *Streptococcus pyogenes*, has revolutionized genome editing by streamlining the process using Cas9 and a customizable sgRNA (Marraffini, 2016). The sgRNA guides Cas9 to precise locations in the genome, causing DSBs to occur. Repairing these DSBs in human hematopoietic stem cells through DNA sequence insertion or deletion can achieve a success rate of over 20%, thus enabling the advancement of CRISPR-based therapies in early clinical trials (Lomova et al., 2019). However, the complexity of applying CRISPR in clinical settings is heightened by the diversity of mutations. Improving our understanding of Cas9's function, including its binding and DNA interrogation kinetics, could enhance the speed and efficiency of genome searching in clinical contexts. Alternative CRISPR-based technologies like base editing and prime editing offer the advantage of introducing precise mutations without creating DSBs and show promise for gene therapy, although they require further development (Saber Sichani et al., 2023). As CRISPR-Cas9-mediated genome editing advances, developing safe and effective methods for packaging and delivering these tools remains crucial for broader clinical use.

The process of delivering CRISPR-Cas9 encompasses both the CRISPR cargo and the method used to transport this cargo into cells (Xu et al., 2019). Delivery techniques for CRISPR-Cas9 consist of the vehicle, which facilitates entry into cells, and the cargo, comprising the Cas nuclease and guide RNA (Xu et al., 2019). There are three primary categories of CRISPR delivery vehicles: viral, non-viral, and physical (Taghdiri and Mussolino, 2024). The choice of delivery vehicle influences whether the Cas nuclease is introduced as DNA, mRNA, or protein. Gene-editing tools come in various forms, including Cas9-encoding DNA plasmids, mRNA constructs, Cas9 RNP complexes, and Cas9 RNP arrays (Liu et al., 2022). Each type of CRISPR delivery vehicle, along with its specific cargo combination, presents unique advantages and disadvantages, impacting the overall safety and effectiveness of the CRISPR application or treatment. **Figure 11.1** shows the types of CRISPR cargoes.

11.2.1 Plasmid DNA

Plasmid DNA (pDNA) is a widely used tool in transfection and gene-editing research, with current efforts aimed at enhancing pDNA constructs to boost transfection efficiency and minimize cytotoxicity. Due to their larger size and weight, pDNA constructs can be prone to off-target effects caused by Cas9 endonucleases. Minicircle DNA vectors (mcDNA) are emerging as a promising alternative to pDNA due to their smaller size and greater capacity for carrying large gene expression cassettes (Gaspar et al., 2015). mcDNA vectors have shown effectiveness in both HDR and HITI, as evidenced by the successful insertion of a tdTomato reporter into three human cell lines (Foley et al., 2022). Combining the CRISPR-Cas9 system with non-viral vectors improves the delivery of nucleic acid–based cargoes, offering better resistance to degradation compared to linear DNA under hydrodynamic shear forces.

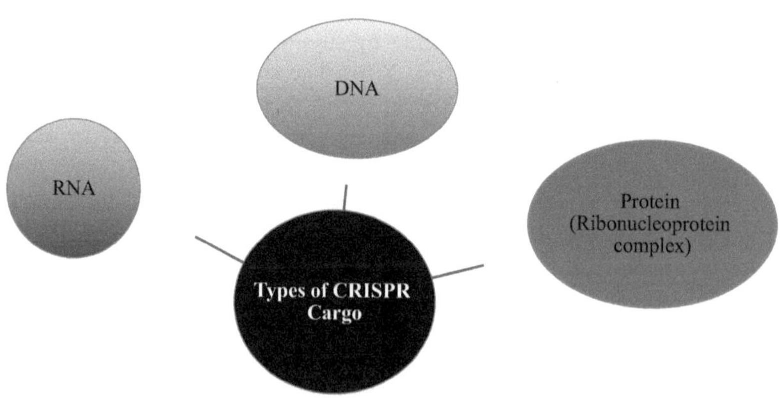

FIGURE 11.1 Types of CRISPR cargoes.

11.2.2 Single Guide RNA

Recent advancements have concentrated on enhancing the guide RNAs (gRNAs) used with Cas9 nucleases (Fu et al., 2014). Researchers have integrated two distinct RNA strands into programmable sgRNAs, leveraging knowledge about the interactions between tracrRNA, crRNA, and Cas9 (Fu et al., 2014). RNA molecules frequently contain hairpin loops as secondary structures that can affect gene expression both locally (cis) and distantly (trans). Modifications to sgRNAs are designed to increase their stability against ribonuclease degradation and to enhance their binding affinity to target sequences. By adjusting the terminal three nucleotides on both ends of three different sgRNA molecules, researchers have improved insertion-deletion (indel) editing in primary human T cells. Furthermore, the presence of a 5′-triphosphate group on sgRNAs can activate innate immune responses through type 1 interferons. Phosphatase treatment to remove this triphosphate group has been found to improve cell viability (Ryan et al., 2023).

11.2.3 Messenger RNA

Messenger RNA (mRNA) is commonly used to deliver Cas9-encoding payloads to cells where genomic integration is not required. mRNA-based systems typically have fewer off-target effects and are suitable for clinical applications. One notable advancement is the use of uridine-depleted transcripts with 5-methoxyuridine, which is currently being tested in both laboratory and clinical settings. mRNA transcripts in CRISPR-Cas9 applications help reduce patient immunogenicity by allowing the rapid clearance of Cas9 from corrected cells before treatment, thus minimizing the need for extended immunosuppressive therapies (Mu and Hu, 2021). Additionally, programmable RNA-binding proteins that respond to small molecules could provide better control over transient mRNA expression. However, mRNA constructs face challenges, such as high molecular weight, hydrophilicity, anionicity, and low stability, which contribute to their lower transfection efficiencies.

11.2.4 Ribonucleoprotein

A RNP complex consists of an RNA molecule and an RNA-binding protein, is critical for CRISPR-Cas9-mediated genome and epigenome alterations. RNPs have several advantages over conventional delivery techniques, including rapid activation, reduced adverse effects, and less toxicity to cells (Wang et al., 2019). They effectively merged multiple Cas9 orthologs, modified versions, and Cas9 fusion proteins, enabling a wide range of editing options. These RNPs are especially valuable for genome and epigenome editing in clinically relevant cell types (Wang et al., 2019).

11.3 DELIVERY METHODS FOR CRISPR-CAS9

Cas9 RNPs, which are composed of recombinant Cas9 protein and a sgRNA, are efficient for accurate genome manipulation (Kim et al., 2014). Nevertheless, their ability to be used in in-depth research is frequently restricted due to their expensive price. To improve RNP effectiveness in clinical settings, scientists have created two main approaches: adding shortened Cas9 target sequences to HDR templates and employing polyglutamic acid nanoparticles to support Cas9 RNPs (Yang et al., 2022). Moreover, changes made to donor DNA, like phosphorothioate modifications and 3′ overhangs, have been demonstrated to enhance HDR efficiency. Choosing the right cargoes and delivery methods is vital in reducing immune responses and off-target effects when applying the CRISPR-Cas9 system to treat genetic disorders (Chen et al., 2020). Cas9 RNPs can be delivered into target cells through various techniques, including viral and non-viral (physical and chemical) methods.

11.3.1 Viral Vectors for Gene Delivery

Over millions of years, viruses have evolved various mechanisms to survive and replicate within host cells, and these characteristics are now being harnessed in gene therapy research. These viruses, with their diverse genetic compositions, are used as vectors to deliver genes, nucleic acids, and other genetic materials into specific cells. Despite inherent risks associated with viral vectors, they are generally considered safe for human use, and research continues to enhance their safety (Butt et al., 2022). Viruses offer significant advantages in gene therapy due to their diverse forms, types of genetic material, and natural ability to target specific cells (Butt et al., 2022). The selection of a viral vector for gene therapy depends on these factors, demonstrating the versatility and adaptability of viruses for therapeutic purposes (Bulcha et al., 2021).

11.3.1.1 Adenoviral Vectors

Adenoviruses were initially viewed as promising tools for gene therapy due to their nonenveloped structure and large, double-stranded DNA genome, which is about 35

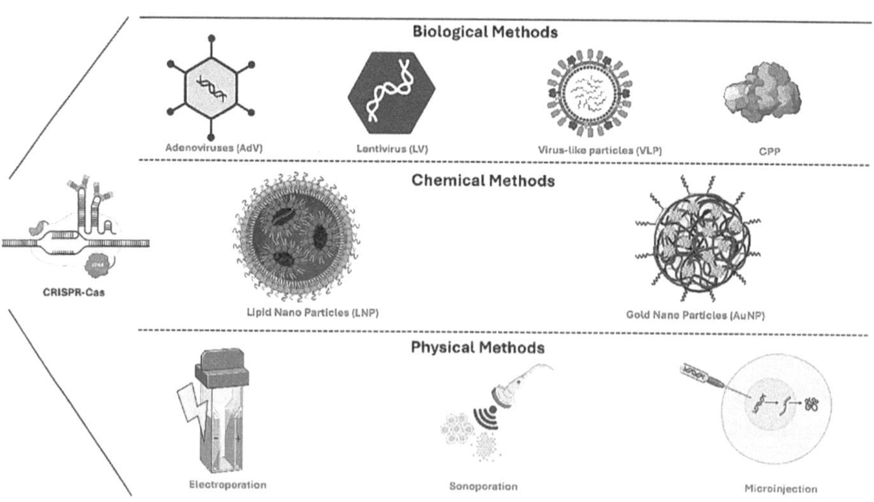

FIGURE 11.2 Different vectors used for delivery of CRISPR-Cas.

kb (Ghosh et al., 2006). However, issues related to their toxicity and interactions with blood components posed challenges. Recent research has focused on improving their gene transfection efficiency through several approaches (Coughlan et al., 2010). These include extensive genome modifications, targeted deletions of non-essential genes while preserving those vital for replication and packaging, and the development of vector hybrids and combined immunotherapy strategies (Coughlan et al., 2010). These efforts reflect ongoing progress in refining adenoviral vectors to enhance their safety and effectiveness for gene therapy, showcasing the continuous advancement in viral vector technology (Lu, 2004).

11.3.1.2 Retroviruses

Retroviruses convert their RNA genome into DNA for integration into the host's genome and have been utilized in gene therapy for conditions like severe combined immunodeficiency due to adenosine deaminase deficiency. Their ability to integrate permanently into the host genome ensures stable gene expression, a notable advantage over other vectors. However, risks, such as insertional mutagenesis, oncogene activation, and issues related to cell tropism and cross-species transmission, pose significant concerns. To mitigate these risks, research is focused on developing replication-deficient retroviral vectors to minimize unwanted genetic modifications and prevent broad host range transmission. Efforts are ongoing to enhance the safety of lentiviral vectors in clinical applications, reflecting the ongoing advancement in improving their safety and therapeutic potential.

11.3.1.3 Adeno-Associated Viruses (AAVs)

AAVs are highly esteemed in gene therapy due to their small size, nonenveloped structure, and lack of integration into the host cell genome (Kotin, 1994). Once inside the human body, AAVs remain inactive, which makes them particularly suitable for delivering genes to slowly dividing cells. They are recognized for their low levels of immunogenicity and toxicity, and their durable capsid allows them to persist in the bloodstream. These characteristics make AAVs some of the safest viral vectors available for gene therapy. Their continued use and enhancement in clinical settings underscore their effectiveness in gene delivery while minimizing immune responses and toxicity, supporting their ongoing development for treating a variety of genetic disorders. The CRISPR-Cas9 system, valued for its accuracy and speed, locates particular DNA sequences by pairing bases and has progressed quickly because of constant work to improve, boost, and broaden the CRISPR toolbox. Although CRISPR-based treatments show promising potential in treating different illnesses, the delivery of these treatments into the human body is still a difficult task. There are two primary ways to administer CRISPR-based therapeutics: *ex vivo* and *in vivo*. *Ex vivo* delivery comprises taking human cells from a patient, altering them in a laboratory, and then returning the modified cells back into the body. This technique allows for a complete analysis of the transformed cells, expanding the spectrum of cell types that gene-editing tools can target. In contrast, *in vivo* administration necessitates the introduction of genome editing tools directly into the patient, either via the bloodstream or at a specific site. Currently, AAV vectors are the primary vehicle for *in vivo* gene therapy delivery (Zhang et al., 2020), particularly in obtaining regulatory approval to deliver AAV-CRISPR directly to the eye. AAV is a small, non-enveloped virus, approximately 25 nm in diameter, with a single-stranded DNA genome of 4.7 kb that includes inverted terminal repeats (ITRs). Recombinant AAV (rAAV) can be engineered by replacing most of its genome with a transgene cassette. Using AAV to deliver smaller Cas proteins allows for more flexible promoter options and simplifies the delivery process. Recent research has improved base editors for the reconstitution of split inteins and AAV vector delivery in mice, achieving genome editing efficiencies between 9% and 59% across different tissues (Ma and Ashnest, 2020).

11.3.1.4 Poxviruses

The smallpox virus, a prominent member of the Poxviridae family, exemplifies its group with two distinct infectious forms: the intracellular mature virus (IMV) and the extracellular enveloped virus (EEV). These characteristics make it a valuable model for studying viral replication and pathogenesis. However, the virus's complexity poses challenges for practical applications, such as gene therapy. The closely related vaccinia virus, despite its efficient life cycle and ample cloning capacity, encounters difficulties due to its intricate nature, complicating its use as a gene therapy vector. Nonetheless, research continues to explore and modify poxviruses for therapeutic purposes, aiming to harness their unique features while overcoming their complexities to ensure safe and effective clinical applications.

11.3.1.5 Lentiviral Vector (LV Vector)

LV are enveloped viruses ranging from 80 to 120 nm in size, and they possess two sets of single-stranded genomic RNA, each measuring 9 kb. They are commonly utilized in ex vivo gene therapy, for example, by stably incorporating CAR cDNA into T cells. Nevertheless, lentiviruses are not suitable for transporting CRISPR-Cas9 because of the potential dangers associated with off-target mutations and the random integration of viral DNA. LV vectors have been utilized for delivering genome editing tools to address genotoxicity concerns. However, the possibility of DNA integration at the Cas9 cleavage site continues to be worrisome. In order to lower the chances of creating replication-competent lentiviruses (RCLs), the essential genes for making LV vectors are spread out among several plasmids to reduce homologous sequences.

11.3.2 Non-Viral Vectors for Gene Delivery

While viral vectors are effective for prolonged gene expression, their development can be complex and may carry risks of toxicity. To overcome these challenges, researchers are exploring non-viral systems as safer alternatives, focusing on their lower toxicity profiles. Non-viral gene delivery methods are generally categorized into physical and chemical approaches. Physical methods involve techniques such as electroporation, sonoporation, magnetoporation, microinjection, needle injection, and gene guns, which use mechanical or energy-based means to facilitate gene transfer. Chemical methods include polymer- and lipid-based systems, along with various inorganic materials, designed to encapsulate and deliver genetic material efficiently. These non-viral strategies provide diverse and promising options for gene therapy, aiming to mitigate safety concerns related to viral vectors while enhancing the versatility and effectiveness of gene delivery systems (Butt et al., 2022).

11.3.2.1 Physical and Energetic Methods of Delivery

Successfully delivering CRISPR-Cas9 cargoes to targeted cells and tissues remains a significant challenge for clinical gene therapies. Various intracellular delivery methods are available, including mechanoporation, electroporation, and acoustoporation.

11.3.2.1.1 Microinjection

Microinjection, a technique that uses small capillaries to deliver gene-editing materials directly into cells, has been crucial in creating genetically modified animals, like sand flies, sheep embryos, and zebrafish. Nevertheless, it has occasionally led to the development of mutant embryos. This technique is especially successful in transporting DNA and mRNA, enabling their unhindered transcription and translation in the nucleus or cytoplasm. Although it requires a lot of time and effort, microinjection is most appropriate for conducting small-scale *in vitro* research.

11.3.2.1.2 Mechanoporation

Mechanoporation can also be achieved by guiding cells toward tiny structures that can penetrate cells, such as nanoneedle arrays, which are more efficient for transferring genetic material into cells that are attached to a surface rather than floating freely.

11.3.2.1.3 Cell Squeezing

Cell squeezing, a mechanoporation method using microfluidics, involves quickly pushing cells through tiny constrictions to break the plasma membrane and permit the entry of different cargoes into the cytosol of various cell types. Han employed a microfluidic tool to mechanically deliver an RNP cargo to silence enhanced green fluorescent protein (EGFP) in SK-BR-3 cells, medical discovery association mammary breakthrough 231 (MDA-MB-231) cells, Strain U-Diffuse Histiocytic Lymphoma 1 (SU-DHL-1) cells, and human primary T cells that were expressing EGFP. Delivery efficiencies of approximately 40% were attained for both RNPs and Cas9 plasmid constructs during the process. Although cell squeezing through microfluidic devices is efficient in transfecting different cell types, it faces obstacles like device blockage, the need to maintain a balance between cell viability and transfection efficiency, and constraints in nucleic acid delivery (Ding et al., 2017).

11.3.2.1.4 Hydrodynamic Manipulation

Hydrodynamic manipulation is a microfluidic-based mechanoporation technique used for the *in vivo* delivery of large macromolecules through the intravenous injection of liquid solutions at high volumes and pressures. Although it often complements other delivery methods, it does not directly penetrate cell membranes. Deng et al. (2018) investigated an inertial microfluidic cell hydroporator (iMCH) for delivering various cargoes to different cell types, achieving high cell viability and over 80% efficiency in collagen type XI alpha 1 chain (COL11A1) gene knockdown in cisplatin-resistant A2780 cells using the CRISPR-Cas9 system. Recent advancements in this platform, including the integration of droplet microfluidics, aim to reduce cargo consumption, enhance throughput, and minimize clogging issues when treating primary human T cells.

11.3.2.1.5 Electroporation

Electroporation is a non-viral technique used to introduce large biomolecules into cells by creating nanometer-sized pores in the cell membrane. It is particularly effective for immune and stem cells, enabling the delivery of RNPs, DNA, and mRNA in both *in vitro* and *ex vivo* settings. However, using large plasmids can reduce cell viability, and uneven electric field distribution in cuvettes can further lower transfection efficiency. Comparisons between electroporation and microfluidic cell squeezing have shown that microfluidic cell squeezing causes minimal transcriptional changes and preserves effector responses. To address issues such as bubble formation during electroporation, innovations like tube-shaped cuvettes have been introduced, enhancing the technique's application for knocking out 2-microglobulin in primary human mesenchymal stem cells.

11.3.2.1.6 Acoustoporation and Sonoporation

These techniques make use of ultrasound to create pores in cell membranes, enabling the delivery of genetic material. The effectiveness of these methods depends on the properties of the cells, the type of acoustic stimulation, and the characteristics of microbubbles involved. While acoustoporation can work without microbubbles, the collapse of microbubbles under high pressure may lead to cell rupture and decreased efficiency in cargo delivery. Recent innovations include cost-effective ultrasound-microfluidic devices for acoustoporating cardiac myoblasts without the use of contrast agents. These devices employ surface acoustic waves to create temporary membrane pores, improving molecular uptake through extracellular streaming.

Hypersonic poration, which uses a nanoelectromechanical resonator to form 200 nm pores and high-frequency bulk acoustic waves (150 MHz), has shown up to 40% efficiency in delivering CRISPR plasmids to HeLa and HEK 293 cells. The choice of delivery method must consider cell viability, and acoustic devices can be adjusted for optimal input voltage and flow rate, achieving throughputs as high as 200,000 cells per minute. A microfluidic platform using lateral cavity acoustic transducers (LCATs) has shown potential for delivering plasmid DNA and sgRNA into cells, offering promise for therapeutic applications. Additionally, acoustofluidic tweezers allow precise genetic material delivery, enabling single-cell-resolution editing in HEK 293 and HeLa cells using CRISPR-Cas9. While particle-based delivery methods discussed earlier are unsuitable for in vivo gene editing due to their reliance on external applications, synthetic carriers present a possible solution to these challenges (Aghaamoo et al., 2022).

11.3.2.2 Chemical Methods for Non-Viral Gene Delivery

Non-viral gene delivery through chemical methods utilizes various compounds or materials to enable the introduction of a genetic material, including plasmid DNA, RNA, or CRISPR-Cas9 components, into target cells. These techniques are prevalent in both research and therapeutic contexts, primarily because of their reduced immunogenicity, straightforward production processes, and adaptability. Chemical non-viral vectors are increasingly important in gene delivery as they can overcome obstacles like cellular uptake, endosomal escape, and targeting the cytoplasm or nucleus (Butt et al., 2022). Following are some key chemical methods employed for non-viral gene delivery.

11.3.2.2.1 Nanocarriers

The term *nanocarrier* refers to a nanoscale structure that facilitates the delivery of therapeutic agents, including drugs, genes, and various other molecules, to targeted sites within the body. These delivery systems are designed to enhance bioavailability, improve targeting precision, and increase the efficacy of therapeutic interventions. A variety of materials, such as lipids, polymers, proteins, and inorganic compounds, can be employed in the construction of nanocarriers. In what follows, the main categories of nanocarriers are presented.

11.3.2.2.1.1 Lipid Nanocarriers

Lipid nanoparticles (LNPs) are frequently used to transport Cas9 RNPs into cells due to their biological compatibility, long circulation period, and low toxicity (Freund et al., 2020). LNPs can be made from commercially available or synthetic lipids, allowing for easy customization and large-scale manufacture (Hashimoto et al., 2016). They have effectively delivered RNPs to several organs in living creatures, including muscles, the brain, the liver, and the lungs (Hung et al., 2018). Furthermore, gold nanoparticles (AuNPs) have been used to transfer RNPs to multiple cell types, revealing high efficacy in genome and epigenome editing while maintaining low toxicity and the capability of packaging donor DNA. Cell-penetrating peptides (CPPs) help RNPs flow through cell membranes via passive or active endocytosis, making them a simple and efficient delivery method (Kholosy et al., 2021).

Lipid nanocarriers are widely used for delivering nucleic acids *in vivo*, particularly in CRISPR-Cas9 therapies (Lee et al., 2017). Despite significant progress, challenges such as toxicity and limited targeting specificity persist. DLin-MC3-DMA has become the material of choice for nucleic acid delivery, replacing traditional lipofection reagents, even in *in vivo* applications. However, its non-degradability and less-than-ideal performance for mRNA or RNP delivery limit its effectiveness and safety. Advances in rational design have led to the development of combinatorial libraries of synthetic lipids and lipid-like materials, which optimize the composition of LNPs to enhance organ selectivity. LNPs modified with tumor-targeting agents have shown improved gene editing specifically in tumors, achieving up to 80% editing *in vivo*. While delivering RNPs via LNP platforms remains challenging, the use of a pH-neutral buffer and permanently cationic lipid components has successfully encapsulated RNPs, resulting in 19.2% HDR *in vitro* (Walther et al., 2022).

LNPs can be directly injected into specific tissues to target gene-editing efforts at precise locations, especially in immunoprivileged areas like the eyes and central nervous system. For example, Lipofectamine-encapsulated LNPs have been employed to modify the inner ear and retina in mice. However, the cytotoxicity of transfection agents may restrict this delivery method's applicability to models other than mice.

Specialized synthetic nanoparticles like nanocapsules and DNA nanoclews are designed for genome editing and have been tested in tumor-bearing mice, where they induced indel formations after intramuscular injection into the tibialis anterior. The formulation of LNPs is crucial for protein cargo delivery, influencing both packaging efficiency and targeting. By adjusting components or their ratios, LNPs can be redirected from the liver to accumulate selectively in other organs, such as the lungs or spleen (Yin et al., 2014).

11.3.2.2.1.2 Polymeric and Supramolecular Nanocarriers

These nanocarriers have long utilized cationic polymers, like chitosan and poly(ethyleneimine) (PEI), for biomolecule delivery, though they face certain challenges. Researchers have improved integration rates by combining different components into functionalized particles, leading to advancements in nanoparticle efficacy. These advancements include the development of custom-designed polymers and modifications to existing ones. For example, dendrimers have been modified with adamantane, and PEI has been enhanced with hyaluronic acid (HA) and mannose, to increase delivery efficiency. Grafting techniques are commonly employed to create supramolecular polymeric nanocarriers, which are essential for supramolecular nanoparticles (SMNPs) used in delivering CRISPR-Cas9 reagents. Additionally, new nanoparticle materials designed for physiological degradation, such as disulfides, esters, and amino esters, have been effective in delivering CRISPR-Cas9 genome editing reagents and achieving high gene knockout rates *in vitro*. Custom polymers are being engineered to improve the packaging and delivery of Cas9 RNPs, such as those modified with adamantane, which have demonstrated gene-editing capabilities in both *in vitro* and *in vivo* settings. These polymers, including Poly(β-amino ester)s (PBAEs), are advantageous for CRISPR-Cas9 delivery due to their versatile biochemical properties, which enhance packaging and delivery efficiency and may provide additional therapeutic benefits (Walther et al., 2022).

Polymer-based nanovectors, crafted from repeating monomers, are intricate, sizable compounds and can be categorized into natural and synthetic types. These polymers can be engineered with specific properties and modified with targeting ligands and shielding agents to enhance their circulation time in the bloodstream (Butt et al., 2022). Despite their potential, cationic polymers used to form polyplexes for gene delivery have faced limitations due to cytotoxicity. As a result, there is increasing interest in biodegradable polymers, which reduce cytotoxicity while effectively delivering genetic material, thus improving the safety and efficacy of gene therapy.

Biodegradable polymers have been used in medical science for over two decades, across various fields, including drug delivery, gene therapy, regenerative medicine, biotechnology, genetic engineering, and tissue engineering. They can be classified into natural or synthetic categories. Natural biodegradable polymers offer several advantages but face challenges related to precise extraction, purification, and batch-to-batch variability. In contrast, synthetic biodegradable polymers provide benefits, such as controlled chemical structures, natural inertness, and consistent batch uniformity, although their biological inertness might limit their use in some therapies. Protein-based vectors, including gelatin and albumin, are notable examples of biodegradable polymers. These proteins are used in gene delivery due to their biocompatibility and biodegradability. Albumin, when combined with other cationic polymers, forms effective non-viral gene delivery nanovectors capable of transfecting HeLa cells. Polysaccharide-based vectors, such as chitosan and its derivatives, are also preferred for gene delivery because their glucose units linked by glycosidic bonds make them suitable for cationic polymer-based gene therapy. Synthetic biodegradable polymers, like polyesters, are useful in medical applications due to their hydrolysable chemical bonds, which are beneficial for sutures, genetic engineering, drug delivery carriers, and gene therapy. Polyphenylene ether (PPE) polymers, which are compatible with biological enzymes and similar to nucleic acids, are particularly suitable for gene delivery, with various techniques developed to enhance their transfection efficiency and minimize toxicity (Butt et al., 2022).

Polycarbonates and polyurethanes are effective gene delivery vectors due to their biocompatibility and adjustable properties. Polycarbonates can be enhanced through chemical modifications, while polyurethanes, combined with cationic polymers, are useful for miRNA delivery, especially in brain tumors. Polymeric-inorganic nanovectors are created to transport Cas9 DNA plasmids to tumors. Hybrid nanovectors incorporate photothermal triggers to release the plasmid and effectively disrupt the Polo-like kinase 1 (Plk1) gene *in vitro* and *in vivo* using CRISPR-Cas9 technology. Polymeric-inorganic hybrid nanoparticles present superior colloidal stability and improved loading and delivery abilities for Cas9 RNPs. These nanoparticles can be designed with precise structural or compositional characteristics to firmly attach to Cas9 RNPs. Metal-organic frameworks (MOFs), known for their diverse microstructures and pore sizes, are highly efficient for delivering Cas9 RNP at the nano level.

Non-viral nanovectors are being more commonly utilized to safely and efficiently deliver CRISPR-Cas9 genome editing systems. These vectors are capable of transporting Cas9 DNA plasmids, Cas9 mRNA, or Cas9 RNPs. At first, lipid-based nanovectors were utilized for delivering CRISPR-Cas9 because of their well-known function in gene delivery. Polymer-based nanovectors have gained popularity for this application due to their excellent colloidal stability. Many non-viral vectors are designed to transport CRISPR-Cas9 systems in order to target and remove specific oncogenes using NHEJ in cancer therapy. New developments involve innovative non-viral carriers for targeting cancer cells and triggering anti-tumor immune response. Although non-viral vectors hold promise for treating genetic diseases, advancements in gene editing or insertion for lengthy therapeutic genes are still constrained. Local or systemic administration of these vectors will result in concentrated delivery, improved gene-editing efficiency, decreased systemic toxicity, and reduced off-target effects. Nanovectors allow for precise targeting in tissues, less unintended effects, and regulated delivery of CRISPR-Cas9 components.

11.3.2.2.2 Inorganic Nanocarriers

Inorganic nanocarriers still face challenges in delivering complex cargoes, such as Cas9 RNPs. Recent developments include core-shell structures that combine inorganic cores with organic components using engineered approaches for targeted delivery or responsiveness to specific stimuli. These advancements aim to improve the precision of gene-editing cargo localization.

11.3.2.2.2.1 Gold Nanoparticles

Gold nanoparticles leverage their plasmonic properties to enhance gene-editing delivery systems. For instance, arginine-functionalized and colloidal gold nanoparticles are designed to co-deliver guide RNA and Cas9, with or without single-stranded oligonucleotides (ssODN). Clinical trials have explored their use for both delivery and diagnostic purposes, utilizing the endosomal-disruptive polymer PAsp (DET) to coat gold nanoparticles and protect the CRISPR cargo. Additionally, functionalized gold nanoparticles, such as those modified with cationic trans-activator of transcription (TAT) peptides, can complex with Cas9-encoding plasmids and guide RNA, enabling targeted gene editing and tumor inhibition when activated by near-infrared (NIR) light. This approach utilizes the plasmonic properties of gold nanoparticles for controlled release and precise gene editing following systemic administration.

11.3.2.2.2.2 Silica Nanoparticles

Silica nanoparticles show considerable potential for CRISPR-Cas9 gene therapies due to their porous structure and ability to respond to various stimuli. However, they face challenges with stability and controlling the release of their cargo. Mesoporous silica nanoparticles are frequently coated with lipids and polymers to enhance the delivery of Cas9-encoding plasmids and RNPs. Chemical functionalization techniques, such as linking with hepatocyte-targeting N-acetylgalactosamine (GalNAc), are used to minimize accumulation in the liver. Moreover, silica-based up-conversion nanoparticles (UCNPs) offer stimulus-responsive delivery of gene-editing agents and have proven effective in reducing tumor size in animal models.

11.3.2.2.3 Liposomes for CRISPR-Cas9 Delivery

Liposomes, including lipoplexes and lipopolyplexes, are emerging as effective carriers for delivering CRISPR-Cas9. Lipoplexes are created by combining cationic liposomes with CRISPR-Cas9 to form multilamellar complexes. Their preparation is straightforward, involving mixing cationic liposomes with CRISPR-Cas9 in specific proportions. Lipopolyplexes, on the other hand, are formed by coating anionic polyplexes with cationic lipid bilayers or vice versa through electrostatic interactions. To improve the stability of these carriers for both *in vitro* and *in vivo* applications, PEGylation is commonly used (Awad et al., 2024). The surface charge of liposomes plays a critical role in CRISPR-Cas9 delivery, influencing each stage of the delivery process. Neutral liposomes tend to offer better stability and longer circulation times in the bloodstream compared to cationic and anionic liposomes. Different surface charges of liposomes come with their own advantages and drawbacks. While monovalent and multivalent liposomes provide good tumor penetration and effective endosomal escape, they often suffer from high cellular absorption and shorter blood circulation times.

Four primary categories of cationic liposomes are utilized for delivering CRISPR-Cas9: monovalent cationic liposomes, pH-responsive cationic liposomes, lipidoids, and multivalent cationic liposomes. Although they have potential, their ability to work effectively in the body may be restricted by limited penetration into tumors. Univalent cationic lipids, such as 1,2-di-O-octadecenyl-3-trimethylammonium propane (DOTMA) and 1,2-dioleoyl-3-trimethylammonium propane (DOTAP), are utilized in the formation of lipoplexes, lipopolyplexes, or membrane/core nanoparticles (MCNPs). Liposomal complexes created from these liposomes typically exhibit strong gene knockout performance in laboratory settings as a result of their efficient absorption into cells and escape from endosomes through the ion-pair mechanism. Lipopolyplexes, which contain cationic polymers, provide increased stability and greater gene transfection effectiveness than lipoplexes through enhanced endosomal escape of CRISPR-Cas9. Multivalent cationic lipids, which originate from monovalent cationic lipids, are utilized in creating lipoplexes or lipopolyplexes. These complexes demonstrate superior gene transfection efficacy and lower toxicity when compared to those constructed with DOTAP. Anionic liposomes, however, face challenges, like poor tumor penetration, short circulation times, high macrophage uptake, and low cellular uptake, due to repulsion by the anionic membrane. Lipidoids, similar to pH-sensitive lipids but with additional protonation sites and tails, form stable nucleic-acid–lipid particles (SNALP) structures for CRISPR-Cas9 delivery. Neutral liposomes, composed of neutral lipids, offer extended blood circulation but struggle with efficient CRISPR-Cas9 encapsulation due to limitations in passive loading. Their success in systemic CRISPR-Cas9 delivery is highly dependent on loading efficiency and cellular uptake (Jayakumar et al., 2010). PEGylation of liposomes can be achieved either by mixing preformed liposomes with PEG-lipid aqueous dispersions or by incorporating PEG-conjugated lipids during liposome synthesis. While PEGylation extends the circulation time of liposomal carriers, it often reduces cellular uptake and endosomal escape, potentially diminishing gene silencing effectiveness (Yin et al., 2017).

pH-sensitive cationic liposomes with a SNALP structure are effective *in vivo* for gene silencing due to their efficient CRISPR-Cas9 encapsulation, neutral surface charge, extended circulation time, and effective cellular uptake through clathrin-mediated endocytosis and micropinocytosis (Sato et al., 2012). PEGylated lipidoid nanoparticles also achieve significant gene knockout efficiency, primarily via macropinocytosis, though 70% are exocytosed from late endosomes. PEGylation can impair the cellular uptake of multivalent cationic liposomes, making those with only 1–2% PEG more suitable for CRISPR-Cas9 delivery. While PEGylating neutral liposomes improves tumor accumulation, it reduces cellular uptake, making PEG modification less beneficial for these liposomes. PEGylated anionic liposomes can enhance circulation time, tumor penetration, and cellular uptake, but their synthesis is more complex, and they show lower CRISPR-Cas9 loading efficiency (Allahham et al., 2022).

11.3.2.2.4 Inorganic Material–Based Vectors

Quantum dots are used in gene delivery as the central component of gene-carrying vectors, enabling the monitoring of the delivered nucleic acids. Carbon nanotubes, constructed from graphene sheets in cylindrical form, encounter obstacles in biological use because of their limited compatibility and inability to dissolve in water. Silica-based systems have been used for a significant amount of time in delivering genes and nucleic acids. These systems can be designed to provide functions like prolonged blood flow, effective delivery of genetic material, minimal harm, increased ability to carry medications, and better absorption by cells. Through the inclusion of cationic substances on the surface of silica-based systems, these carriers are able to engage with negatively charged DNA, thereby enabling the non-invasive transportation of genes. Furthermore, CRISPR-based technologies have transformed the field of life sciences by offering unique treatment approaches to target a wide array of diseases (Butt et al., 2022).

11.4 ADVANTAGES AND DISADVANTAGES OF COMMONLY USED VECTORS

Oncoretroviral Vectors

Advantages

- Effective and reliable gene transfer.
- In non-human primates, transduction rates of HSCs can reach 40%.
- High-fidelity gene transfer, since there are no chromosomal rearrangements and integration is intact.
- Several clinical studies indicate that they are generally safe.

Disadvantages

- Minimal expression.
- A single provirus copy or fewer per cell
- Chromosomal position effects are sensitive to it.
- Sensitive to DNA repeats, introns.
- Restrictions on the therapeutic gene's size (<7 kb) insert.
- *In vivo* delivery is challenging due to low titers (<10^7 TU/ml).

Lentiviral Vectors

Advantages

- Effective and reliable gene transfer.
- Over 90% of HSCs are successfully transduced.
- Results in high expression levels of the transgene.
- Achieves high titers (>10^8 TU/ml).
- Extensive clinical background as a result of AIDS.
- No chromosomal rearrangements and complete integration result in high-fidelity gene transfer.

Disadvantages

- Responsive to changes in chromosomal position.
- Limits on the size of the therapeutic gene (<10 kb) insert.

Adenoviral Vectors

Advantages

- Efficient gene transfer.
- Delivery of several copies of the genome to each target cell, which usually results in extremely high expression.
- Using vectors from the most recent generation, rather large therapeutic genes (25–30 kb).
- High fidelity of gene transfer.
- Genetic stability exists in vector genomes.
- *In vivo* administration.
- Broad clinical competence.
- No issues with insertional mutagenesis.

Disadvantages

- Transfer and expression are transient.
- Transgene activity usually lasts for 1–2 months in cells that are not dividing, but it is briefer in dividing cells because adenoviruses do not integrate into the cells.
- Individuals with pre-existing immunity to adenoviruses may exhibit poor transgene delivery and expression levels.
- Since the viral capsid and residual viral proteins induce inflammation, vectors are immunogenic.
- Safety issues: High doses of vectors can be fatal, and they spread quickly.

(Continued)

Oncoretroviral Vectors	
Advantages	**Disadvantages**
Adeno-Associated Vectors (AAV)	
Advantages	**Disadvantages**
• Efficient gene transfer. • Several genome copies per target cell. • High-fidelity gene transfer. • Under certain conditions, stable expression. • Vector genomes show evidence of genetic stability. • Capable of integrating into the genome of the target cell *in vivo* administration. • At tested doses, generally safe.	• Small transgenes (approx. 4.7 kb). • Because the virus doesn't always integrate, transfer and expression aren't always stable. • Questionable tropism for HSC. • Frequently immunogenic. • Individuals may have low amounts of transgene delivery and expression due to pre-existing immunity.

Source: Thomas et al. (2003).

11.5 POTENTIAL USES OF RNP-MEDIATED GENOME AND EPIGENOME EDITING IN MAMMALIAN CELLS

11.5.1 Hematopoietic Stem Cells

Hematopoietic cells offer significant potential for treating a range of conditions, from cancer and infectious diseases to autoimmune disorders and genetic immunodeficiencies. Current research focuses on gene-edited T cells for cancer immunotherapy and HIV treatment, with Cas9 RNPs enhancing editing success while preserving the cells' growth capability for scalable treatments (Li et al., 2019). Base editing is used to create off-the-shelf T cells by reducing alloreactivity and modifying checkpoint inhibitors, with BE4 mRNA delivery proving more effective than BE4 RNP delivery for multiplex knockouts (Webber et al., 2019). B cells, due to their long-term persistence in the body, are also a promising target for gene therapy, with Cas9 RNP showing advantages in editing for disorders related to protein deficiencies (Eyquem et al., 2017).

HSPCs are crucial focal points for genome editing uses (Cornu et al., 2021). RNP complexes help improve genome editing by decreasing cellular toxicity, minimizing overall changes in gene expression, and reducing off-target effects. Furthermore, RNPs do not show higher toxicity levels when compared to mock-treated controls and continued to exhibit the ability to differentiate into all progenitor cells (Lattanzi et al., 2019).

11.5.2 Human Pluripotent Stem Cells (hPSCs)

Human pluripotent stem cells (hPSCs) are greatly valued for their potential in genetic modification and transplantation treatments because they can continuously self-renew and transform into any specific cell type (Yu et al., 2015). Cas9 plasmids and ssODN donor templates have effectively corrected mutations in the collagen type VII alpha 1 chain (COL7A1) gene, while RNPs have fixed homozygous frameshift mutations in iPSCs (Berthault et al., 2024). Moreover, the combination of Cas9 RNPs with AAV donor templates has shown effective transgene knock-in and gene correction in both iPSCs and ESCs (Gaj et al., 2017). The modified cells maintained their ability to develop into various types of cells and showed higher amounts of corrected adult hemoglobin genes, leading to a significant decrease in sickle hemoglobin levels (Gaj et al., 2017).

11.5.3 Zygotes

Editing zygotes using RNP is essential for studying mammalian embryogenesis and creating genetically modified animals for disease modeling. Initial CRISPR trials involved the introduction of Cas9 mRNA into mouse zygotes via either microinjection or electroporation, resulting in frequently low editing efficacy and significant mosaicism (Hashimoto et al., 2015). RNPs have been shown to be a better approach, decreasing mosaicism and enhancing editing efficiency to improve embryo viability. A new method, called improved genome editing via oviductal nucleic acids delivery (i-GONAD), has been created to edit mouse zygotes in their original location using RNPs (Sato et al., 2023). This technique leads to 57% mosaicism, improved editing success rates, and 97% HDR knock-in of a reporter cassette (Sato et al., 2023). Moreover, RNPs have been employed for generating specific point mutations in mouse zygotes using the BE3 base editor, resulting in accurate C-to-T conversions at the dystrophin (DMD) locus in 56% of blastocysts, with some unintended modifications observed (Xie et al., 2019). Using both 2PN and 3PN zygotes, RNP delivery alongside ssODNs effectively fixed mutations related to β-thalassemia in human embryos, making use of the healthy oocyte allele for HDR repair (Bloomer et al., 2022). Editing bovine zygotes using RNP resulted in a 92% indel formation at the octamer-binding transcription factor 4 (Oct4) gene, with only 1 out of 87 gene-corrected embryos successfully developing to the blastocyst stage (Doudna, 2015).

11.5.4 Retinal Cells

Due to the non-dividing nature of retinal cells, achieving high HDR efficiencies has historically been difficult, limiting the success of CRISPR-mediated gene correction (Yu and Wu, 2018). However, recent studies have shown that inhibiting p53 can improve HDR gene correction rates in retinal pigment epithelial-1 (RPE1) cells (Conti and Di Micco, 2018). Although HDR editing in retinal cells has shown potential, Cas9 delivery has predominantly relied on AAV vectors. To minimize the risk of inflammatory responses, non-viral delivery methods are currently being explored. For instance, researchers investigated the use of RNPs to target and disrupt the vascular endothelial growth factor A (VEGFA) gene in RPE cells (Ameri et al., 2020). Cells isolated from the injection site demonstrated a 25% indel formation rate at the VEGFA locus. Furthermore, knocking out VEGFA resulted in a reduction of CNV in a mouse model (Benati et al., 2020).

11.5.5 Cochlear Cells

CRISPR-Cas9 gene editing has shown potential in treating sensorineural hearing loss in animal models. Techniques to boost inner-ear cell proliferation could address progressive hearing loss from injury. Specific mutations in the TMC1 gene are linked to post-lingual sensorineural hearing loss in humans. Researchers employed RNP-based gene therapy in mice to mitigate hearing loss, and RNP-mediated base editing was used on inner ear sensory cells, including the organ of Corti, stria vascularis, and modiolus, enabling precise C-to-T conversions (Pan et al., 2017). These genetic modifications reduce β-catenin breakdown, enhance Wnt signaling, and promote hair cell growth, potentially reversing hearing loss due to environmental damage (Pan et al., 2017).

11.6 CHALLENGES TO *IN VIVO* DELIVERY OF CRISPR-CAS9 SYSTEM

Delivering CRISPR-Cas9 therapeutics *in vivo* presents unique challenges compared to traditional protein and gene therapies. Unlike single-component treatments, CRISPR-Cas9 therapy requires the simultaneous delivery of multiple functional components—namely, the Cas9 nuclease, guide RNA, and potentially additional donor DNA—into the same target cell nucleus. This complexity affects the overall efficiency of delivery and introduces logistical issues in synchronizing the timing and localization of these components within target cells (Lin et al., 2014). As a result, achieving effective *in vivo* CRISPR-Cas9 therapy requires overcoming these specific delivery challenges, alongside addressing important considerations related to the underlying biology, ethical concerns, and costs of the therapy (Wilson and Carroll, 2019).

11.6.1 Biochemical and Immunogenic Instability

The inherent biochemical properties of key CRISPR-Cas9 components—such as Cas9 protein, plasmid DNA, mRNA, sgRNA, and donor DNA—create significant challenges for their stability and delivery *in vivo*. The Cas9 protein is prone to degradation by proteases and hydrolysis within cells and tissues. Its positive charge can also lead to rapid clearance from the bloodstream due to binding with negatively charged blood and immune system components when administered systemically. Similarly, DNA and RNA molecules are unstable *in vivo*, quickly degraded by the body's nucleases. These exogenous molecules—including plasmid DNA, RNA, and protein—are also immunogenic and may provoke immune responses when introduced into the body. Although strategies like removing CpG motifs from plasmid DNA and chemically modifying RNA molecules have helped reduce immunogenicity, the use of Cas9 nucleases from *S. aureus* and *Streptococcus pyogenes* remains problematic due to pre-existing anti-Cas9 antibodies from these common pathogens (Hendel et al., 2015). To address these issues, CRISPR-Cas9 components need to be completely encapsulated within effective delivery vectors to maintain their stability and reduce immune responses. Additionally, research is ongoing to engineer bacterial Cas9 nucleases to eliminate immunogenic epitopes and to explore new CRISPR nucleases that provide enhanced genome editing efficiency and accuracy (Klinman et al., 1997).

11.6.2 Biodistribution of Delivery Vectors to Target Tissues after Systemic Injection

AAVs are highly esteemed in the viral vectors field for their ability to infect a wide range of cells in various organs and tissues, with their targeting specificity determined by the serotype. In order to improve AAV targeting to particular tissues, techniques like directed evolution, finding new AAVs from ancient species, and modifying capsid proteins are used. When considering non-viral vectors, the size of particles plays a crucial role (Cheng et al., 2021). After being injected intravenously, nanoparticles over several hundred nanometers tend to become stuck in lung capillary beds. However, nanoparticles under 100 nm are better at spreading throughout the body. Cationic components are frequently employed to give a positive surface charge to vectors in order to make it easier to deliver negatively charged DNA and RNA molecules (Guo and Huang, 2012). Nevertheless, this favorable electrical charge may result in quick attachment to negatively charged elements in the bloodstream, creating sizable complexes that often get stuck in the lungs, limiting their delivery to other specific tissues. Although advantageous for lung-focused therapies, this trait must be sufficiently concealed to target different tissues. Moreover, non-viral vectors circulating in the blood are promptly recognized as foreign by the host immune system, leading to swift elimination and decreased tissue-specific delivery (Cheng et al., 2021). PEGylation is often utilized to hide the surface charge and safeguard non-viral vectors from opsonization, thus improving their stability when stored (Capasso et al., 2013). Nevertheless, it is important to carefully manage the density of the PEG chains, as an excess amount can impede cellular uptake. Additionally, PEG can induce an immune reaction, causing the body to create anti-PEG IgM antibodies that help speed up the removal of PEGylated vectors when used repeatedly (Yang et al., 2015).

11.6.3 Delivery Specifically Targeted to Certain Types of Cells

Large negatively charged hydrophilic molecules like DNA and RNA find it difficult to get through the negatively charged hydrophobic cell membrane on their own. Although both viral and non-viral vectors may shield molecules against degradation in the bloodstream and tissues, they frequently have difficulty precisely targeting certain cell types after intravascular injection (Yin et al., 2014). To transport genes to certain cell types, non-viral vectors' exteriors can be modified with targeting agents, such as small molecule ligands, aptamers, peptides, and antibodies. To reduce off-target effects, certain cell types must be targeted in diseases that need systemic delivery of CRISPR-Cas9 systems. Yet specific conditions may still achieve success in genome editing therapies clinically, even in the absence of perfectly cell type–specific vectors. This is due to the fact that focusing on and fixing the genetic material of an ample number of cells in the impacted area using local delivery can result in substantial medical enhancements while reducing safety risks. The effectiveness of local administration for CRISPR-Cas9 genome editing therapy is mainly influenced by factors like disease stage, severity, target gene properties, and tissue type (Milton Harris and Chess, 2003).

11.6.4 Effective Release of CRISPR-Cas9 Cargoes into Cells

Delivery vectors for CRISPR are guided into early endosomes through receptor-mediated endocytosis, where they then move on to late endosomes and lysosomes (Rouet et al., 2018). Within these sections, the CRISPR-Cas9 components can be broken down by the acidic surroundings and enzymes. Releasing the cargo into the cytosol before endosomes mature into late endosomes or lysosomes is essential in order to avoid degradation, which poses a significant challenge for non-viral vectors as opposed to viral vectors (Rouet et al., 2018). Different approaches have been explored to enhance endosomal release, like employing cationic lipids to disrupt endosomal membranes or harnessing the "proton sponge" capability of PEI polymers. However, the effectiveness of these approaches remains limited. Membrane fusion peptides and cell-penetrating peptides show potential in disrupting endosomal and lysosomal membranes, which could enhance the escape of the cargo. Additionally, directly delivering nanoparticles to the cytosol through membrane fusion can bypass the endosomal pathway, thereby increasing the availability of the cargo in the cytoplasm. In summary, overcoming endosomal escape is a significant challenge for non-viral CRISPR delivery vectors, and ongoing research is focused on developing effective strategies to improve this critical step in the delivery process (Hoffman and Lai, 2020).

11.6.5 Efficient Nuclear Entry

The *nucleus* is the primary site where CRISPR-Cas9 components operate for the majority of genome editing purposes. During cell division, the nuclear envelope disintegrates, enabling the entry of large molecules such as Cas9 nucleases, plasmid DNA, mRNA, sgRNA, and donor DNA into the nucleus. Nevertheless, cells that do not divide, like neurons, myocytes, and adipocytes, present a difficulty because of their unbroken nuclear membranes. Further investigation is needed to explore the effectiveness of nuclear localization sequences in facilitating the entry of plasmid DNA into the nucleus *in vivo*. While CRISPR-Cas9 genome editing has been successful in both dividing and non-dividing cells in preclinical and clinical studies, the exact methods of how Cas9 nucleases move across the nuclear envelope are still not understood. It is essential to further comprehend this process, especially in non-dividing cells, to progress CRISPR-Cas9 technology and create better delivery systems in the future (Zhao et al., 2020).

11.7 CONCLUSION AND FUTURE PERSPECTIVES

CRISPR-Cas9 genome editing is a groundbreaking technology with tremendous potential for treating a wide array of inherited and acquired diseases. Despite its potential, most clinical trials have predominantly utilized *ex vivo* delivery methods, highlighting the significant challenge of achieving safe and effective *in vivo* delivery. To fully capitalize on the capabilities of CRISPR-Cas9, it is crucial to develop innovative delivery systems, as traditional gene and protein delivery methods often fall short. While progress has been made with both viral and non-viral vectors for *in vivo* applications, comprehensive data on their long-term efficacy and loading efficiency remains scarce. Advances in engineered Cas9 nucleases, novel biomaterials, and integrated vector systems capable of delivering all CRISPR-Cas9 components could propel the field forward. Challenges still exist in achieving systemic and cell-specific delivery, as well as in editing the mitochondrial genome. Innovative delivery systems that provide spatial and temporal control over CRISPR-Cas9 activity can further reduce off-target mutations. By integrating advancements in delivery science and fostering ongoing collaborations, CRISPR-Cas9 therapeutics are positioned to reach their full potential in the near future.

REFERENCES

Aghaamoo, M., Y. H. Chen, X. Li, N. Garg, R. Jiang, J. T. H. Yun, & A. P. Lee. (2022). "High-throughput and dosage-controlled intracellular delivery of large cargos by an acoustic-electric micro-vortices platform." *Advanced Science* 9(1).

Allahham, N., A. Buanz, S. Brocchini, & G. R. Williams. (2022). "Histidylated poly(lysine) dendrimers for siRNA delivery." *British Journal of Pharmacy* 7(2).

Ameri, Hossein, Christopher Murat, Amirmohsen Arbabi, Wei Jiang, Srikanth R. Janga, Peter Zhifeng Qin, & Sarah F. Hamm-Alvarez. (2020). "Reduced expression of vegf-a in human retinal pigment epithelial cells and human muller cells following CRISPR-Cas9 ribonucleoprotein-mediated gene disruption." *Translational Vision Science & Technology* 9(8): 23–23.

Amirkhanov, R. N., & G. A. Stepanov. (2019). "Systems of delivery of CRISPR/Cas9 ribonucleoprotein complexes for genome editing." *Russian Journal of Bioorganic Chemistry* 45: 431–437.

Awad, A. K., A. Ahmed, O. A. A. Mohamed, & M. A. Rais. (2024). "A healthy heart for all: Boosting cardiac surgery access in low-income countries." *International Journal of Surgery* 110(6).

Benati, D., C. Patrizi, & A. Recchia. (2020). "Gene editing prospects for treating inherited retinal diseases." *Journal of Medical Genetics* 57(7).

Berthault, Camille, Sonia Gaucher, Olivier Gouin, Alain Schmitt, Mei Chen, David Woodley, Matthias Titeux, Alain Hovnanian, & Araksya Izmiryan. (2024). "Highly efficient ex vivo correction of COL7A1 through Ribonucleoprotein-based CRISPR/Cas9 and homology-directed repair to treat recessive dystrophic epidermolysis bullosa." *Journal of Investigative Dermatology* 144(6): 1322–1333.

Betermier, Mireille, Pascale Bertrand, & Bernard S. Lopez. (2014). "Is non-homologous end-joining really an inherently error-prone process?." *PLoS Genetics* 10(1): e1004086.

Bloomer, Hanan, Jennifer Khirallah, Yamin Li, & Qiaobing Xu. (2022). "CRISPR/Cas9 ribonucleoprotein-mediated genome and epigenome editing in mammalian cells." *Advanced Drug Delivery Reviews* 181: 114087.

Bulcha, J. T., Y. Wang, H. Ma, P. W. L. Tai, & G. Gao. (2021). "Viral vector platforms within the gene therapy landscape." *Signal Transduction and Targeted Therapy* 6(1).

Butt, M. H., M. Zaman, A. Ahmad, R. Khan, T. H. Mallhi, Hasan, M. M., Y. H. Khan, S. Hafeez, E. E. S. Massoud, M. H. Rahman, & Cavalu, S. (2022). "Appraisal for the potential of viral and nonviral vectors in gene therapy: A review." *Genes* 13(8).

Capasso, Cristian, Mari Hirvinen, & Vincenzo Cerullo. (2013). "Beyond gene delivery: Strategies to engineer the surfaces of viral vectors." *Biomedicines* 1(1): 3–16.

Chen, X., Y. Chen, H. Xin, T. Wan, & Y. Ping. (2020). "Near-infrared optogenetic engineering of photothermal nanoCRISPR for programmable genome editing." *Proceedings of the National Academy of Sciences of the United States of America* 117(5).

Cheng, Hao, Feng Zhang, & Yang Ding. (2021). "CRISPR/Cas9 delivery system engineering for genome editing in therapeutic applications." *Pharmaceutics* 13(10): 1649.

Chu, Van Trung, Timm Weber, Benedikt Wefers, Wolfgang Wurst, Sandrine Sander, Klaus Rajewsky, & Ralf Kühn. (2015). "Increasing the efficiency of homology-directed repair for CRISPR-Cas9-induced precise gene editing in mammalian cells." *Nature Biotechnology* 33(5): 543–548.

Conti, Anastasia, & Raffaella Di Micco. (2018). "p53 activation: A checkpoint for precision genome editing?." *Genome Medicine* 10: 1–4.

Cornu, T. I., C. Mussolino, M. C. Müller, C. Wehr, W. V. Kern, & T. Cathomen. (2021). "HIV gene therapy: An update." *Human Gene Therapy* 32(1–2).

Coughlan, Lynda, Raul Alba, Alan L. Parker, Angela C. Bradshaw, Iain A. McNeish, Stuart A. Nicklin, & Andrew H. Baker. (2010). "Tropism-modification strategies for targeted gene delivery using adenoviral vectors." *Viruses* 2(10): 2290.

Davis, Luther, & Nancy Maizels. (2014). "Homology-directed repair of DNA nicks via pathways distinct from canonical double-strand break repair." *Proceedings of the National Academy of Sciences* 111(10): E924–E932.

Deng, Y., M. Kizer, M. Rada, J. Sage, X. Wang, D. J. Cheon, & A. J. Chung. (2018). "Intracellular delivery of nanomaterials via an inertial microfluidic cell hydroporator." *Nano Letters* 18(4).

Ding, Xiaoyun, Martin P. Stewart, Armon Sharei, James C. Weaver, Robert S. Langer, and Klavs F. Jensen. 2017. "High-throughput nuclear delivery and rapid expression of DNA via mechanical and electrical cell-membrane disruption." *Nature biomedical engineering* 1(3): 1–7.

Doudna, J. (2015). "Perspective: Embryo editing needs scrutiny." *Nature* 528(7580).

Eyquem, J., J. Mansilla-Soto, T. Giavridis, S. J. C. Van Der Stegen, M. Hamieh, K. M. Cunanan, A. Odak, M. Gönen, & M. Sadelain. (2017). "Targeting a CAR to the TRAC locus with CRISPR/Cas9 enhances tumour rejection." *Nature* 543(7643).

Foley, Ruth A., Ruby A. Sims, Emily C. Duggan, Jessica K. Olmedo, Rachel Ma, & Steven J. Jonas. (2022). "Delivering the CRISPR/Cas9 system for engineering gene therapies: Recent cargo and delivery approaches for clinical translation." *Frontiers in Bioengineering and Biotechnology* 10: 973326.

Freund, E. C., J. Y. Lock, J. Oh, T. Maculins, L. Delamarre, C. J. Bohlen, B. Haley, & A. Murthy. (2020). "Efficient gene knockout in primary human and murine myeloid cells by non-viral delivery of CRISPR-Cas9." *Journal of Experimental Medicine* 217(7).

Fu, Yanfang, Jeffry D. Sander, Deepak Reyon, Vincent M. Cascio, & J. Keith Joung. (2014). "Improving CRISPR-Cas nuclease specificity using truncated guide RNAs." *Nature Biotechnology* 32(3): 279–284.

Gaj, Thomas, Brett T. Staahl, Gonçalo M. C. Rodrigues, Prajit Limsirichai, Freja K. Ekman, Jennifer A. Doudna, & David V. Schaffer. (2017). "Targeted gene knock-in by homology-directed genome editing using Cas9 ribonucleoprotein and AAV donor delivery." *Nucleic Acids Research* 45(11): e98.

Gaspar, Vítor, Duarte de Melo-Diogo, Elisabete Costa, André Moreira, João Queiroz, Chantal Pichon, Ilídio Correia, & Fani Sousa. (2015). "Minicircle DNA vectors for gene therapy: Advances and applications." *Expert Opinion on Biological Therapy* 15(3): 353–379.

Ghosh, Siddhartha S., Packirisamy Gopinath, & A. Ramesh. "Adenoviral vectors: A promising tool for gene therapy." *Applied Biochemistry and Biotechnology* 133 (2006): 9–29.

Guo, X., & L. Huang. (2012). "Recent advances in nonviral vectors for gene delivery." *Accounts of Chemical Research* 45(7): 971–979.

Hashimoto, M., Y. Yamashita, & T. Takemoto. (2016). "Electroporation of Cas9 protein/sgRNA into early pronuclear zygotes generates non-mosaic mutants in the mouse." *Developmental Biology* 418(1).

Hashimoto, Masakazu, & Tatsuya Takemoto. (2015). "Electroporation enables the efficient mRNA delivery into the mouse zygotes and facilitates CRISPR/Cas9-based genome editing." *Scientific Reports* 5(1): 11315.

Hendel, A., R. O. Bak, J. T. Clark, A. B. Kennedy, D. E. Ryan, S. Roy, I. Steinfeld, B. D. Lunstad, R. J. Kaiser, A. B. Wilkens, R. Bacchetta, A. Tsalenko, D. Dellinger, L. Bruhn, & M. H. Porteus. (2015). "Chemically modified guide RNAs enhance CRISPR-Cas genome editing in human primary cells." *Nature Biotechnology* 33(9).

Hoffman, A. S., & J. J. Lai. (2020). "Three significant highlights of controlled drug delivery over the past 55 years: PEGylation, ADCs, and EPR." *Advanced Drug Delivery Reviews* 158.

Hung, K. L., I. Meitlis, M. Hale, C. Y. Chen, S. Singh, S. W. Jackson, C. H. Miao, I. F. Khan, D. J. Rawlings, & R. G. James. (2018). "Engineering protein-secreting plasma cells by homology-directed repair in primary human B cells." *Molecular Therapy* 26(2).

Jayakumar, R., K. P. Chennazhi, S. V. Nair, Tetsuya Furuike, & Hiroshi Tamura. (2010). "26 Chitosan-Conjugated DNA Nanoparticle Delivery." In *Chitin, Chitosan, Oligosaccharides and Their Derivatives: Biological Activities and Applications*, 357–366. CRC Press.

Kholosy, W. M., M. Visscher, K. Ogink, H. Buttstedt, K. Griffin, A. Beier, J. P. Gerlach, J. J. Molenaar, N. Geijsen, M. de Boer, & A. Chatsisvili. (2021). "Simple, fast and efficient iTOP-mediated delivery of CRISPR/Cas9 RNP in difficult-to-transduce human cells including primary T cells." *Journal of Biotechnology* 338.

Kim, S., D. Kim, S. W. Cho, J. Kim, & J. S. Kim (2014). "Highly efficient RNA-guided genome editing in human cells via delivery of purified Cas9 ribonucleoproteins." *Genome Research* 24(6), 1012–1019.

Klinman, D. M., G. Yamshchikov, & Y. Ishigatsubo. (1997). "Contribution of CpG motifs to the immunogenicity of DNA vaccines." *The Journal of Immunology* 158(8).

Kotin, R. M. (1994). "Prospects for the use of adeno-associated virus as a vector for human gene therapy." *Human Gene Therapy* 5(7).

Kowalski, P. S., A. Rudra, L. Miao, & D. G. Anderson. (2019). "Delivering the messenger: Advances in technologies for therapeutic mRNA delivery." *Molecular Therapy* 27(4).

Lattanzi, Annalisa, Vasco Meneghini, Giulia Pavani, Fatima Amor, Sophie Ramadier, Tristan Felix, Chiara Antoniani, et al. 2019. "Optimization of CRISPR/Cas9 delivery to human hematopoietic stem and progenitor cells for therapeutic genomic rearrangements." *Molecular Therapy* 27(1): 137–150.

Lee, K., M. Conboy, H. M. Park, F. Jiang, H. J. Kim, M. A. Dewitt, V. A. Mackley, K. Chang, A. Rao, C. Skinner, T. Shobha, M. Mehdipour, H. Liu, W. C. Huang, F. Lan, N. L. Bray, S. Li, J. E. Corn, K. Kataoka . . . N. Murthy. (2017). "Nanoparticle delivery of Cas9 ribonucleoprotein and donor DNA *in-vivo* induces homology-directed DNA repair." *Nature Biomedical Engineering* 1(11).

Li, D., X. Li, W. L. Zhou, Y. Huang, X. Liang, L. Jiang, X. Yang, J. Sun, Z. Li, W. D. Han, & W. Wang. (2019). "Genetically engineered t cells for cancer immunotherapy." *Signal Transduction and Targeted Therapy* 4(1).

Lin, S., B. T. Staahl, R. K. Alla, & J. A. Doudna. (2014). "Enhanced homology-directed human genome engineering by controlled timing of CRISPR/Cas9 delivery." *ELife* 3.

Lino, C. A., J. C. Harper, J. P. Carney, & J. A. Timlin. (2018). "Delivering CRISPR: A review of the challenges and approaches." *Drug Delivery* 25(1).

Liu, Guanwen, Qiupeng Lin, Shuai Jin, & Caixia Gao. "The CRISPR-Cas toolbox and gene editing technologies." *Molecular Cell* 82, no. 2 (2022): 333–347.

Liu, Mingjie, Saad Rehman, Xidian Tang, Kui Gu, Qinlei Fan, Dekun Chen, & Wentao Ma. "Methodologies for improving HDR efficiency." *Frontiers in Genetics* 9 (2019): 691.

Lomova, A., D. N. Clark, Campo-Fernandez, B., Flores-Bjurström, C., M. L. Kaufman, Fitz-Gibbon, S., X. Wang, E. Y. Miyahira, D. Brown, M. A. DeWitt, J. E. Corn, R. P. Hollis, Z. Romero, & D. B. Kohn. (2019). "Improving gene editing outcomes in human hematopoietic stem and progenitor cells by temporal control of DNA repair." *Stem Cells* 37(2): 284–294.

Lu, Y. (2004). "Recombinant adeno-associated virus as delivery vector for gene therapy - a review." *Stem Cells and Development* 13(1).

Ma, X., & J. R. Ashnest. (2020). "Genome editing: Small-size CasΦ is shining." *Molecular Plant* 13(9).

Marraffini, L. A. (2016). "The CRISPR-Cas system of Streptococcus pyogenes: Function and applications." In *Streptococcus pyogenes: Basic Biology to Clinical Manifestations,* 2nd edition. University of Oklahoma Health Sciences Center.

Milton Harris, J., & R. B. Chess. (2003). "Effect of pegylation on pharmaceuticals." *Nature Reviews Drug Discovery* 2(3).

Mu, Xin, & Sun Hur. (2021). "Immunogenicity of in vitro-transcribed RNA." *Accounts of Chemical Research* 54(21): 4012–4023.

Pan, B., C. Askew, A. Galvin, Heman-Ackah, S., Y. Asai, A. A. Indzhykulian, F. M. Jodelka, M. L. Hastings, J. J. Lentz, L. H. Vandenberghe, J. R. Holt, & Géleóc, G. S. (2017). "Gene therapy restores auditory and vestibular function in a mouse model of Usher syndrome type 1c." *Nature Biotechnology* 35(3).

Porto, E. M., A. C. Komor, I. M. Slaymaker, & G. W. Yeo. (2020). "Base editing: Advances and therapeutic opportunities." *Nature Reviews Drug Discovery* 19(12). https://doi.org/10.1038/s41573-020-0084-6

Rouet, Romain, Benjamin A. Thuma, Marc D. Roy, Nathanael G. Lintner, David M. Rubitski, James E. Finley, Hanna M. Wisniewska et al. (2018). "Receptor-mediated delivery of CRISPR-Cas9 endonuclease for cell-type-specific gene editing." *Journal of the American Chemical Society* 140(21): 6596–6603.

Ryan, D. E., Diamant-Levi, T., I. Steinfeld, D. Taussig, Visal-Shah, S., S. Thakker, B. D. Lunstad, R. J. Kaiser, R. McCaffrey, Ortiz, M., J. Townsend, W. R. W. Welch, M. Singh, Curry, B., D. J. Dellinger, & Bruhn, L. (2023). "Phosphonoacetate modifications enhance the stability and editing yields of guide RNAs for Cas9 editors." *Biochemistry* 62(24).

Saber Sichani, Ali, Maryam Ranjbar, Maryam Baneshi, Farid Torabi Zadeh, & Jafar Fallahi. (2023). "A review on advanced CRISPR-based genome-editing tools: Base editing and prime editing." *Molecular Biotechnology* 65(6): 849–860.

Sato, Masahiro, Kazunori Morohoshi, Masato Ohtsuka, Shuji Takabayashi, Emi Inada, Issei Saitoh, Satoshi Watanabe, & Shingo Nakamura. (2023). "Recent advances in the production of genome-edited animals using i-GONAD, a novel in vivo genome editing system, and its possible use for the study of female reproductive systems." *OBM Genetics* 7(4): 1–30.

Sato, Yusuke, Hiroto Hatakeyama, Yu Sakurai, Mamoru Hyodo, Hidetaka Akita, & Hideyoshi Harashima. (2012). "A pH-sensitive cationic lipid facilitates the delivery of liposomal siRNA and gene silencing activity in vitro and in vivo." *Journal of Controlled Release* 163(3): 267–276.

Shen, B., J. Zhang, Wu, H., J. Wang, Ma, K., Z. Li, Zhang, X., P. Zhang, & Huang, X. (2013). "Generation of gene-modified mice via Cas9/RNA-mediated gene targeting." *Cell Research* 23(5).

Taghdiri, Maryam, & Claudio Mussolino. (2024). "Viral and non-viral systems to deliver gene therapeutics to clinical targets." *International Journal of Molecular Sciences* 25(13): 7333.

Thomas, C. E., A. Ehrhardt, & M. A. Kay. (2003). "Progress and problems with the use of viral vectors for gene therapy." *Nature Reviews Genetics* 4(5).

Walther, J., D. Wilbie, V. S. J. Tissingh, Öktem, M., Van Der Veen, H., B. Lou, & Mastrobattista, E. (2022). "Impact of formulation conditions on lipid nanoparticle characteristics and functional delivery of CRISPR RNP for gene knock-out and correction." *Pharmaceutics* 14(1).

Wang, Y., B. Wang, Xie, H., Q. Ren, Liu, X., F. Li, Lv, X., X. He, Cheng, C., R. Deng, Li, J., J. Zhao, Song, Z., & Gu, F. (2019). "Efficient human genome editing using SaCas9 Ribonucleoprotein Complexes." *Biotechnology Journal* 14(7).

Webber, Beau R., Cara-lin Lonetree, Mitchell G. Kluesner, Matthew J. Johnson, Emily J. Pomeroy, Miechaleen D. Diers, Walker S. Lahr et al. (2019). "Highly efficient multiplex human T cell engineering without double-strand breaks using Cas9 base editors." *Nature Communications* 10(1): 5222.

Wilson, R. C., & Carroll, D. (2019). "The daunting economics of therapeutic genome editing." *The CRISPR Journal* 2(5).

Xie, Jingke, Weikai Ge, Nan Li, Qishuai Liu, Fangbing Chen, Xiaoyu Yang, Xingyun Huang et al. (2019). "Efficient base editing for multiple genes and loci in pigs using base editors." *Nature Communications* 10(1): 2852.

Xu, Xiaojie, Tao Wan, Huhu Xin, Da Li, Hongming Pan, Jun Wu, & Yuan Ping. (2019). "Delivery of CRISPR/Cas9 for therapeutic genome editing." *The Journal of Gene Medicine* 21(7): e3107.

Yang, Qi, & Samuel K. Lai. (2015). "Anti-PEG immunity: Emergence, characteristics, and unaddressed questions." *Wiley Interdisciplinary Reviews: Nanomedicine and Nanobiotechnology* 7(5): 655–677.

Yang, Wu, Jiaqi Yan, Pengzhen Zhuang, Tao Ding, Yu Chen, Yu Zhang, Hongbo Zhang, & Wenguo Cui. (2022). "Progress of delivery methods for CRISPR-Cas9." *Expert Opinion on Drug Delivery* 19(8): 913–926.

Yin, H., R. L. Kanasty, A. A. Eltoukhy, A. J. Vegas, J. R. Dorkin, & D. G. Anderson. (2014). "Non-viral vectors for gene-based therapy." *Nature Reviews Genetics* 15(8).

Yin, H., K. J. Kauffman, & D. G. Anderson. (2017). "Delivery technologies for genome editing." *Nature Reviews Drug Discovery* 16(6).

Yu, C., Y. Liu, Ma, T., K. Liu, Xu, S., Y. Zhang, Liu, H., La Russa, M., M. Xie, Ding, S., & L. S. Qi. (2015). "Small molecules enhance CRISPR genome editing in pluripotent stem cells." *Cell Stem Cell* 16(2).

Yu, W., & Wu, Z. (2018). "*In-vivo* applications of CRISPR-based genome editing in the retina." *Frontiers in Cell and Developmental Biology* 6(May).

Zaboikin, Michail, Tatiana Zaboikina, Carl Freter, and Narasimhachar Srinivasakumar. (2017). "Non-homologous end joining and homology directed DNA repair frequency of double-stranded breaks introduced by genome editing reagents." *PLoS One* 12(1): e0169931.

Zhang, S., J. Shen, Li, D., & Cheng, Y. (2020). "Strategies in the delivery of Cas9 ribonucleoprotein for CRISPR/Cas9 genome editing." *Theranostics* 11(2).

Zhao, Z., A. Ukidve, Kim, J., & Mitragotri, S. (2020). "Targeting strategies for tissue-specific drug delivery." *Cell* 181(1).

Zuris, J. A., D. B. Thompson, Y. Shu, J. P. Guilinger, J. L. Bessen, J. H. Hu, M. L. Maeder, J. K. Joung, Z. Y. Chen, & D. R. Liu. (2015). "Cationic lipid-mediated delivery of proteins enables efficient protein-based genome editing *in-vitro* and *in-vivo*." *Nature Biotechnology* 33(1).

12 Ethical Issues in Genome Editing

Huma Hameed, Muhammad Zaman, Saleha Faheem, Mahtab Ahmad Khan, Shazia Akram Ghumman, Tooba Mehboob, and Naila Tabassam

12.1 INTRODUCTION

After the discovery of Deoxyribonucleic acid (DNA), efforts were undertaken to modify the DNA sequence at specific locations (Ramezani & Dietz, 2020; Schloss et al., 2020). Since the 1990s, advances in genome engineering technology have led to the development of the incredibly complex, straightforward, inexpensive, and user-friendly clustered regularly interspaced palindromic repeats (CRISPR) and CRISPR-associated protein 9 (Cas9) system of genome editing (Giani et al., 2020; Mattei et al., 2022). CRISPR technology is not a breakthrough, although other genome editing platforms, such as transcription activator-like effector nucleases (TALENS) (Harale et al., 2022) and zinc finger nucleases (ZFNs) (Harmatz et al., 2022; Katayama et al., 2024), were used for a while before falling out of favor because of their complexity, expense, and time consumption. Unlike CRISPR, which needs a guide RNA (gRNA) molecule capable of identifying the target sequence, the aforementioned methods include protein engineering (Gerashchenkov et al., 2020; Kose et al., 2024). The field of genome engineering has enormous potential to advance agriculture, medicine, food, and health care (Ahmar et al., 2021; Harish et al., 2022). Thanks to CRISPR-Cas9, biological advancements have seen a revolution (Javaid et al., 2022; Ou et al., 2021). Through the investigation of genomic rearrangements and the progression of cancer, this method holds promise for the treatment of hereditary disorders (Chakravarty & Solit, 2021; Claringbould & Zaugg, 2021). Despite its wide range of applications, ethicists have serious worries about CRISPR's ability to rapidly modify human germline cells in a way that would be passed on to progeny (Doss, 2022).

CRISPR was initially described as short straight repeats scattered with short sequences when it was discovered in the *Escherichia coli* (*E. coli*) genome by Japanese researchers in 1987 (El-Ashry, 2023). CRISPR was eventually discovered to play a role in gene regulation and DNA repair in a wide range of bacteria and archaea (Vigouroux & Bikard, 2020). A significant portion of the spacer sequences linked to CRISPRs were discovered in 2005 to be derived from plasmid and viral sequences (Mosterd et al., 2021; Shmakov et al., 2020). Numerous research groups have hypothesized that adaptive immunity may be influenced by CRISPR-Cas9 (Buquicchio &

DOI: 10.1201/9781003537151-12

Satpathy, 2021; Garcia-Robledo et al., 2020). This idea was later verified in 2007 by researchers looking into *Streptococcus thermophilus* (Hou et al., 2024; Proust et al., 2020). Cas proteins stop a virus from replicating inside prokaryotes (*E. coli*), according to a study published in *Science* (Makarova et al., 2020). The ability of CRISPR-Cas9 to target DNA was demonstrated the same year (Jacinto et al., 2020).

This book chapter seeks to provide a concise analysis and road map of all the ethical concerns brought up by the CRISPR technology. There is already a wealth of literature on the ethical difficulties raised by CRISPR, but most of the works that have been published up to this point have tended to concentrate on a small number of these problems (and particularly on those that are technical in nature) (Lange & Kappel, 2022). Our goal is to provide a comprehensive assessment of any subject that comes up, taking into account sociological, anthropological, and technical factors. Though they are not unique to CRISPR, some of them appear in nearly every genome editing approach, yet CRISPR may make them more obvious, faster, or more intense (Lorenzo et al., 2022). Overall, this chapter covers the following subjects in providing this worldwide overview: the features of the technology and its application, associated ethical concerns, the technique's applications and goals, and lastly, the social justice dilemma.

12.2 IMPORTANCE OF CRISPR AND CAS9 APPROACH

The CRISPR approach is a genome editing method that was initially employed in 2012 by a team of University of Berkeley academics (Wang & Doudna, 2023). Additionally, there are other, comparable recombinant DNA editing methods, notably ZFN and TALEN; however, CRISPR offers a superior compromise of three important aspects: affordability, accessibility, and accuracy (Bisht et al., 2024; Panigrahi et al., 2023). In addition, it is simpler to utilize than the alternatives because the latter need for more workers with specialized training and more time (Mindell & Reynolds, 2023).

It is a process that begins with bacterial DNA, which acts as an immune system to fight viruses (Bernheim & Sorek, 2020). These sequences are able to identify viruses that infiltrate bacteria and "activate" an enzyme that degrades the bacterium (Grabowski et al., 2021). The enzyme uses the pieces that are produced as a result to immunize the bacteria against the virus (Seeberger, 2021). This procedure is made feasible by CRISPR (the RNA molecule that transfers the biological information found in DNA for protein synthesis) and Cas9, a specialized bacterial enzyme that is capable of precisely repairing DNA breaks (Schwartz et al., 2021). Prokaryotic organisms have an innate defense mechanism, called CRISPR-Cas9, that protects them against invading genetic material, like plasmids and bacteriophages.

Short DNA repeats with plasmid or viral origin or CRISPR sequences are a type of repeats found in bacterial genomes that are consistently spaced and clustered together (Ganger et al., 2022; Pinilla-Redondo et al., 2022). CRISPR-Cas9 are the genes that encode helicase plus nuclease proteins that are connected to CRISPR repeat sequences that cut or disintegrate DNA (Mohamadi et al., 2020). Cas9 is an endonuclease that belongs to the Cas gene family and has the ability to cut DNA from two active cut sites on either end of the DNA double helix (Kantor et al., 2020;

Khanzadi & Khan, 2020). The source was found to be *Streptococcus pyogenes*. The CRISPR-Cas system recognizes a virus or bacteria's DNA when it enters the body and destroys it (Nazir et al., 2024). It was later demonstrated that genome editing may be accomplished by instructing the CRISPR-Cas system to find as well as cut specific target DNA regions (Li et al., 2023a; Malinin et al., 2021). The discovery that CRISPR-Cas9 can modify the human genome, which can be utilized for therapeutic and diagnostic applications, signaled the start of genetic engineering's new era (Cao et al., 2021; Zaib et al., 2022). After 2020, an excess of CRISPR-Cas9-based tools for DNA and ribonucleic acid (RNA) gene editing had been developed due to the technology's fast progress (Batool et al., 2021; Shanmugam et al., 2020).

12.3 ETHICAL ISSUE LINKED TO CRISPR-CAS9 TECHNOLOGY

The ethical implications of human genetic alterations have long been a topic of discussion; CRISPR-Cas9-mediated genome editing has offered a fresh viewpoint. A careful analysis of the ethical and societal ramifications of this technology is necessary, given the unpredictable nature of its uses and their far-reaching effects. Fundamental beliefs are held by the public as well as religious scholars, among other segments of society (Singh, 2024). A rapidly developing tool in molecular biology, CRISPR-Cas9 enables researchers to modify the DNA of a variety of species to their desired specifications. Since its introduction in 2012, its simplicity and convenience of use have led to its rapid adoption. Applications in clinical medicines, agriculture, and human genetic modifications to treat hereditary illnesses are also being explored. Though not entirely new, the ethical discussion around human genetic alteration has gained momentum thanks to CRISPR-Cas9-mediated genome editing. Owing to its erratic and extensive effects, a thorough discussion of its moral and societal ramifications is required. This study addresses the moral issues raised by this new technology, highlighting the necessity of in-depth discussion among scientists, ethicists, businesspeople, and decision-makers, as well as the views of the general public and religious scholars (Gostimskaya, 2022).

12.4 WHY ETHICAL CONCERNS ARISE

The discovery of CRISPR-Cas9 has transformed the field of molecular biology research by opening up new avenues for genetic alteration, cell mutation, and gene editing (Kropf, 2024). It may be used to change the genes of plants and animals to produce drought-resistant crops or powerful police dogs, as well as to turn on or off genes to investigate their functions and cause mutations in cells to investigate cell cancer. Additionally, CRISPR has been used to permanently change the human genome in order to remove mutations that lead to sickness or improve desired traits in young people (Valdés & Lecaros, 2023). Due to the fact that genetic modifications produced in non-reproductive cells are not heritable but can be passed on to the progeny, technology has made genetically modified people and designer babies a reality. This has sparked a new discussion over human germline changes and prompted a wave of genome editing research throughout the kingdom of life, primarily limited to non-human species. Chinese researchers used CRISPR-based gene editing in human embryos for a study they did in 2015, which sparked discussion about the technology's potential to create designer babies and ethical issues. Despite using non-viable embryos for the experiment, it proved the principle and might be used to live embryos in the future. Concerns about the environment, particularly those pertaining to human genomes, have been raised by CRISPR-mediated alterations in plants, such as improved crop yield (Souza et al., 2002).

12.5 ETHICAL ISSUES OF CRISPR-CAS TECHNOLOGIES ON MAMMALIAN SYSTEMS

Concerns about ethics include whether or not CRISPR would be utilized to improve human traits like memory, learning ability, and muscular bulk, and if access to the technology would be equal for all people. There are ethical concerns regarding the possibility of editing human germline (eggs, sperm, and very early embryos) using CRISPR (Gonzalez-Avila et al., 2021). For instance, how much may genetic alterations impact next generations, and is it appropriate to do research on human embryos? Eighteen scientists from seven nations—the United States, Canada, France, and China—called for a voluntary stop to any research involving the altering of the human germline in March 2018 (Khambhati et al., 2020).

12.5.1 Rationale for Hazards and Outcomes

The need that benefits outweigh hazards is a crucial ethical consideration in research. Risks need to be given more consideration since they could harm the environment or living things. The use of the CRISPR-Cas9 technology carries some risk since it might result in harmful target alterations. Human cells have a high frequency of off-target effects, but mice and zebrafish have modest frequencies. One issue is that several DNA sequences that are very similar or identical to the intended target DNA sequence can be found in huge genomes (Li et al., 2023b). These unwanted sequences may also be cut by CRISPR-Cas9, resulting in alterations that might lead to cell death or transformation. Attempts have been performed to decrease off-target mutations; nonetheless, further development is required, particularly for the exact adjustments required for therapeutic treatments, where improvement is required. The safe distribution of CRISPR-Cas9 to difficult-to-transfect or infect cell types or tissues is a significant concern as well (Uddin et al., 2020).

12.5.2 A State of Ecological Imbalance

When conducting RNA-guided gene drives with the CRISPR-Cas9 technology, it is imperative to examine specificity while taking into account aim effects. Since the human gene drive is still active, there is always a chance that the target may mutate, and this chance might rise with each generation. If there's a chance that genes will be transferred to other species, there's also a chance that altered sequences will be transferred, which might spread the undesirable characteristic over national boundaries to related creatures. Controlling the distribution of the gene drive characteristic might be challenging. Moreover, the loss of an entire population that is the focus of a gene drive might have a significant impact on the balance of the environment. For instance, new plagues could emerge. He insists on a thorough evaluation of every possible use and the requirement for regulatory standards. Safety precautions are required to prevent the spread of organisms that might harm the environment or have an adverse effect on human health (Hemalatha et al., 2023).

12.5.3 Regulations Linked to Consumer

The ability of the CRISPR-Cas9 system to produce precise genetic alterations makes it more challenging to control genetically modified organisms in the marketplace and to detect them once they are outside of the lab. Any genetically modified organism intended for consumer use must be approved by regulatory agencies, like the US Food and Drug Administration. It is unclear how these agencies will respond to the prospect of an enlarged CRISPR-Cas9 market. Patent regulation is another problem. There are several financial interests at play here (Sarma et al., 2021). For an extended period, transgenic organisms and human gene sequences intended for therapeutic application have been protected by patents, allowing for the substantial advancement of biotechnology. On the other hand, patenting may lead to legal disputes. Biotech companies have already clashed and disagreed on patenting CRISPR-Cas9 for human medicinal applications. Researchers Doudna and Charpentier squared off against Zhang in a patent battle over a method for using human cells. The scope of patenting might likewise be overly wide. Patenting, on the one hand, presents an ethical dilemma that prioritizes business and gives biotechnology enterprises, while patenting might contribute to industry regulation. However, the focus ought to be on releasing or commercializing only safe goods (Mali, 2022).

12.5.4 Modifying the Genome for Improvement

The potential use of genome editing for non-therapeutic treatments raises another ethical concern. For safety considerations, its usage in germlines is prohibited. However, the effectiveness of the CRISPR-Cas9 method raises the prospect of manipulating somatic cells to tailor genetics to our desired lives. A genetic component contributes toward many phenotypic traits, independent of environment, which is modifiable. The method may be used, for instance, to improve the performance of athletes, deter aggressive behavior, or to reduce addiction. Gene therapies often aim to enhance a patient's health for its own advantage. However, in the future, the criminal justice system may require violent, dangerous repeat offenders or those with violent genes to have their genomes edited (Barman et al., 2020). Informed consent issues arise when an intervention is carried out during a child's developmental stage, since it is debatable whether parents or guardians should have the authority to make decisions about their future on behalf of their offspring for reasons unrelated to health. If certain groups or people have hereditary advantages over others, such as superior intelligence, this will cause social problems (Mushtaq et al., 2021).

12.5.5 Creation of Animal Chimeras in Preparation for Organ Replacement

Many people who must wait a long time for a human organ donor may find hope in the development of human/animal chimeras for organ donation. However, the development of these chimeras could contain human germ and brain cells. Because of their danger and the way, they defy the natural order, chimpanzees have sparked ethical debate over whether to treat them like animals or like humans (Jamil et al., 2023). Some believe that chimera embryos have the capacity to produce beings with tissue or cells obtained from humans, which might have an impact on the dignity and identity of the human species. However, the presence of human cells in an organism does not make it human, nor does it diminish its dignity. The chimera's human-like traits are purely biological in origin; they have no bearing on the animal's moral standing or ability to become aware (Babochkina et al., 2020).

12.5.6 Ethical Standpoints of CRISPR-Cas9

Concerns regarding human safety, dignity, and the possibility of eugenics exploitation have been raised by the CRISPR-based gene-editing project for human germline alteration. A stakeholder group demanded that human genome research be put on hold until a consensus was established on society's acceptance of it. The purpose of the Napa, California, moratorium was to spark a public discussion on the moral and societal implications of the technology. The technology's cost–benefit analysis has divided scientists, with Janssens and Cecile voicing worries about its possible application to human improvement (Doss, 2022).

Although CRISPR can improve desired qualities, handling numerous variations and complicated attributes like intellect limits its technological viability. Editing encompasses more than simply adjusting a single gene; it also includes merging many genes and seeing how they interact with one another and the environment (Panuntun, 2020). Future editing will be uncertain as a result. Rather than treating complicated diseases like diabetes or cancer, initial trials on human embryos concentrated on rectifying gene mutations in disorders like β-thalassemia, a recessive genetic ailment caused by a single mutation. It is predicted that more complicated circumstances are not viable to improve with current technology; however, polygenic features or illnesses show considerable potential (Seiter & Fuselier, 2021).

Considering, the possibility of off-target mutations and the challenge of fully comprehending the effects of gene editing on subsequent generations, CRISPR technology is not yet developed enough to regulate human heredity. However, these technological problems may be resolved with little optimization (Tang et al., 2021). Sharma et al. (2023) reported that ethical norms require that this kind of study be limited to embryos that are less than 14 days old. They thus endorse the appropriate use of germline editing technologies in human embryos (Sharma et al., 2023). Lander et al. (2020) reported that until there are severe genetic illnesses and no other viable treatment options, genome editing techniques used to human defective zygotes should be prohibited. Moreover, this study also drew attention to the ethical issues raised by human genome editing, such as the possibility of utilizing the finest genomes for the wealthiest people in order to create hereditary class differences. The use of genome editing technology does not raise any significant ethical issues, notwithstanding these worries (Lander et al., 2020; Santiago-Frangos et al., 2023).

12.6 POSSIBLE RISKS RELATED TO CRISPR-CAS9 TECHNOLOGY

Using CRISPR may have unforeseen repercussions. It may, for instance, accidentally target a particular place in the DNA and alter it, perhaps leading to illness or other negative effects. Although some studies suggested that mistakes might not happen often, the full effects of these mistakes are yet unknown (Rafii et al., 2022). Furthermore, CRISPR activity may cause cell stress and inhibit editing. Furthermore, after having their DNA changed, some cells can recover, but not all of them. Acceptance of germline editing technology depends critically on patient safety, especially in clinical settings, where genetic abnormalities may be avoided (Rasul et al., 2022). On the other hand, care must be taken to prevent negative outcomes. By effectively modifying the Cas9 enzyme to decrease the quantity of off-target mutations, Feng Zhang, a synthetic biologist at Harvard and Massachusetts Institute of Technology, may allay worries about the safety of CRISPR-mediated germline editing. Even with the speed at which science is progressing, the therapeutic use of this technology is still a long way off (Cheng et al., 2021). Possible risk factors associated to CRISPR-Cas9 applications have been explained in Table 12.1.

12.6.1 Patient's Safety Issues

Clinical studies with CRISPR-Cas-based gene therapies can ensure patient safety by closely monitoring dosage administration and making sure the treatments wear off after a certain amount of time. Since germline modifications are irreversible, it could be necessary to create new policies or make adjustments to safeguard patients from unfavorable side effects (Yang et al., 2022). The risk potential of CRISPR-Cas technology varies depending on the application; certain applications may not be feasible anytime soon, while others may be acceptable or expected to be relevant in the near future. Positively, CRISPR-Cas technology has the potential to significantly improve the environment and human health, although this will mostly rely on how the technology is applied (Cheng et al., 2020).

12.6.2 Immune Reaction to the Natural Cas9 Protein

One of the three primary parts of the CRISPR system's structure, the Cas9 protein is necessary for binding the double-stranded DNA, coupled-cutting it precisely three nucleotides before the protospacer adjacent motif (PAM) sequence using the mRNA guide. This protein comes from the bacteria *Streptococcus pyogenes*, which causes various common diseases in people. The body reacts to it by recognizing it as an antigen and mounting an immunological defense (Mehta & Merkel, 2020). Similarly, it has been found that *Staphylococcus aureus* has an established immune response to the homologous Cas9 protein. Indeed, the two bacteria (*Staphylococcus aureus* and *Streptococcus pyogenes*) produce the two primary Cas9 proteins i.e. *Staphylococcus aureus* Cas9 (SaCas9) and *Streptococcus pyogenes* Cas9 (SpCas9). However, because the human immune system perceives these proteins as alien, it produces an immune reaction to them after injection, which causes the Cas9 protein to quickly degrade and stop being able to modify genes (Behr et al., 2021).

TABLE 12.1
Possible Risk Factors Associated with CRISPR-Cas9 Applications

Organism	Risk Factor	CRISPR-Cas Application	References
Natural plants	Comparatively little to no danger (fewer off-target effects, such as in zebrafish)	Enhancing crops, adding disease resistance, eradicating dangerous invading species, and using gene drive to reverse the resistance of pesticides as well as herbicides in weeds and insects.	Zhu et al., 2020
Animal models	Quite high (target mutations, characteristics transferred to unrelated creatures in the environment, ecological imbalance brought about by the eradication of a population targeted by gene drive)	Animal models are used in research to use gene silencing or mutation to analyze illnesses or the development process. For instance, using a mouse model to research how mutations affect cancer.	Mehra & Kumar, 2022
Humans	Moderate (mutations of the target) Extremely high (informed consent, uncertain effects on future generations, and the requirement to reorganize clinical trial protocols)	Treat illnesses by introducing new genes with protective properties, repairing mutations that cause disease, or replacing endogenous genes that cause disease (somatic gene therapies). Genetic modification to treat complicated genetic illnesses, like diabetes, or to introduce desirable qualities, like intellect, etc.	Hendriks et al., 2020

12.7 CHALLENGES LINKED TO CRISPR-CAS9 TECHNOLOGY

Concerns regarding safety, ethics, and regulations have also been raised by CRISPR gene editing. As examples, consider the following.

12.7.1 Off-Target Effects

With CRISPR-Cas9 genome editing, off-target mutations are a big worry, since they may hurt both people and the environment. Because bigger genomes potentially have numerous similar sites for cleavage, which can result in aberrant transformations and cell death, research shows that humans have a higher frequency of off-target mutations than zebrafish. In cells that are difficult to infect, CRISPR-Cas9-based therapies are challenging to implement. Acknowledgment of the technique hinges on accurate editing using CRISPR-Cas9 (Guo et al., 2023). The integrity of the environment may be threatened by off-target mutations; thus, it is important to weigh the advantages and hazards. Gene drive is a continuous process that can move genes between organisms, changing the surrounding environment. Certain experts alert about the potential dangers of unintentionally releasing genetically modified organisms into the environment. The reason CRISPR is so effective is that it can modify DNA rapidly—only takes a few days to do so. But there's also concern about the possibility that CRISPR may go awry and cause plants to become invasive weeds or create infectious viruses that attack people or crops (Newman et al., 2020). In order to regulate organisms that might cause environmental damage, safety and security measures should be created together with regulatory criteria for the deployment of CRISPR. The primary methods that have been effectively used in earlier research may be divided into three categories: (1) use of bioinformatics tools to develop more precise gRNA and targeting prediction, (2) use of Cas9 nickases, and (3) inclusion of anti-CRISPR proteins (Khoshandam et al., 2024).

12.7.2 CRISPR-Cas9 Delivery Challenges

Targeting the appropriate sequence inside the tumor and selecting an appropriate, safe, and accurate delivery method to introduce the CRISPR system into the tumor location, particularly in vivo, one further problem to take into account is the nucleus. Delivery methods for CRISPR-Cas9 technologies include physical, viral, and extracellular vesicle-based system delivery methods. Every technique has its limits and is employed for a certain reason (Son & Park, 2022). The primary obstacles in selecting the appropriate vector, however, are those related to packing, distribution, and site selection. For instance, there are several restrictions on the use of viral vectors, including insertional constraint and immune response, despite their widespread usage in both *in vivo* and *in vitro* settings. For example, it raises the danger of mutations and mistargeting after assuming viral vector *in vivo* and exposing it constantly for an extended period of time (Yip, 2020).

12.7.3 Regulatory Issues

The ease with which CRISPR-Cas9 may cause genomic alterations renders it equally challenging to detect the same altered creature outside of the lab, which prompts questions about their legal framework. We may anticipate an enlarged market for genetically modified organisms, which will raise concerns about legislation, if CRISPR-Cas9 continues to spread over the world. Furthermore, patenting concerns must be settled. There is disagreement among scientific organizations over the patenting of genetically altered organisms for medical purposes. There are a lot of financial interests related to CRISPR-Cas9 (Wani et al., 2022). It is anticipated that CRISPR-Cas9 technology patents will grant the relevant businesses a great deal of influence. Numerous countries are currently grappling with the question of how to regulate CRISPR and other gene-editing technologies. In the United States, for example, it is currently unclear whether crops that have undergone gene editing would fall under the same legal framework as conventional genetically modified (GM) plants (Gupta et al., 2021). Federal Drug Administration (FDA) began soliciting public input on its regulatory policy in 2017, but as of March 2020, the agency had not issued any more industry guidance about its current regulatory strategy. Genetically engineered crops will be subject to the same regulations as genetically modified plants, according to a 2018 ruling by a European court (Barman et al., 2020).

12.8 IMPROVEMENTS IN ETHICAL CONCERNS LINKED TO CRISPR-CAS9 TECHNOLOGY

12.8.1 Role of Biohacker Communities

Biohacker communities are dedicated toward research, development, as well as innovation in scientific and technological activities. They argue that pharmaceutical companies benefit from developing methods that can be performed at home without any clinical controls or medical supervision and at an affordable cost (Lim, 2021). The main task of these communities is to generate treatments to increase lifespan, cure the diseases, as well as increase human capacities. These communities support ethical arguments and philosophical theories, such as DIYBio, or "do it yourself biology," which aims to increase human capacity within the framework of freedom (Meyer & Vergnaud, 2020). Since it was brought up at CodeCon in 2005, DIYBio has become more and more relevant. Free research in DNA biology was encouraged, which sparked debates over the morality of synthetic biology and the alteration of eukaryotic creatures. FDA of the United States has identified biohackers operating inside its borders who are experimenting with therapies such as intradermal injections of CRISPR-Cas-modified DNA molecules (Voort, 2022).

12.8.2 Legal Aspects in CRISPR-Cas9

Even though DNA technology is lawful, concerns over the usage of recombinant DNA have been expressed since the 1975 Asilomar meeting. The bioethical arguments surrounding genetic engineering are still being discussed; in the 1960s, ethical concerns raised doubts about gene therapy. Human gene therapy was allowed up to 1990, but dependability data was not discovered until after 2000. The first well-known bioethical question surrounding genetic alteration is whether or not to permit CRISPR-Cas technology for gene editing, given that human application of this technology is improbable. Concerns over gene-editing regulation have been voiced in a number of international declarations; the Hinxton Group and the Academy of Sciences of the United States of North America contend that clinical research should not be restricted because to abuse worries. The Academy of Sciences of the United States of North America and the Hinxton Group met in the UK in 2015 and voiced concerns on the regulation of gene editing (Dayan, 2020).

Gene-editing research on embryos were permitted by the UK's Authority for Human Embryology and Fertilization, which sparked debates over the possibility of creating HIV-resistant individuals by CRISPR-Cas modification. Mexico is a center for stem cell therapies, but because of its lack of scientific and legal oversight and easy access to untested uses, it is prone to bioethical issues. The General Health Law, Regulation of the Law on Research Matters, Official Mexican Standard NOM-012-SSA3–2012, and the Declaration of Helsinki govern research involving human subjects or human tissues in Mexico (Ahmad et al., 2021).

Because they are simple to conceive but difficult to implement, CRISPR-Cas-based genetic alteration approaches are new because they enable the genetic manipulation of people. Although there are legitimate concerns about its usage, since it may change an organism's physiology, if health issues are resolved, it is still seen as advantageous. The scientific community is concerned, meanwhile, about the absence of defined legal boundaries for scientists, which leaves room for interpretation about the extent of genetic alteration and its implications for bioethical behavior (D'Souza et al., 2023).

Not just scientists and medical professionals should be able to examine the state of these technologies, evaluate the advantages and disadvantages of therapeutic genome editing, and scrutinize existing information about them. A public dialogue that takes into account both scientific and ethical realities should be encouraged, and bioethicists are qualified to mediate and are knowledgeable about the advantages and disadvantages of these new technologies. Assuring the moral application of these systems to close gaps and social inequities while taking sociopolitical, economic, and historical considerations into account should be the government's duty in enforcing rules.

To sum up, the goal of this chapter is to dispel the concerns and fears that the world's population has created by starting a conversation about the true problems with these technologies (Janik et al., 2020).

12.9 CONCLUSION

Biotechnology has advanced significantly as a result of the use of CRISPR-Cas, which is the most notable discovery of the twenty-first century. The outlook for what lies ahead is that anthropocentric expectations for the CRISPR-Cas system include the treatment of uncommon or fatal illnesses, including cancer, diabetes, and congenital anemia, among others. The use of this technology in the medical field will enable the battle against illnesses like SARS-CoV-2, HIV, malaria, dengue, and the Zika virus. Apart from the aforementioned benefits, there exists a chance to impede the progression of antibiotic-resistant bacteria and lower the pathogenicity of pathogenic bacterial isolates. The goal is to constantly work toward improving human health. However, caution must be used to avoid making the patient change unintentionally (Allemailem et al., 2023).

The CRISPR-Cas system has a wide range of applications, though, and agriculture stands to gain greatly by using it to enhance food quality and increase supply in order to fight hunger and food shortages. According to Nidhi et al. (2021), CRISPR-Cas has been shown to be useful in the biotechnology sector for installation or modification of metabolic pathways. This could boost product yields and optimize procedures for getting biologically derived goods. The application of CRISPR-Cas entails legal as well as bioethical guidelines, which at first should preserve patient integrity, human dignity, and the contents of their genetic data to prevent improper application (Sen & Thummer, 2022). Addressing the inherent risk of receiving these treatments—which may or may not have short-term negative effects—should also be taken into account. Legal law and bioethics will need to collaborate in order to control how patient data is used while defending basic rights like health. Given that certain illnesses may be eliminated with the use of these approaches, it is important to properly assess and monitor them rather than outright forbid them (Tong et al., 2024).

REFERENCES

Ahmad, Aftab, Muhammad Zubair Ghouri, Nayla Munawar, Muhammad Ismail, Sidra Ashraf, and Syed Ovais Aftable 2021. "Regulatory, ethical, and social aspects of CRISPR crops." *CRISPR Crops: The Future of Food Security* 261–287.

Ahmar, Sunny, Tahir Mahmood, Sajid Fiaz, Freddy Mora-Poblete, Muhammad Sohaib Shafique, Muhammad Sohaib Chattha, and Ki-Hung Jung. 2021. "Advantage of nanotechnology-based genome editing system and its application in crop improvement." *Frontiers in Plant Science* 12: 663849.

Allemailem, Khaled S, Saleh A Almatroodi, Ahmad Almatroudi, Faris Alrumaihi, Waleed Al Abdulmonem, Wafa Abdullah I Al-Megrin, Adel Nasser Aljamaan, Arshad Husain Rahmani, and Amjad Ali Khan. 2023. "Recent advances in genome-editing technology with CRISPR/Cas9 variants and stimuli-responsive targeting approaches within tumor cells: A future perspective of cancer management." *International Journal of Molecular Sciences* 24 (8): 7052.

Babochkina, TI, LA Gerlinskaya, and MP Moshkin. 2020. "Generation of donor organs in chimeric animals via blastocyst complementation." *Vavilov Journal of Genetics and Breeding* 24 (8): 913.

Barman, Antara, Bornali Deb, and Supriyo Chakraborty. 2020. "A glance at genome editing with CRISPR–Cas9 technology." *Current Genetics* 66: 447–462.

Batool, Asiya, Fayaz Malik, and Khurshid Iqbal Andrabi. 2021. "Expansion of the CRISPR/Cas genome-sculpting toolbox: Innovations, applications and challenges." *Molecular Diagnosis & Therapy* 25: 41–57.

Behr, Matthew, Jing Zhou, Bing Xu, and Hongwei Zhang. 2021. "In vivo delivery of CRISPR-Cas9 therapeutics: Progress and challenges." *Acta Pharmaceutica Sinica B* 11 (8): 2150–2171.

Bernheim, Aude, and Rotem Sorek. 2020. "The pan-immune system of bacteria: Antiviral defence as a community resource." *Nature Reviews Microbiology* 18 (2): 113–119.

Bisht, Deepanker, Sagar Salave, Nimeet Desai, Purnima Gogoi, Dhwani Rana, Prachurya Biswal, Gautami Sarma, Derajram Benival, Nagavendra Kommineni, and Dhruv Desai. 2024. "Genome editing and its role in vaccine, diagnosis, and therapeutic advancement." *International Journal of Biological Macromolecules* 131802.

Buquicchio, Frank A, and Ansuman T Satpathy. 2021. "Interrogating immune cells and cancer with CRISPR-Cas9." *Trends in Immunology* 42 (5): 432–446.

Cao, Genmao, Xuezhen Xuan, Ruijing Zhang, Jie Hu, and Honglin Dong. 2021. "Gene therapy for cardiovascular disease: Basic research and clinical prospects." *Frontiers in Cardiovascular Medicine* 8: 760140.

Chakravarty, Debyani, and David B Solit. 2021. "Clinical cancer genomic profiling." *Nature Reviews Genetics* 22 (8): 483–501.

Cheng, Hao, Feng Zhang, and Yang Ding. 2021. "CRISPR/Cas9 delivery system engineering for genome editing in therapeutic applications." *Pharmaceutics* 13 (10): 1649.

Cheng, Xing, Shaoyi Fan, Chengcai Wen, and Xianfa Du. 2020. "CRISPR/Cas9 for cancer treatment: Technology, clinical applications and challenges." *Briefings in Functional Genomics* 19 (3): 209–214.

Claringbould, Annique, and Judith B Zaugg. 2021. "Enhancers in disease: Molecular basis and emerging treatment strategies." *Trends in Molecular Medicine* 27 (11): 1060–1073.

D'Souza, Russell, Mary Mathew, and Krishna Mohan Surapaneni. 2023. "A scoping review on the ethical issues in the use of CRISPR-Cas9 in the creation of human disease models." *Journal of Clinical & Diagnostic Research* 17 (12).

Dayan, Fazli. 2020. "Ethico-legal aspects of CRISPR Cas-9 genome editing: A balanced approach." *Bangladesh Journal of Medical Science* 19 (1): 11.

Doss, Tarun M. 2022. "The ethical concerns raised by applications of CRISPR-Cas9 in gene drives and human germline editing." *The National High School Journal of Science* 1–15.

Ganger, Suman, Geetanjali Harale, and Parama Majumdar. 2022. "Clustered regularly interspaced short palindromic repeats/CRISPR-associated (CRISPR/Cas) systems: Discovery, structure, classification, and general mechanism." In *CRISPR/Cas-Mediated Genome Editing in Plants*, 65–97. Apple Academic Press.

Garcia-Robledo, Juan Esteban, María Claudia Barrera, and Gabriel J Tobón. 2020. "CRISPR/Cas: From adaptive immune system in prokaryotes to therapeutic weapon against immune-related diseases: CRISPR/Cas9 offers a simple and inexpensive method for disease modeling, genetic screening, and potentially for disease therapy." *International Reviews of Immunology* 39 (1): 11–20.

Gerashchenkov, GA, NA Rozhnova, BR Kuluev, O Yu Kiryanova, GR Gumerova, AV Knyazev, ZR Vershinina, EV Mikhailova, DA Chemeris, and RT Matniyazov. 2020. "Design of guide RNA for CRISPR/Cas plant genome editing." *Molecular Biology* 54: 24–42.

Giani, Alice Maria, Guido Roberto Gallo, Luca Gianfranceschi, and Giulio Formenti. 2020. "Long walk to genomics: History and current approaches to genome sequencing and assembly." *Computational and Structural Biotechnology Journal* 18: 9–19.

Gonzalez-Avila, Luis Uriel, Juan Manuel Vega-López, Leda Ivonne Pelcastre-Rodríguez, Omar Alejandro Cabrero-Martínez, Cecilia Hernández-Cortez, and Graciela Castro-Escarpulli. 2021. "The Challenge of CRISPR-Cas toward bioethics." *Frontiers in Microbiology* 12: 657981.

Gostimskaya, Irina. 2022. "CRISPR–Cas9: A history of its discovery and ethical considerations of its use in genome editing." *Biochemistry (Moscow)* 87 (8): 777–788.

Grabowski, Łukasz, Krzysztof Łepek, Małgorzata Stasiłojć, Katarzyna Kosznik-Kwaśnicka, Karolina Zdrojewska, Monika Maciąg-Dorszyńska, Grzegorz Węgrzyn, and Alicja Węgrzyn. 2021. "Bacteriophage-encoded enzymes destroying bacterial cell membranes and walls, and their potential use as antimicrobial agents." *Microbiological Research* 248: 126746.

Guo, Congting, Xiaoteng Ma, Fei Gao, and Yuxuan Guo. 2023. "Off-target effects in CRISPR/Cas9 gene editing." *Frontiers in Bioengineering and Biotechnology* 11: 1143157.

Gupta, Shweta, Adarsh Kumar, Rupali Patel, and Vinay Kumar. 2021. "Genetically modified crop regulations: Scope and opportunity using the CRISPR-Cas9 genome editing approach." *Molecular Biology Reports* 48 (5): 4851–4863.

El-Ashry, Amira H. 2023. "The CRISPR/Cas system: Gene Editing by bacterial defense." *Novel Research in Microbiology Journal* 7 (5): 2101–2115.

Harale, Geetanjali, Sheetal Pardeshi, Parama Majumdar, and Suman Ganger. 2022. "Transcription activator-like effector nucleases (TALENs): Genome editing tool to explore enhanced activity of antidiabetic plants." In *Antidiabetic Potential of Plants in the Era of Omics*, 403–428. Apple Academic Press.

Harish, Vancha, Devesh Tewari, Manish Gaur, Awadh Bihari Yadav, Shiv Swaroop, Mikhael Bechelany, and Ahmed Barhoum. 2022. "Review on nanoparticles and nanostructured materials: Bioimaging, biosensing, drug delivery, tissue engineering, antimicrobial, and agro-food applications." *Nanomaterials* 12 (3): 457.

Harmatz, Paul, Carlos E Prada, Barbara K Burton, Heather Lau, Craig M Kessler, Liching Cao, Marina Falaleeva, Andres G Villegas, Jennifer Zeitler, and Kathleen Meyer. 2022. "First-in-human in vivo genome editing via AAV-zinc-finger nucleases for mucopolysaccharidosis I/II and hemophilia B." *Molecular Therapy* 30 (12): 3587–3600.

Hemalatha, Palanivel, Ebrahim M Abda, Shipra Shah, S Venkatesa Prabhu, M Jayakumar, N Karmegam, Woong Kim, and M Govarthanan. 2023. "Multi-faceted CRISPR-Cas9 strategy to reduce plant based food loss and waste for sustainable bio-economy–a review." *Journal of Environmental Management* 332: 117382.

Hendriks, Delilah, Hans Clevers, and Benedetta Artegiani. 2020. "CRISPR-Cas tools and their application in genetic engineering of human stem cells and organoids." *Cell Stem Cell* 27 (5): 705–731.

Hou, Chengjie, Xin Song, Zhiqiang Xiong, Guangqiang Wang, Yongjun Xia, and Lianzhong Ai. 2024. "Investigating the role of β-disodium glycerophosphate and urea in promoting growth of Streptococcus thermophilus from omics-integrated genome-scale models." *Foods* 13 (7): 1006.

Jacinto, Filipe V, Wolfgang Link, and Bibiana I Ferreira. 2020. "CRISPR/Cas9-mediated genome editing: From basic research to translational medicine." *Journal of Cellular and Molecular Medicine* 24 (7): 3766–3778.

Jamil, Arif, Shamima Parvin Lasker, Ahmed Ragib Chowdhury, and See Ming Seow. 2023. "Gene modification in non-human animal for developing human compatible organs: Ethical, legal, clinical and societal issues." *Bangladesh Journal of Bioethics* 14 (2): 13–18.

Janik, Edyta, Marcin Niemcewicz, Michal Ceremuga, Lukasz Krzowski, Joanna Saluk-Bijak, and Michal Bijak. 2020. "Various aspects of a gene editing system—CRISPR–Cas9." *International Journal of Molecular Sciences* 21 (24): 9604.

Javaid, Darakhshan, Shahid Yousuf Ganie, Younis Ahmad Hajam, and Mohd Salim Reshi. 2022. "CRISPR/Cas9 system: A reliable and facile genome editing tool in modern biology." *Molecular Biology Reports* 49 (12): 12133–12150.

Kantor, Ariel, Michelle E McClements, and Robert E MacLaren. 2020. "CRISPR-Cas9 DNA base-editing and prime-editing." *International Journal of Molecular Sciences* 21 (17): 6240.

Katayama, Shota, Masahiro Watanabe, Yoshio Kato, Wataru Nomura, and Takashi Yamamoto. 2024. "Engineering of zinc finger nucleases through structural modeling improves genome editing efficiency in cells." *Advanced Science* 2310255.

Khambhati, Khushal, Nisarg Gohil, Gargi Bhattacharjee, and Vijai Singh. 2020. "Development and challenges of using CRISPR-Cas9 system in mammalians." In *Genome Engineering via CRISPR-Cas9 System*, 83–93. Elsevier.

Khanzadi, Manzoor N, and Abid A Khan. 2020. "CRISPR/Cas9: Nature's gift to prokaryotes and an auspicious tool in genome editing." *Journal of Basic Microbiology* 60 (2): 91–102.

Khoshandam, Mohadeseh, Hossein Soltaninejad, Marziyeh Mousazadeh, Amir Ali Hamidieh, and Saman Hosseinkhani. 2024. "Clinical applications of the CRISPR/Cas9 genome-editing system: Delivery options and challenges in precision medicine." *Genes & Diseases* 11 (1): 268–282.

Kose, Ali Mertcan, Ozan Kocadagli, Cihan Taştan, Cagdas Aktan, Onur Mert Ünaldı, Elanur Güzenge, and Hamza Emir Erdil. 2024. "Unveiling off-target mutations in CRISPR guide RNAs: Implications for gene region specificity." *The CRISPR Journal* 7 (3): 168–178.

Kropf, Mario. 2024. "The ethically significant difference between dual use and slippery slope arguments, in relation to CRISPR-Cas9: Philosophical considerations and ethical challenges." *Research Ethics* 17470161241240587.

Lander, Noelia, Teresa Cruz-Bustos, and Roberto Docampo. 2020. "A CRISPR/Cas9-riboswitch-based method for downregulation of gene expression in Trypanosoma cruzi." *Frontiers in Cellular and Infection Microbiology* 10: 68.

Lange, Victor, and Klemens Kappel. 2022. "CRISPR gene-therapy: A critical review of ethical concerns and a proposal for public decision-making." *Canadian Journal of Bioethics* 5 (2): 78–87.

Li, Chao, Wen Chu, Rafaqat Ali Gill, Shifei Sang, Yuqin Shi, Xuezhi Hu, Yuting Yang, Qamar U Zaman, and Baohong Zhang. 2023a. "Computational tools and resources for CRISPR/Cas genome editing." *Genomics, Proteomics & Bioinformatics* 21 (1): 108–126.

Li, Tianxiang, Yanyan Yang, Hongzhao Qi, Weigang Cui, Lin Zhang, Xiuxiu Fu, Xiangqin He, Meixin Liu, Pei-feng Li, and Tao Yu. 2023b. "CRISPR/Cas9 therapeutics: Progress and prospects." *Signal Transduction and Targeted Therapy* 8 (1): 36.

Lim, Yong-Bee. 2021. "For liberty and accessible science for all-building a better understanding of US community laboratories and the do-it-yourself (DIY) biology movement." George Mason University.

Lorenzo, David, Montse Esquerda, Francesc Palau, and Francisco J Cambra. 2022. "Ethics and genomic editing using the CRISPR-Cas9 technique: Challenges and conflicts." *Nanoethics* 16 (3): 313–321.

Makarova, Kira S, Albertas Timinskas, Yuri I Wolf, Ayal B Gussow, Virginijus Siksnys, Česlovas Venclovas, and Eugene V Koonin. 2020. "Evolutionary and functional classification of the CARF domain superfamily, key sensors in prokaryotic antivirus defense." *Nucleic Acids Research* 48 (16): 8828–8847.

Mali, Franc. 2022. "Key socio-economic and (bio) ethical challenges in the CRISPR-Cas9 patent landscape." *Genome Editing in Drug Discovery* 315–327.

Malinin, Nikolay L, GaHyun Lee, Cicera R Lazzarotto, Yichao Li, Zongli Zheng, Nhu T Nguyen, Matthew Liebers, Ved V Topkar, A John Iafrate, and Long P Le. 2021. "Defining genome-wide CRISPR–Cas genome-editing nuclease activity with GUIDE-seq." *Nature Protocols* 16 (12): 5592–5615.

Mattei, Alexandra L, Nina Bailly, and Alexander Meissner. 2022. "DNA methylation: A historical perspective." *Trends in Genetics* 38 (7): 676–707.

Mehra, Vinay Kumar, and Satish Kumar. 2022. "The application of CRISPR/Cas9 technology for farm animals: A review." *Agricultural Reviews* 43 (1): 54–61.

Mehta, Aditi, and Olivia M Merkel. 2020. "Immunogenicity of Cas9 protein." *Journal of Pharmaceutical Sciences* 109 (1): 62–67.

Meyer, Morgan, and Frédéric Vergnaud. 2020. "The rise of biohacking: Tracing the emergence and evolution of DIY biology through online discussions." *Technological Forecasting and Social Change* 160: 120206.

Mindell, David A, and Elisabeth Reynolds. 2023. *The Work of the Future: Building Better Jobs in an Age of Intelligent Machines*. MIT Press.

Mohamadi, Sahar, Saeed Zaker Bostanabad, and Reza Mirnejad. 2020. "CRISPR arrays: A review on its mechanism." *Journal of Applied Biotechnology Reports* 7 (2): 81–86.

Mosterd, Cas, Geneviève M Rousseau, and Sylvain Moineau. 2021. "A short overview of the CRISPR-Cas adaptation stage." *Canadian Journal of Microbiology* 67 (1): 1–12.

Mushtaq, Muntazir, Aejaz Ahmad Dar, Milan Skalicky, Anshika Tyagi, Nancy Bhagat, Umer Basu, Basharat Ahmad Bhat, Abbu Zaid, Sajad Ali, and Tanvir-Ul-Hassan Dar. 2021. "CRISPR-based genome editing tools: Insights into technological breakthroughs and future challenges." *Genes* 12 (6): 797.

Nazir, Taha, Hameed Mirza, and Nida Taha. 2024. "History, development and characteristics of CRISPR-Cas system."

Newman, Anthony, Lora Starrs, and Gaetan Burgio. 2020. "Cas9 cuts and consequences; detecting, predicting, and mitigating CRISPR/Cas9 on-and off-target damage: Techniques for detecting, predicting, and mitigating the on-and off-target effects of Cas9 editing." *Bioessays* 42 (9): 2000047.

Nidhi, Sweta, Uttpal Anand, Patrik Oleksak, Pooja Tripathi, Jonathan A Lal, George Thomas, Kamil Kuca, and Vijay Tripathi. 2021. "Novel CRISPR–Cas systems: An updated review of the current achievements, applications, and future research perspectives." *International Journal of Molecular Sciences* 22 (7): 3327.

Ou, Xuejin, Qizhi Ma, Wei Yin, Xuelei Ma, and Zhiyao He. 2021. "CRISPR/Cas9 gene-editing in cancer immunotherapy: Promoting the present revolution in cancer therapy and exploring more." *Frontiers in Cell and Developmental Biology* 9: 674467.

Panigrahi, Kaushik Kumar, Ayesha Mohanty, Smruti Ranjan Padhan, Priya brata Bhoi, Syed Mohammad Bashir Ali, and Purandar Mandal. 2023. "Recent advances in genome editing towards sustainable agriculture." *Genome Editing and Global Food Security* 155–193.

Panuntun, Daniel Fajar. 2020. "Christian ethics toward CRISPR Cas-9 gene editing for human being." *ICCIRS 2019: Proceedings of the First International Conference on Christian and Inter Religious Studies, ICCIRS 2019*, December 11–14. 2019, Manado, Indonesia.

Pinilla-Redondo, Rafael, Jakob Russel, David Mayo-Muñoz, Shiraz A Shah, Roger A Garrett, Joseph Nesme, Jonas S Madsen, Peter C Fineran, and Søren J Sørensen. 2022. "CRISPR-Cas systems are widespread accessory elements across bacterial and archaeal plasmids." *Nucleic Acids Research* 50 (8): 4315–4328.

Proust, Lucas, Eloi Haudebourg, Alain Sourabié, Martin Pedersen, Iris Besançon, Véronique Monnet, and Vincent Juillard. 2020. "Multi-omics approach reveals how yeast extract peptides shape Streptococcus thermophilus metabolism." *Applied and Environmental Microbiology* 86 (22): e01446-20.

Rafii, Saeed, Emad Tashkandi, Nedal Bukhari, and Humaid O Al-Shamsi. 2022. "Current status of CRISPR/Cas9 application in clinical cancer research: Opportunities and challenges." *Cancers* 14 (4): 947.

Ramezani, Hamid, and Hendrik Dietz. 2020. "Building machines with DNA molecules." *Nature Reviews Genetics* 21 (1): 5–26.

Rasul, Mohammed Fatih, Bashdar Mahmud Hussen, Abbas Salihi, Bnar Saleh Ismael, Paywast Jamal Jalal, Anna Zanichelli, Elena Jamali, Aria Baniahmad, Soudeh Ghafouri-Fard, and Abbas Basiri. 2022. "Strategies to overcome the main challenges of the use of CRISPR/Cas9 as a replacement for cancer therapy." *Molecular Cancer* 21 (1): 1–30.

Santiago-Frangos, Andrew, William S Henriques, Tanner Wiegand, Colin C Gauvin, Murat Buyukyoruk, Ava B Graham, Royce A Wilkinson, Lenny Triem, Kasahun Neselu, and Edward T Eng. 2023. "Structure reveals why genome folding is necessary for site-specific integration of foreign DNA into CRISPR arrays." *Nature Structural & Molecular Biology* 30 (11): 1675–1685.

Sarma, Hemen, N. F. Islam, Ram Prasad, M. N. V. Prasad, Lena Q Ma, and Jörg Rinklebe. 2021. "Enhancing phytoremediation of hazardous metal (loid) s using genome engineering CRISPR–Cas9 technology." *Journal of Hazardous Materials* 414: 125493.

Schloss, Jeffery A, Richard A Gibbs, Vinod B Makhijani, and Andre Marziali. 2020. "Cultivating DNA sequencing technology after the human genome project." *Annual Review of Genomics and Human Genetics* 21 (1): 117–138.

Schwartz, Matthew L, M Wayne Davis, Matthew S Rich, and Erik M Jorgensen. 2021. "High-efficiency CRISPR gene editing in C. elegans using Cas9 integrated into the genome." *PLoS Genetics* 17 (11): e1009755.

Seeberger, Peter H. 2021. "Discovery of semi-and fully-synthetic carbohydrate vaccines against bacterial infections using a medicinal chemistry approach: Focus review." *Chemical Reviews* 121 (7): 3598–3626.

Seiter, Katie M, and Linda Fuselier. 2021. "Content knowledge and social factors influence student moral reasoning about CRISPR/Cas9 in humans." *Journal of Research in Science Teaching* 58 (6): 790–821.

Sen, Tirthankar, and Rajkumar P Thummer. 2022. "CRISPR and iPSCs: Recent developments and future perspectives in neurodegenerative disease modelling, research, and therapeutics." *Neurotoxicity Research* 40 (5): 1597–1623.

Shanmugam, Sabarathinam, Huu-Hao Ngo, and Yi-Rui Wu. 2020. "Advanced CRISPR/Cas-based genome editing tools for microbial biofuels production: A review." *Renewable Energy* 149: 1107–1119.

Sharma, Akshay, Jaap-Jan Boelens, Maria Cancio, Jane S Hankins, Prafulla Bhad, Marjohn Azizy, Andrew Lewandowski, Xiaojun Zhao, Shripad Chitnis, and Radhika Peddinti. 2023. "CRISPR-Cas9 editing of the HBG1 and HBG2 promoters to treat sickle cell disease." *New England Journal of Medicine* 389 (9): 820–832.

Shmakov, Sergey A, Irina Utkina, Yuri I Wolf, Kira S Makarova, Konstantin V Severinov, and Eugene V Koonin. 2020. "CRISPR arrays away from cas genes." *The CRISPR Journal* 3 (6): 535–549.

Singh, Sanjay. 2024. "Gene editing technologies: CRISPR/Cas9 and beyond for genetic disease therapy and research." *Universal Research Reports* 11 (3): 1–7.

Son, Seungmin, and Sang Ryeol Park. 2022. "Challenges facing CRISPR/Cas9-based genome editing in plants." *Frontiers in Plant Science* 13: 902413.

Souza, FSD, MCSN Botelho, and RS Lisbôa. 2002. "Hepatology: Hepatology principles and practice: History, morphology, biochemistry, diagnostics, clinic, therapy." *Journal of American Medical Association* 288: 1–15.

Tang, Yanan, Lu Gao, Wei Feng, Chen Guo, Qianfan Yang, Feng Li, and X Chris Le. 2021. "The CRISPR–Cas toolbox for analytical and diagnostic assay development." *Chemical Society Reviews* 50 (21): 11844–11869.

Tong, Ling-Wu, Yong-Shan Hu, Shi-Jing Yu, Cheng-Lei Li, and Jing-Wei Shao. 2024. "Current application and future perspective of CRISPR/Cas9 gene editing system mediated immune checkpoint for liver cancer treatment." *Nanotechnology* 35(40): 402002.

Uddin, Fathema, Charles M Rudin, and Triparna Sen. 2020. "CRISPR gene therapy: Applications, limitations, and implications for the future." *Frontiers in Oncology* 10: 1387.

Valdés, Erick, and Juan Alberto Lecaros. 2023. "Ethical issues in genetically modified foods: From transgenesis to CRISPR-Cas9 genome editing technology." In *Handbook of Bioethical Decisions. Volume I: Decisions at the Bench*, 723–736. Springer.

Vigouroux, Antoine, and David Bikard. 2020. "CRISPR tools to control gene expression in bacteria." *Microbiology and Molecular Biology Reviews* 84 (2).

Voort, Thomas van de. 2022. "Self-made success, achieved together. How and why the DIY-bio community engages in DIYbio projects and what makes them successful." Master's thesis. Utrecht University Student Theses Repository.

Wang, Joy Y, and Jennifer A Doudna. 2023. "CRISPR technology: A decade of genome editing is only the beginning." *Science* 379 (6629): eadd8643.

Wani, Atif Khurshid, Nahid Akhtar, and Saurabh Shukla. 2022. "CRISPR/Cas9: Regulations and challenges for law enforcement to combat its dual-use." *Forensic Science International* 334: 111274.

Yang, Wu, Jiaqi Yan, Pengzhen Zhuang, Tao Ding, Yu Chen, Yu Zhang, Hongbo Zhang, and Wenguo Cui. 2022. "Progress of delivery methods for CRISPR-Cas9." *Expert Opinion on Drug Delivery* 19 (8): 913–926.

Yip, Bon Ham. 2020. "Recent advances in CRISPR/Cas9 delivery strategies." *Biomolecules* 10 (6): 839.

Zaib, Sumera, Mushtaq A Saleem, and Imtiaz Khan. 2022. "CRISPR-Cas9 genome engineering: Trends in medicine and health." *Mini Reviews in Medicinal Chemistry* 22 (3): 410–421.

Zhu, Haocheng, Chao Li, and Caixia Gao. 2020. "Applications of CRISPR–Cas in agriculture and plant biotechnology." *Nature Reviews Molecular Cell Biology* 21 (11): 661–677.

Index

Note: Tables are indicated by **bold**.